Monographs in Computer Science

T0191552

Editors

David Gries
Fred B. Schneider

Annabelle McIver
Carroll Morgan

Abstraction, Refinement and Proof for Probabilistic Systems

With 63 Figures

 Springer

Annabelle McIver
Department of Computing
Macquarie University
Sydney, NSW 2109
Australia
anabel@ics.mq.edu.au

Carroll Morgan
School of Computer Science
   and Engineering
University of New South Wales
Sydney, NSW 2052
Australia
carrollm@cse.unsw.edu.au

Series Editors:
David Gries
Department of Computer Science
Cornell University
Upson Hall
Ithaca, NY 14853-7501

Fred B. Schneider
Department of Computer Science
Cornell University
Upson Hall
Ithaca, NY 14853-7501

Cover illustration: Herman's-graph variant. See Figure 2.8.3 on p. 59.

ISBN   978-1-4419-2312-7   e-ISBN   978-0-387-27006-7

Printed on acid-free paper.

9 8 7 6 5 4 3 2 1

springeronline.com

Preface

Probabilistic techniques in computer programs and systems are becoming more and more widely used, for increased efficiency (as in random algorithms), for symmetry breaking (distributed systems) or as an unavoidable artefact of applications (modelling fault-tolerance). Because interest in them has been growing so strongly, stimulated by their many potential uses, there has been a corresponding increase in the study of their correctness — for the more widespread they become, the more we will depend on understanding their behaviour, and their limits, exactly.

In this volume we address that last concern, of understanding: we present a method for *rigorous reasoning about probabilistic programs and systems*. It provides an operational model — "how they work" — and an associated program logic — "how we should reason about them" — that are designed to fit together. The technique is simple in principle, and we hope that with it we will be able to increase dramatically the effectiveness of our analysis and use of probabilistic techniques in practice.

Our contribution is a probabilistic calculus that operates at the level of the program text, and it is *light-weight* in the sense that the amount of reasoning is similar in size and style to what standard assertional techniques require. In the fragment at right, for example, each potential loop entry occurs with probability $1/2$; the resulting iteration establishes $x \geq 1/2$ with probability exactly p for any $0 \leq p \leq 1$. It is thus an implementation of the general operation *choose with probability p*, but it uses only simple tests of unbiased random bits (to implement the loop guard). It should take only a little quantitative logic to confirm that claim, and indeed we will show that just four lines of reasoning suffice.

```
x := p;
while 1/2 do
    x := 2x;
    if x ≥ 1
        then x := x - 1
    fi
od
```

Economy and precision of reasoning are what we have come to expect for standard programs; there is no reason we should accept less when they are probabilistic.

The cover illustration comes from page 59.
The program fragment is adapted from Fig. 7.7.10 on page 210.

Scope and applicability

Methods for the analysis of probabilistic systems include automata, labelled transition systems, model checking and logic (*e.g.* dynamic or temporal). Our work falls into the last category: we overlay the Hoare-logic paradigm with probabilistic features imported from Markov processes, taking from each the essential characteristics required for a sound mathematical theory of refinement and proof. The aim is to accommodate modelling and analysis of both sequential and distributed probabilistic systems, and to allow — even encourage — movement between different levels of abstraction.

Our decision to focus on *logic* — and a proof system for it — was motivated by our experience with logical techniques more generally: they impose a discipline and order which promotes clarity in specifications and design; the resulting proofs can often be carried out, and checked, with astonishing conciseness and accuracy; and the calculation rules of the logic lead to an algebra that captures useful equalities and inequalities at the level of the programs themselves.

Although we rely ultimately on an operational model, we use it principally to validate the logic (and that, in turn, justifies the algebra) — direct reliance on the model's details for individual programs is avoided if possible. (However we do not hesitate to use such details to support our intuition.) We feel that operational reasoning is more suited to the algorithmic methods of verification used by model checkers and simulation tools which can, for specific programs, answer questions that are impractical for the general approach that a logic provides.

Thus the impact of our approach is most compelling when applied to programs which are intricate either in their implementation or their design, or have generic features such as undetermined size or other parameters. They might appear as probabilistic source-level portions of large sequential programs, or as abstractions from the probabilistic modules of a comprehensive system-level design; we provide specific examples of both situations. In the latter case the ability to abstract modules' properties has a significant effect on the overall verification enterprise.

Technical features

Because we generalise the well-established assertional techniques of specifications, pre- and postconditions, there is a natural continuity of reasoning style evident in the simultaneous use of the new and the familiar approaches: the probabilistic analysis can be deployed more, or less, as the situation warrants.

A major feature is that we place probabilistic choice and abstraction together, in the same framework, without having to factor either of them out for separate treatment unless we wish to (as in fact we do in Chap. 11). This justifies the *abstraction and refinement* of our title, and is what gives us access to the stepwise-development paradigm of standard programming where systems are "refined" from high levels of abstraction towards the low levels that include implementation detail.

As a side-effect of including abstraction, we retain its operational counterpart *demonic choice* as an explicit operator ⊓ in the cut-down probabilistic programming language *pGCL* which we use to describe our algorithms — that is, the new probabilistic choice operator $_p\oplus$ refines demonic choice rather than replacing it. In Chap. 8 we consider angelic choice ⊔ as well, which is thus a further refinement.

Probabilistic and demonic choice together allow an elementary treatment of the hybrid that selects "with probability *at least p*" (or similarly "*at most p*"), an abstraction which accurately models our unavoidable ignorance of exact probabilities in real applications. Thus in our mathematical model we are able to side-step the issue of "approximate refinement."

That is, rather than saying "this coin refines a fair coin with probability 95%," we would say "this coin refines one which is within 5% of being fair." This continues the simple view that either an implementation refines a specification or it does not, which simplicity is possible because we have retained the original treatment in terms of sets of behaviours: abstraction is inclusion; refinement is reverse inclusion; and demonic choice is union. In that way we maintain the important relationship between the three concepts. (Section 6.5 on pp. 169ff illustrates this geometrically.)

Organisation and intended readership

The material is divided into three major parts of increasing specialisation, each of which can to a large extent be studied on its own; a fourth part contains appendices. We include a comprehensive index and extensive cross-referencing.

Definitions of notation and explanations of standard mathematical techniques are carefully given, rather than simply assumed; they appear as footnotes at their first point of use and are made visually conspicuous by using SMALL CAPITALS for the defined terms (where grammar allows). Thus in many cases a glance should be sufficient to determine whether any footnote contains a definition. In any case all definitions, whether or not in footnotes, may be retrieved by name through the index; and those with numbers are listed in order at page xvii.

Because much of the background material is separated from the main text, the need for more advanced readers to break out of the narrative should be reduced. We suggest that on first reading it is better to consult the footnotes only when there is a term that appears to require definition — otherwise the many cross-references they contain may prove distracting, as they are designed for "non-linear" browsing once the main ideas have already been assimilated.

Part I, *Probabilistic guarded commands*, gives enough introduction to the probabilistic logic to prove properties of small programs such as the one earlier, for example at the level of an undergraduate course for Formal-Methods-inclined students that explains "what to do" but not necessarily "why it is correct to do that." These would be people who need to understand how to reason about programs (and why), but would see the techniques as intellectual tools rather than as objects of study in their own right.

We have included many small examples to serve as models for the approach (they are indexed under *Programs*), and there are several larger case studies (for example in Chap. 3).

Part II, *Semantic structures*, develops in detail the mathematics on which the probabilistic logic is built and with which is it justified. That is, whereas the earlier sections present and illustrate the new reasoning techniques, this part shows where they have come from, why they have the form they do and — crucially — why they are correct.

That last point is especially important for students intending to do research in logic and semantics, as it provides a detailed and extended worked example of the fundamental issue of proving reasoning techniques *themselves* to be correct (more accurately, "valid"), a higher-order concept than the more familiar theme of the previous part in which we presented the techniques *ex cathedra* and used them to verify particular programs.

This part would thus be suitable for an advanced final-year undergraduate or first-year graduate course, and would fit in well with other material on programming semantics. It defines and illustrates the use of many of the standard tools of the subject: lattices, approximation orders, fixed points, semantic injections and retractions *etc.*

Part III, *Advanced topics*, concentrates on more exotic methods of specification and design, in this case probabilistic temporal/modal logics. Its final chapter, for example, contains material only recently discovered and leads directly into an up-to-date research area. It would be suitable for graduate students as an introduction to this specialised research community.

Part IV includes appendices collecting material that either leads away from the main exposition — *e.g.* alternative approaches and why we have not taken them — or supports the text at a deeper level, such as some of the more detailed proofs.

It also contains a short list of algebraic laws that demonic/probabilistic program fragments satisfy, generated mainly by our needs in the examples and proofs of earlier sections. An interesting research topic would be a more systematic elaboration of that list with a view to incorporating it into probabilistic Kleene- or omega algebras for distributed computations.

Overall, readers seeking an introduction to probabilistic formal methods could follow the material in order from the beginning. Those with more experience might instead sample the first chapter from each part, which would give an indication of the scope and flavour of the approach generally.

Original sources

Much of the material is based on published research, done with our colleagues, in conference proceedings and journal articles; but here it has been substantially updated and rationalised — and we have done our best to bring the almost ten years' worth of developing notation into a uniform state.

For self-contained presentations of the separate topics, and extra background, readers could consult our earlier publications as shown overleaf.

At the end of each chapter we survey the way in which our ideas have been influenced by — and in some cases adopted from — the work of other researchers, and we indicate some up-to-date developments.

Acknowledgements

Our work on probabilistic models and logic was carried out initially at the University of Oxford, together with Jeff Sanders and Karen Seidel and with the support of the UK's *Engineering and Physical Sciences Research Council* (the EPSRC) during two projects led by Sanders and Morgan over the years 1994–2001.

Morgan spent sabbatical semesters in 1995–6 at the University of Utrecht, as the guest of S. Doaitse Swierstra, and at the University of Queensland and the Software Verification and Research Centre (SVRC), as the guest of David Carrington and Ian Hayes. The foundational work the EPSRC projects produced during that period — sometimes across great distances — benefited from the financial support of those institutions but especially from the academic environment provided by the hosts and by the other researchers who were receptive to our initial ideas [MMS96].

Ralph Back at Åbo Akademi hosted our group's visit to Turku for a week in 1996 during which we were able to explore our common interests in refinement and abstraction as it applied to the new domain; that led later to a three-month visit by Elena Troubitsyna from the *Turku Center for Computer Science* (TUCS), to our group in Oxford in 1997, and contributed to what has become Chap. 4 [MMT98].

David Harel was our host for a two-week visit to Israel in 1996, during which we presented our ideas and benefited from the interaction with researchers there.

Chapters' dependence on original sources

Chapter 1	see	[MM99b, SMM, MMS00]
Chapter 2	see	[Mor96, MMS00]
Chapter 3	see	[MM99b]
Chapter 4	see	[MMT98]
Chapter 5	see	[MMS96]
Chapter 6		*is new material*
Chapter 7	see	[Mor96, MM01b]
Chapter 8	see	[MM01a]
Chapter 9	see	[MM97]
Chapter 10	see	[MM99a]
Chapter 11	see	[MM02]

The sources listed opposite are in chronological order of writing, thus giving roughly the logical evolution of the ideas.

Subsequently we have continued to work with Sanders and with Ken Robinson, Thai Son Hoang and Zhendong Jin, supported by the *Australian Research Council* (ARC) over the (coming) years 2001–8 in their *Large Grant* and *Discovery* programmes, at the Universities of Macquarie and of New South Wales.

Joe Hurd from the Computer Laboratory at Cambridge University visited us in 2002, with financial assistance from Macquarie University; and Orieta Celiku was supported by TUCS when she visited in 2003. Both worked under McIver's direction on the formalisation of *pGCL*, and its logic, in the mechanised logic *HOL*.

Hoang, Jin and especially Eric Martin have helped us considerably with their detailed comments on the typescript; also Ralph Back, Ian Hayes, Michael Huth, Quentin Miller and Wayne Wheeler have given us good advice. Section B.1 on the algebraic laws satisfied by probabilistic programs has been stimulated by the work (and the critical eyes) of Steve Schneider and his colleagues at Royal Holloway College in the U.K.

We thank the members of IFIP Working Groups 2.1 and 2.3 for their many comments and suggestions.

LRI Paris   Annabelle McIver
May 2004   Carroll Morgan

In memoriam AJMvG

List of sources in order of writing

[MMS96] C.C. Morgan, A.K. McIver, and K. Seidel.   Probabilistic predicate transformers.   *ACM Transactions on
Programming Languages and Systems*, 18(3):325–53, May
1996.

[Mor96]   C.C. Morgan.   Proof rules for probabilistic loops.   In
He Jifeng, John Cooke, and Peter Wallis, editors, *Pro-
ceedings of the BCS-FACS 7th Refinement Workshop*,
Workshops in Computing. Springer-Verlag, July 1996.

[MM01b]   A.K. McIver and C.C. Morgan.   Partial correctness for
probabilistic programs.   *Theoretical Computer Science*,
266(1–2):513–41, 2001.

[SMM]   K. Seidel, C.C. Morgan, and A.K. McIver.   Probabilistic
imperative programming: a rigorous approach.   Extended
abstract appears in Groves and Reeves [GR97], pages 1–2.

[MMT98]   A.K. McIver, C.C. Morgan, and E. Troubitsyna.   The
probabilistic steam boiler: a case study in probabilistic data
refinement.   In J. Grundy, M. Schwenke, and T. Vickers,
editors, *Proc. International Refinement Workshop, ANU,
Canberra*, Discrete Mathematics and Computer Science,
pages 250–65. Springer-Verlag, 1998.

[MM01a]   A.K. McIver and C.C. Morgan.   Demonic, angelic and unbounded probabilistic choices in sequential programs.   *Acta
Informatica*, 37:329–54, 2001.

[MM99b]   C.C. Morgan and A.K. McIver.   *pGCL*: Formal reasoning
for random algorithms.   *South African Computer Journal*,
22, March 1999.

[MM97]   C.C. Morgan and A.K. McIver.   A probabilistic temporal calculus based on expectations.   In Groves and Reeves
[GR97], pages 4–22.

[MM99a]   C.C. Morgan and A.K. McIver.   An expectation-based
model for probabilistic temporal logic.   *Logic Journal of
the IGPL*, 7(6):779–804, 1999.

[MMS00]   A.K. McIver, C.C. Morgan, and J.W. Sanders.   Probably
Hoare? Hoare probably! In J.W. Davies, A.W. Roscoe, and
J.C.P. Woodcock, editors, *Millennial Perspectives in Com-
puter Science*, Cornerstones of Computing, pages 271–82.
Palgrave, 2000.

[MM02]   A.K McIver and C.C. Morgan.   Games, probability and the
quantitative μ-calculus qMu. In *Proc. LPAR*, volume 2514
of *LNAI*, pages 292–310. Springer-Verlag, 2002.

Contents

List of definitions *etc.*

Definitions and notations for standard mathematical concepts are given in footnotes, rather than in the main text, so that they do not interrupt the flow. They can be found directly, via the index, where they are indicated by bold-face page references. Thus for example at "fixed point, least" we find that the *least fixed-point* and its associated "μ notation" are defined in Footnote 33 on p. 21, and a second form of the μ notation is defined in Footnote 32 on p. 102.

Part I

Probabilistic guarded commands and their refinement logic

1
Introduction to *pGCL*:
Its logic and its model

1.1   Sequential program logic

Since the mid-1970's, any serious student of rigorous program development will have encountered "assertions about programs" — they are predicates which, when inserted into program code, are supposed to be "true at that point of the program." Formalised — *i.e.* made into a logic — they look like either

$$\{pre\}\ prog\ \{post\} \qquad\qquad \text{Hoare-style}$$
$$\text{or} \qquad pre \Rrightarrow wp.prog.post\ , \qquad\qquad \text{Dijkstra-style} \qquad\left.\right\}\ (1.1)$$

in each case meaning "from any state satisfying precondition *pre*, the sequential program *prog* is guaranteed to terminate in a state satisfying postcondition *post*." [1] Formulae *pre* and *post* are written in first-order predicate logic over the program variables, and *prog* is written in a sequential programming language. Often Dijkstra's *Guarded Command Language* [Dij76], called *GCL*, is used in simple expositions like this one, since it contains just the essential features, and no clutter.

A conspicuous feature of Dijkstra's original presentation of guarded commands was the novel "demonic" choice. He explained that it arose naturally if one developed programs hand-in-hand with their proofs of correctness: if a single specification admitted say two implementations, then a third possibility was program code that seemed to choose unpredictably between the two. Yet in its pure form, where for example

$$prog \quad \sqcap \quad prog' \qquad\qquad (1.2)$$

is a program that can unpredictably behave either as *prog* or as *prog'*, this "demonic" nondeterminism seemed at first — to some — to be an unnecessary and in fact gratuitously confusing complication. Why would anyone ever want to introduce unpredictability *deliberately*? Programs are unpredictable enough already.

If one really wanted programs to behave in some kind of "random" way, then more useful surely would be a construction like the

$$prog \quad {}_{\frac{1}{2}}\!\oplus \quad prog' \qquad\qquad (1.3)$$

that behaves as *prog* on half of its runs, and as *prog'* on the other half. Of course on any *particular* run the behaviour is unpredictable, and even over many runs the proportions will not necessarily be exactly "50/50" — but over a long enough period one will find approximately equal evidence of each behaviour.

A logic and a model for programs like (1.3) was in fact provided in the early 1980's [Koz81, Koz85], where in the "Kozen style" the pre- and post-formulae became real- rather than Boolean functions of the state, and $\sqcap$ was replaced by ${}_{p}\!\oplus$ in the programming language. Those logical statements

---

[1] We will use the Dijkstra-style.

(1.1) now took on a more general meaning, that "if program *prog* is run many times from the same initial state, the average value of *post* in the resulting final states is at least the actual value that *pre* had in the initial state." Naturally we are relying on the expressions' *pre* and *post* having real- rather than Boolean type when we speak of their average, or *expected* value.

The original — *standard*, we call it — Boolean logic was still available of course via the embedding false, true $\mapsto 0, 1$.

Dijkstra's demonic ⊓ was not so easily discarded, however. Far from being "an unnecessary and confusing complication," it is the very basis of what is now known as *refinement* and *abstraction* of programs. (The terms are complementary: an implementation refines its specification; a specification abstracts from its implementation.) To specify "set r to a square-root of s" one could write *directly in the programming language GCL*

$$r := -\sqrt{s} \quad \sqcap \quad r := \sqrt{s}, \quad {}^2 \tag{1.4}$$

something that had never been possible before. This explicit, if accidental, "programming feature" caught the tide that had begun to flow in that decade and the following: the idea that specifications and code were merely different ways of describing the same thing (as advocated by Abrial, Hoare and others; making an early appearance in Back's work [Bac78] on what became the Refinement Calculus [Mor88b, Bac88, Mor87, Mor94b, BvW98]; and as found at the heart of specification and development methods such as *Z* [Spi88] and *VDM* [Jon86]).

Unfortunately, *probabilistic* formalisms were left behind, and did not embrace the new idea: *replacing* ⊓ *by* ${}_p\oplus$, they lost demonic choice; without demonic choice, they lost abstraction and refinement; and without those, they had no nontrivial path from specification to implementation, and no development calculus or method.

---

²Admittedly this is a rather clumsy notation when compared with those designed especially for specification, *e.g.*

$r: [r^2 = s]$	a specification statement (Back, Morgan, Morris)
$(r')^2 = s$	(the body of) a *Z* schema (Abrial, *Oxford*)
$\overrightarrow{r}^2 = \overleftarrow{s}$	*VDM* (Bjørner, Jones)
any r' **with** $(r')^2 = s$ **then** $r := r'$ **end**	a generalised substitution (Abrial)

But the point is that the specification could be written in a "programming language" at all: it was beginning to be realised that there was no reason to distinguish the meanings of specifications and of programs (a point finally crystallised in the subtitle *Assigning Programs to Meanings* of Abrial's book [Abr96a], itself a reference 30 years further back to Floyd's paper [Flo67] where it all began).

To have a probabilistic *development method*, we need both $\sqcap$ and $_p\oplus$ — we cannot abandon one for the other. Using them together, we can for example describe "flip a nearly fair coin" as

$$c:= \mathsf{heads}\,_{0.49}\oplus\mathsf{tails} \quad \sqcap \quad c:= \mathsf{heads}\,_{0.51}\oplus\mathsf{tails} \; .$$

What we are doing here is specifying a coin which is within 1% of being fair — just as well, since perfect $_{0.5}\oplus$ coins do not exist in nature, and so we could never implement a specification that required one.[3] This program abstracts, slightly, from the precise probability of heads or tails.

In this introduction we will see how the seminal ideas of Floyd, Hoare, Dijkstra, Abrial and others can be brought together and replayed in the probabilistic context suggested by Kozen, and how the milestones of sequential program development and refinement — the concepts of

- program assertions;

- loop invariants;

- loop variants;

- program algebra (*e.g.* monotonicity and conjunctivity)

— can be generalised to include probability. Our simple programming language will be Dijkstra's, but with $_p\oplus$ added and — crucially — demonic choice $\sqcap$ retained: we call it *pGCL*.

Section 1.2 gives a brief overview of *pGCL* and its use of so-called *expectations* rather than predicates in its accompanying logic; Section 1.3 then supplies operational intuition by relating *pGCL* operationally to a form of gambling game. (The rigorous operational semantics is given in Chap. 5, and a deeper connection with games is given in Chap. 11.) Section 1.4 completes the background by reviewing elementary probability theory.

Section 1.5 gives the precise syntax and expectation-transformer semantics of *pGCL*, using the infamous "Monty Hall" game as an example. Finally, in Sec. 1.6 we make our first acquaintance with the algebraic properties of *pGCL* programs.

Throughout we write $f.x$ instead of $f(x)$ for *function application* of f to argument x, with left association so that $f.g.x$ is $(f(g))(x)$; and we use ": =" for *is defined to be*. For *syntactic substitution* we write $expr\,\langle var \mapsto term\rangle$

---

[3]That means that probabilistic formalisms without abstraction in their specifications must introduce probability into their refinement operator if they are to be of any practical use: writing for example *prog* $\sqsubseteq_{0.99}$ *prog'* can be given a sensible meaning even if the probability in *prog* is exact [DGJP02, vBMOW03, Yin03]. But we do not follow that path here.

to indicate replacing *var* by *term* in *expr*. We use "overbar" to indicate *complement* both for Booleans and probabilities: thus $\overline{\mathsf{true}}$ is false, and $\bar{p}$ is $1 - p$.

1.2   The programming language *pGCL*

We'll use *square brackets* [·] to convert Boolean-valued predicates to arithmetic formulae which, for reasons explained below, we call *expectations*. Stipulating that [false] is zero and [true] is one makes [*P*] in a trivial sense the probability that a given predicate *P* holds: if false, it holds with probability zero; if true, it holds with probability one.[4]

For our first example, consider the simple program

$$x := -y \quad {}_{\frac{1}{3}}\oplus \quad x := +y \tag{1.5}$$

over integer variables $x, y\colon \mathbb{Z}$, using the new construct ${}_{\frac{1}{3}}\oplus$ which we interpret as "choose the left branch $x := -y$ with probability $1/3$, and choose the right branch with probability $1 - 1/3$."

Recall [Dij76] that for any predicate *post* over *final* states, and a standard command *prog*,[5] the "weakest precondition" predicate *wp.prog.post* acts over *initial* states: it holds just in those initial states from which *prog* is guaranteed to reach *post*. Now suppose *prog* is probabilistic, as Program (1.5) is; what can we say about the *probability* that *wp.prog.post* holds in some initial state?

It turns out that the answer is just *wp.prog.[post]*, once we generalise *wp.prog* to expectations instead of predicates. For that, we begin with the two definitions [6]

$$wp.(x := E).postE \quad := \quad \text{``}postE \text{ with } x \text{ replaced} \tag{1.6}$$
$$\text{everywhere by } E\text{''} \,{}^{7}$$

$$wp.(prog\,{}_{p}\oplus prog').postE \quad := \quad p * wp.prog.postE \tag{1.7}$$
$$+\ \bar{p} * wp.prog'.postE \,,$$

in which *postE* is an expectation, and for our example program we ask *what is the probability that the predicate "the final state will satisfy $x \geq 0$" holds in some given initial state of the program (1.5)?*

To find out, we calculate *wp.prog.[post]* using the definitions above; that is

---

[4]Note that this nicely complements our "overbar" convention, because for any predicate *P* the two expressions $\overline{[P]}$ and $[\overline{P}]$ are therefore the same.

[5]Throughout we use STANDARD to mean "non-probabilistic."

[6]Here we are defining the language as we go along; but all the definitions are collected together in Fig. 1.5.3 (p. 26).

[7]In the usual way, we take account of free and bound variables, and if necessary rename to avoid variable capture.

$$wp.(x:= -y \, {\textstyle\frac{1}{3}}\oplus x:= +y).[x \geq 0]$$

$\equiv^8$
$$(1/3) * wp.(x:= -y).[x \geq 0] \quad\quad\quad\quad \text{using (1.7)}$$
$$+ \quad (2/3) * wp.(x:= +y).[x \geq 0]$$

$\equiv$
$$(1/3)\,[-y \geq 0] + (2/3)\,[+y \geq 0] \quad\quad\quad\quad \text{using (1.6)}$$
$\equiv$
$$[y < 0]\,/3 \; + \; [y = 0] \; + \; 2\,[y > 0]\,/3\,. \quad\quad \text{using arithmetic}$$

Thus our answer is the last arithmetic formula above, which we call a "pre-expectation" — and the probability we seek is found by reading off the formula's value for various initial values of y, getting

when $y < 0$,	$1/3 + 0 + 2(0)/3$	$=$	$1/3$
when $y = 0$,	$0/3 + 1 + 2(0)/3$	$=$	1
when $y > 0$,	$0/3 + 0 + 2(1)/3$	$=$	$2/3$.

Those results indeed correspond with our operational intuition about the effect of ${\textstyle\frac{1}{3}}\oplus$.

For our second example we illustrate abstraction from probabilities: a demonic version of Program (1.5) is much more realistic in that we set its probabilistic parameters only within some tolerance. We say informally (but still precisely) that

- $x:= -y$ is to be executed with probability *at least* 1/3,

- $x:= +y$ is to be executed with probability *at least* 1/4 and   (1.8)

- it is certain that one or the other will be executed.

Equivalently we could say that alternative $x:= -y$ is executed with probability between 1/3 and 3/4, and that otherwise $x:= +y$ is executed (therefore with probability between 1/4 and 2/3).

With demonic choice we can write Specification (1.8) as

$$x:= -y \, {\textstyle\frac{1}{3}}\oplus x:= +y \quad \sqcap \quad x:= -y \, {\textstyle\frac{3}{4}}\oplus x:= +y\,, \quad\quad (1.9)$$

because we do not know or care whether the left or right alternative of $\sqcap$ is taken — and it may even vary from run to run of the program, resulting in an "effective" $_p\oplus$ with p somewhere between the two extremes.[9]

---

[8] Later we explain the use of "$\equiv$" rather than "$=$".

[9] We will see later that a convenient notation for (1.9) uses the abbreviation

$$prog \;{}_p\oplus_q\; prog' \quad := \quad prog \,{}_p\oplus\, prog' \;\sqcap\; prog' \,{}_q\oplus\, prog\,;$$

we would then write it $x:= -y \; {}_{\frac{1}{3}}\oplus_{\frac{1}{4}} \; x:= +y$, or even $x:= \; -y \, {}_{\frac{1}{3}}\oplus_{\frac{1}{4}} +y$.

To treat Program (1.9) we need a third definition,

$$wp.(prog \sqcap prog').postE \quad := \quad wp.prog.postE \text{ min } wp.prog'.postE \,, \quad (1.10)$$

using min because we regard demonic behaviour as attempting to make the achieving of *post* as <u>im</u>probable as it can. Repeating our earlier calculation (but more briefly) gives this time

$$wp.(\text{ Program } (1.9) \text{ }).[x \geq 0]$$

$$\equiv \qquad \begin{array}{l} [y \leq 0]/3 + 2[y \geq 0]/3 \qquad\qquad \text{using (1.6), (1.7), (1.10)} \\ \text{min} \quad 3[y \leq 0]/4 + [y \geq 0]/4 \end{array}$$

$$\equiv \qquad [y < 0]/3 \; + \; [y = 0] \; + \; [y > 0]/4 \,. \qquad\qquad \text{using arithmetic}$$

Our interpretation has become

- When y is initially negative, a demon chooses the left branch of $\sqcap$ because that branch is more likely (2/3 *vs.* 1/4) to execute $x := +y$ — the best we can say then is that $x \geq 0$ will hold with probability at least 1/3.

- When y is initially zero, a demon cannot avoid $x \geq 0$ — either way the probability of $x \geq 0$ finally is one.

- When y is initially positive, a demon chooses the right branch because that branch is more likely to execute $x := -y$ — the best we can say then is that $x \geq 0$ finally with probability at least 1/4.

The same interpretation holds if we regard $\sqcap$ as abstraction instead of as run-time demonic choice. Suppose Program (1.9) represents some mass-produced physical device and, by examining the production method, we have determined the tolerance (1.8) we can expect from a particular factory. If we were to buy one from the warehouse, all we could conclude about its probability of establishing $x \geq 0$ is just as calculated above.

Refinement is the converse of abstraction: we have

Definition 1.2.1 PROBABILISTIC REFINEMENT   For two programs *prog*, *prog'* we say that *prog'* is a refinement of *prog*, written *prog* $\sqsubseteq$ *prog'*, whenever for all post-expectations *postE* we have

$$wp.prog.postE \quad \Rrightarrow \quad wp.prog'.postE \qquad\qquad (1.11)$$

We use the symbol $\Rrightarrow$ for $\leq$ (extended pointwise) between expectations, which emphasises the similarity between probabilistic- and standard refinement.[10]   □

---

[10]We are aware that "$\Rrightarrow$" looks more like "$\geq$" than it does "$\leq$"; but for us its resemblance to "$\Rightarrow$" is the important thing.   . . .

From (1.11) we see that in the special case when expectation *postE* is an embedded predicate [*post*], the meaning of $\Rrightarrow$ ensures that a refinement *prog'* of *prog* is at least as likely to establish *post* as *prog* is.[11] That accords with the usual definition of refinement for standard programs — for then we know *wp.prog.*[*post*] is either zero or one, and whenever *prog* is certain to establish *post* (whenever *wp.prog.*[*post*] $\equiv$ 1) we know that *prog'* also is certain to do so (because then $1 \Rrightarrow wp.prog'.$[*post*]).

For our third example we prove a refinement: consider the program

$$x := -y \quad {}_{\frac{1}{2}}\oplus \quad x := +y \; , \qquad\qquad (1.12)$$

which clearly satisfies Specification (1.8); thus it should refine Program (1.9), which is just that specification written in *pGCL*. With Definition (1.11), we find for any *postE* that

      $wp.($ Program (1.12) $).postE$

$\equiv$ $wp.(x := -y).postE/2$ definition ${}_p\oplus$, at (1.7)
        + $wp.(x := +y).postE/2$

$\equiv$ $postE^-/2 \quad + \quad postE^+/2$ introduce abbreviations

$\equiv$ $(3/5)(postE^-/3 + 2postE^+/3)$ arithmetic
        + $(2/5)(3postE^-/4 + postE^+/4)$

$\Lleftarrow$ $postE^-/3 + 2postE^+/3$ any linear combination exceeds min
    min $3postE^-/4 + postE^+/4$

$\equiv$ $wp.($ Program (1.9) $).postE$.

The refinement relation (1.11) is indeed established for the two programs.

The introduction of 3/5 and 2/5 in the third step can be understood by noting that demonic choice $\sqcap$ can be implemented by any probabilistic choice whatever: in this case we used ${}_{\frac{3}{5}}\oplus$. Thus a proof of refinement using program algebra might read

      Program (1.12)
$=$ $x := -y \quad {}_{\frac{1}{2}}\oplus \quad x := +y$

---

[10]Similar conflicts of interest arise when logicians use "$\supset$" for *implies* although, interpreted set-theoretically, *implies* is in fact "$\subseteq$". And then there is "$\sqsubseteq$" for refinement, which corresponds to "$\supseteq$" of behaviours.

[11]We see later in this chapter, however, and in Sec. A.1, that it is not sound to consider only post-expectations *postE* of the form [*post*] in Def. 1.2.1: it is necessary for refinement, *but not sufficient*, that *prog'* be at least as likely to establish any postcondition *post* as *prog* is.

$$= \qquad (x:=-y \quad {}_{\frac{1}{3}}\oplus \quad x:=+y)$$
$$\quad {}_{\frac{3}{5}}\oplus \quad (x:=-y \quad {}_{\frac{3}{4}}\oplus \quad x:=+y)$$

$$\sqsupseteq \qquad x:=-y \quad {}_{\frac{1}{3}}\oplus \quad x:=+y \qquad\qquad (\sqcap) \sqsubseteq ({}_p\oplus) \text{ for any } p \text{ }^{12}$$
$$\sqcap \qquad x:=-y \quad {}_{\frac{3}{4}}\oplus \quad x:=+y$$

$$= \qquad \text{Program } (1.9) \;.$$

1.3 An informal computational model: $pGCL$ describes gambling

We now use a simple card-and-dice game as an informal introduction to the computational model for $pGCL$, to support the intuition for probabilistic choice, demonic choice and their interaction. To start with, we consider the simplest case: non-looping programs without $\sqcap$ or ${}_p\oplus$.

1.3.1 The standard game

Imagine we have a board of numbered squares, and a selection of numbered cards laid on it with at most one card per square; winning squares are indicated by coloured markers. The squares are the program states; the program is the pattern of numbered cards; the coloured markers indicate the postcondition.

To play the game

> An initial square is chosen (according to certain rules which do not concern us); subsequently
>
> - if the square contains a card the card is removed, and play continues from the square whose number appeared on the card, and
> - if the square does not contain a card, the game is over.
>
> When the game is over the player has won if his final square contains a marker — otherwise he has lost.

This simple game is deterministic: any initial state always leads to the same final state. And because the cards are removed after use it is also guaranteed to terminate, if the board is finite. It is easily generalised however to include other features of standard programs:

---

[12] By $(\sqcap) \sqsubseteq ({}_p\oplus)$ we mean that for all *prog, prog'* we have

$$prog \sqcap prog' \quad \sqsubseteq \quad prog \text{ }_p\oplus prog',$$

which is an instance of our Law 7 given on p. 323, in Sec. B.1 on program algebra.

looping If the cards are *not* removed after use, the game can "loop." A looping-forever player loses.

aborting If a card reads *go to jail*, the program is said to "abort" and the player can be sent to any square whatever, including a special supplementary "jail" square from which there is no escape. A jailed player loses.

demonic nondeterminism If each square can contain several cards, face-down, and the rules are modified so that the next state is determined by choosing just one of them "blind," then play is nondeterministic. Taking the demonic (pessimistic) view, the player should expect to lose unless he is guaranteed to reach a winning position no matter which blind choices he makes.

In the standard game, for each (initial) square one can examine the cards before playing to determine whether a win is guaranteed from there. But once the game has started, the cards are turned face-down.

The set of squares from which a win is guaranteed is the weakest precondition.[13]

1.3.2 The probabilistic game

Suppose now that each card contains not just one but, rather, a list of successor squares, and the choice from the list is made by rolling a die. In this deterministic game,[14] play becomes a succession of die rolls, taking the player from square to square; termination (no card) and winning (marker) are defined as before.

When squares can contain several cards face down, each with a separate list of successors to be resolved by die roll, we are dealing with probability and demonic nondeterminism together: first the card is chosen "blind" (*i.e.* demonically); the card is turned over and a die roll (probability) determines which of its listed alternatives to take.

In the probabilistic game one can ask for the *greatest guaranteed probability* of winning; as in the standard case, the prediction will vary depending on the initial square. (It's because of demonic nondeterminism, as illustrated below, that the probability might be only a lower bound.)

---

[13] A glance at Fig. 6.7.1 (p. 173) will show where we are headed in the visualisation of probabilistic preconditions!

[14] Note that we still call this game "deterministic," in spite of the probabilistic choices, and there are good mathematical reasons for doing so. (In Chap. 5, for example, we see that such programs are maximal in the refinement order.) An informal justification is that deterministic programs are those with repeatable behaviours and, even for probabilistic programs, the output *distribution* is repeatable (to within statistical confidence measures) provided the program contains no demonic choice; see *e.g.* p. 135.

In Fig. 1.3.1 is an example game illustrating some of the above points. The greatest guaranteed probability of winning from initial state 0 is only $1/2$, in spite of the fact that the player can win every time if he is lucky enough to choose the first card in the pile; but he might be unlucky enough never to choose the first card, and we must assume the worst.

1.3.3 Expected winnings in the probabilistic game

For standard programs, the computational model of execution supports a complementary, "logical" view — given a set of final states (the postcondition) we can examine the program to determine the largest set of initial states (the weakest precondition) from which execution of the program is guaranteed to reach the designated final states. The sets of states are *predicates*, and the program is being regarded as a predicate *transformer*.

Regarding sets of states as characteristic functions (from the state space into $\{0,1\}$), we generalise to "probabilistic predicates" by extending the range of those functions to all of $\mathbb{R}_{\geq}$, the non-negative reals.[15]

Probabilistic programs become functions from probabilistic postconditions to probabilistic weakest preconditions — we call them *post-expectations* and *greatest pre-expectations*. The corresponding generalisation in the game is as follows.

Rather than placing winning markers on the board, we place *money* — rather than strictly winning or losing, the player simply keeps whatever money he finds in his final square. In Fig. 1.3.2 we show the effect of translating our original game. In fact, not much changes: the *probability* of winning (in Fig. 1.3.1) translates into the equivalent *expected payoff* (Fig. 1.3.2) as the corresponding fraction of £1, illustrating this important fact:

> The expected value of a characteristic function over a distribution is the same as the probability assigned to the set that function describes.

Thus using expectations is at least as general as using probabilities explicitly, since we can always restrict ourselves to $\{0,1\}$-valued functions from which probabilities are then recovered.

For probabilistic programs, the operational interpretation of execution thus supports a "logical" view also — given a function from final states to $\mathbb{R}_{\geq}$ (the post-expectation) one can examine the program beforehand to determine for each initial state the minimum expected (or "average") win when the game is played repeatedly from there (the greatest pre-expectation) — also therefore a function from states to $\mathbb{R}_{\geq}$.

---

[15] In later chapters we will be more precise about the range of expectations, requiring them in particular to be *bounded above*.

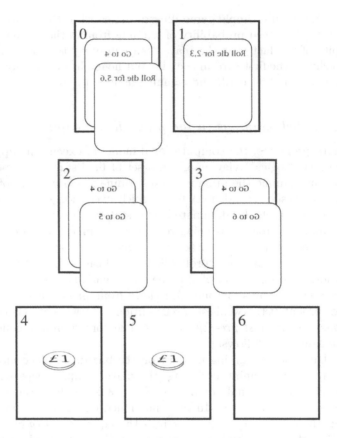

To play from a square, you first pick one of the face-down cards. (In the diagram, we are seeing what's on the cards with our *x-ray* vision.) Then you roll a die to choose one of the alternatives on the card. (In this case the die is two-sided, *i.e.* it is a coin.)

As special cases, a *standard* step (non-probabilistic) has only one alternative per card, but possibly many cards; and a *deterministic* step has only one card, but possibly many alternatives on it. A standard and deterministic step has one card, and only one alternative.

The winning final positions — the postcondition — are the states $\{4, 5\}$, marked with a £1 coin. From initial state 2 a win is guaranteed; from state 0 or 1 the minimum guaranteed probability of winning is 1/2; from state 3 the minimum probability is zero, since the second card might be chosen every time.

The probabilities are summarised in Fig. 1.3.2.

Figure 1.3.1. CARD-AND-DICE GAME OPERATIONAL SEMANTICS FOR *pGCL*

The *post-expectation*:

Final state	0	1	2	3	4	5	6
Payoff awarded if this state reached	0	0	0	0	£1	£1	0

The probability of winning (ending on a £1) (from Fig. 1.3.1):

Initial state	0	1	2	3	4	5	6
Greatest guaranteed probability of winning	1/2	1/2	1	0	1	1	0

The *greatest pre-expectation*:

Initial state	0	1	2	3	4	5	6
Greatest guaranteed expected payoff	50p	50p	£1	0	£1	£1	0

Figure 1.3.2. A PROBABILISTIC AND NONDETERMINISTIC GAMBLING GAME

Since the functions are *expectations*, the program is being regarded as an *expectation transformer*.[16]

We are not limited to £1 coins for indicating postconditions — that is only an artefact of embedding standard postconditions into the probabilistic world. In general any amount of money can be placed in a square, and that is the key to allowing a smooth sequential composition of programs at the logical level — for if the program *game* of Fig. 1.3.2 were executed after some other program *prog*, the precondition of the two together with respect to the postcondition {4, 5} would be calculated by applying *wp.prog* to the *greatest pre-expectation* table for *game*. That is because sequential composition of programs becomes, as usual, functional composition of the corresponding transformers: we have

$$wp.(prog;\, game).\{4,5\} \quad := \quad wp.prog.(\overbrace{wp.game.\{4,5\}}^{\text{expected win table}}),$$

and that table contains non-integer values (for example 50p).

Another reason for allowing arbitrary values in $\mathbb{R}_\geq$ is that using only standard postconditions ({0, 1}-valued) — equivalently, using explicit probabilities (recall the important fact above) — is not discriminating enough when nondeterminism is present: certain programs are identified that should be distinguished, and the semantics becomes *non-compositional*. (See Sec. A.1 for why this happens.)

---

[16]For deterministic (yet probabilistic) programs, the card-game model and the associated transformers are essentially Kozen's original construction [Koz81, Koz85]. We have added demonic (and later angelic) nondeterminism.

1.4   Behind the scenes: elementary probability theory

In probability theory, an *event* is a subset of some given *sample space* S, so that the event is said to have occurred if the sampled value is in that set; a *probability distribution* Pr over the sample space is a function from its events into the closed interval $[0, 1]$, giving for each event the probability of its occurrence. In the general case, for technical reasons, not necessarily all subsets of the sample space are events.[17]

In our case we consider countable sample spaces, and take every (sub-)set of S to be an event — and so we can regard a probability distribution more simply as a function from S directly to probabilities (rather than from its subsets). Thus $\Pr\colon S \to [0, 1]$, and the probability of a more general event is now just the sum of the probabilities of its elements: we are using *discrete* distributions.[18]

A *random variable* X is a function from the sample space to the non-negative reals;[19] and the *expected value* $\mathrm{Exp}.X$ of that random variable is defined in terms of the (discrete) probability distribution Pr; we have the summation

$$\mathrm{Exp}.X \quad := \quad \left(\sum_{s \in S} \Pr.s * X.s \right) . \quad ^{20} \tag{1.13}$$

It represents the "average" value of $X.s$ over many repeated samplings of s according to the distribution $\Pr$.[21]

In fact expected values can also be characterised without referring directly to an underlying probability distribution:

If a function Exp is of type $(S \to \mathbb{R}_{\geq}) \to \mathbb{R}_{\geq}$, and it is

non-negative   so that $\mathrm{Exp}.X \geq 0$ for all $X\colon S \to \mathbb{R}_{\geq}$,
linear   so that for $X, Y\colon S \to \mathbb{R}_{\geq}$ and $c, d\colon \mathbb{R}_{\geq}$ we have

$$\mathrm{Exp}.(c * X + d * Y) \quad = \quad c * \mathrm{Exp}.X + d * \mathrm{Exp}.Y$$

---

[17]This may occur if the sample space is uncountable, for example; the general technique for such cases involves σ-algebras [GS92]. See Footnote 7 on p. 297 for an example.

[18]The price paid for using discrete distributions is that there are some "everyday" situations we cannot describe, such as the uniform "continuous" distribution over the real interval $[0, 1]$ that might be the result of the program "choose a real number x randomly so that $0 \leq x \leq 1$." We get away with it because no such program can be written in *pGCL* — at least, not at this stage.

[19]Footnote 12 on p. 134 gives a more generous definition.

[20]Although the parentheses may look odd around $\sum$ — we write $(\sum \cdots)$ rather than $\sum(\cdots)$ — we always indicate the scope of bound variables (like s) with explicit delimiters, since it helps to avoid errors when doing calculations.

[21]Our "important fact" (p. 13) is now stated "if X is the characteristic function of some event P, then $\mathrm{Exp}.X$ is the probability that event P will occur."

and **normalised**   so that it satisfies Exp.$\underline{1} = 1$, where $\underline{1}$ is the
constant function returning 1 for all arguments in S,

then it is an expectation over some probability distribution: it
can be shown that it is expressible uniquely in the form (1.13)
for some Pr.[22]

The relevance of the above is that our real-valued expressions over the
state — what we are calling "expectations" — are random variables, and
that the expression

$$wp.prog.postE \,, \qquad\qquad (1.14)$$

as a function of *initial* values for the state variables, is a random variable
as well. As a function of state variables, it is the expected value of the
random variable *postE* (also a function of state variables, but those taken
after execution) over the distribution of final states produced by executions
of *prog*, and so

$$preE \quad \Rrightarrow \quad wp.prog.postE \qquad\qquad (1.15)$$

says that *preE* gives in any initial state a lower bound for the expected
value of *postE* in the final distribution reached via execution of *prog* begun
in that initial state.

In general, we call random variables *post-expectations* when they are to
be evaluated in a final state, and we call them *pre-expectations* when they
are calculated as at (1.14). And, like pre- and postconditions in standard
programs, if placed "between" two programs a single random variable is a
post-expectation for the first and a pre-expectation for the second.

But how do *prog* and an initial state determine a distribution? In fact the
underlying distributions are found on the cards of the game from Sec. 1.3
— the sample space is the set of squares, and each card gives an explicit
distribution over that space. If we consider the deterministic game, and
regard "make *one* move in the game" as a program in its own right, then
we have a function from initial state to final distribution — the function
taking a square to the card that square contains.[23] For any postcondition
postE written, say, as an expression over names N of squares, and initial
square N_0, the expression $wp.move.postE \, \langle N \mapsto N_0 \rangle$ is the expectation of
postE over the distribution of square names given on the card found at N_0.

---

[22]It is a special case of the *Riesz Representation Theorem* which states, loosely speaking, that knowledge of the expectation (assumed to be given directly) of every random variable uniquely determines an underlying probability distribution. See for instance Feller [Fel71, p. 135].

[23]For nondeterministic programs we are thus considering a function from state to *sets* of distributions, from a square to the set of cards there; again we see the general computational model underlying the expectation-transformer semantics.

For example, in Figs. 1.3.1 and 1.3.2 we see the above features: program *move* is given by the layout of the cards (Fig. 1.3.1); and the resulting pre- and post-expectations are tabulated in Fig. 1.3.2. All three tables there are random variables over the state space $\{0, \cdots, 6\}$.

When we move to more general programs, we must relax the conditions that characterise expectations. If *prog* is possibly nonterminating — if it is recursive or contains **abort** — then *wp.prog.postE* may violate the normalisation condition Exp.$\underline{1} = 1$. However as a function which satisfies the first two conditions it can still be regarded as an expectation in a weak sense. That was shown by Kozen [Koz81] and later Jones [Jon90], who defined expectations with regard to "probability distributions" which may sum to less than one. Those are in fact a special case of Jones's *evaluations*,[24] and she gave conditions similar to the above for their existence [Jon90, p. 117].

Finally, if program *prog* is not deterministic then we move further away from elementary theory, because *wp.prog.postE* is no longer an expectation even in the weak sense: it not linear. It is still however the minimum of a set of expectations: if *prog* and *prog'* are deterministic programs then *wp.(prog ⊓ prog').postE* is the pointwise minimum of the two expectations *wp.prog.postE* and *wp.prog'.postE*. This definition is one of the main features of this approach.

Thus although linearity is lost, it is not gone altogether: we retain so-called *sub*-linearity,[25] which implies that for any $c_1, c_2 \colon \mathbb{R}_{\geq}$ and any program *prog* we still have

$$wp.prog.(c_1 * postE_1 + c_2 * postE_2)$$
$$\Lleftarrow \quad c_1 * wp.prog.postE_1 \quad + \quad c_2 * wp.prog.postE_2 \ .$$

And clearly non-negativity continues to hold.

The characterisations of expectations given above for the simpler cases might suggest that non-negative and sublinear functionals uniquely determine a *set* of probability distributions — and, in Chap. 5, that is indeed shown to be the case: sublinearity is the key "healthiness condition" for expectation transformers.[26]

1.5   Basic syntax and semantics of *pGCL*

1.5.1   Syntax

Let *prog* range over programs and p over real number expressions taking values between zero and one inclusive; assume that x stands for a list of distinct variables, and *expr* for a list of expressions (of the same length as x

---

[24] She was working in a much more general context.

[25] The actual property is slightly more general than we give here; see Sec. 1.6.

[26] Halpern and Pucella [HP02] have recently studied similar properties.

where appropriate); and let the program *scheme C* be a program in which program *names* like xxx can appear. The syntax of $pGCL$ is as follows:

$$prog \;:=\; \textbf{abort} \mid \textbf{skip} \mid x\!:=\, E \mid prog;\, prog \mid$$
$$prog\,_p{\oplus}\, prog \mid prog \sqcap prog \mid \qquad (1.16)$$
$$(\textbf{mu}\ xxx \cdot C)$$

The first four constructs, namely **abort**, **skip**, assignment and sequential composition, are just the conventional ones [Dij76].

The remaining constructs are for probabilistic choice, nondeterministic choice and recursion: given p in the closed interval $[0,1]$ we write $prog\,_p{\oplus}$ $prog'$ for the probabilistic choice between programs $prog$ and $prog'$; they have probability p and $1-p$ respectively of being selected. In many cases p will be a constant, but in general it can be an expression over the state variables.

1.5.2 Shortcuts and "syntactic sugar"

For convenience we extend our logic and language with the following notations.

Boolean embedding — For predicate *pred* we write $[pred]$ for the expectation "1 **if** *pred* **else** 0".[27]

Conditional — The conditional

$$prog\ \textbf{if}\ pred\ \textbf{else}\ prog'$$

or   **if** *pred* **then** *prog* **else** *prog'* **fi** ,

chooses program *prog* (resp. *prog'*) if Boolean *pred* is true (resp. false). It is defined $prog\ _{[pred]}{\oplus}\ prog'$.

If **else** is omitted then **else skip** is assumed. (See also the "hybrid" conditional of Sec. 3.1.2.)

Implication-like relations — For expectations exp, exp' we write

$$exp \Rightarrow exp' \quad \text{for} \quad exp \text{ is everywhere less than or equal to } exp'$$
$$exp \equiv exp' \quad \text{for} \quad exp \text{ and } exp' \text{ are everywhere equal}$$
$$exp \Leftarrow exp' \quad \text{for} \quad exp \text{ is everywhere greater than or equal to } exp'$$

We distinguish $exp \Rightarrow exp'$ from $exp \leq exp'$ — the former is a statement *about* exp and exp', thus true or false as a whole; the latter is itself a Boolean-valued expression over the state, possibly true in some states and false in others.[28] Similarly we regard $exp = exp'$ as

---

[27] We will not distinguish predicates from Boolean-valued expressions.

[28] Note that $exp \Rightarrow exp'$ is different again, in fact badly typed if exp and exp' are expectations: one real-valued function cannot "imply" another.

true in just those states where *exp* and *exp'* are equal, and false in the rest.

The closest standard equivalent of $\Rightarrow$ is the entailment relation $\models$ between predicates[29] — and in fact *post* $\models$ *post'* exactly when $[post] \Rightarrow [post']$, meaning that the "embedding" of $\models$ is $\Rightarrow$.

Multi-way probabilistic choices — A probabilistic choice over N alternatives can be written horizontally

$$(prog_1 @ p_1 \mid \cdots \mid prog_N @ p_N)$$

or vertically

$$\begin{array}{ll} prog_1 & @ p_1 \\ prog_2 & @ p_2 \\ \quad \vdots \\ prog_N & @ p_N \end{array}$$

in which the probabilities are enumerated and sum to no more than one.[30] We can also write a "probabilistic comprehension" $(\lVert i \colon I \bullet prog_i @ p_i)$ over some countable index set I. In general, we have

$$\begin{array}{ll} & wp.(prog_1 @ p_1 \mid \cdots \mid prog_N @ p_N).postE \\ := & p_1 * wp.prog_1.postE + \cdots + p_N * wp.prog_N.postE \,. \end{array}$$

It means "execute $prog_1$ with probability at least p_1, and $prog_2$ with probability at least $p_2 \ldots$"[31]

If the probabilities sum to 1 exactly, then it is a simple N-way probabilistic branch; if there is a deficit $1 - \Sigma_i p_i$, it gives the probability of aborting.

When all the programs $prog_i$ are assignments with the same left-hand side, say $x := expr_i$, we write even more briefly

$$x := (expr_1 @ p_1 \mid \cdots \mid expr_N @ p_N) \,.$$

Variations on $_p\oplus$ — By $prog \oplus_p prog'$ we mean $prog' \, _p\oplus \, prog$, and in general we write $prog \, _p\oplus_{p'} \, prog'$ for

$$\begin{array}{ll} prog & @ \quad p \\ prog' & @ \quad p' \\ prog \sqcap prog' & @ \quad 1 - (p+p') \,, \end{array}$$

the program that executes *prog* with probability at least p and *prog'*

---

[29]One predicate ENTAILS another, written $\models$, just when it implies the other in all states.

[30]See Sec. 4.3 for an example of the vertical notation.

[31]It is "at least p_i" because if the probabilities sum to less than one there will be an "aborting" component, which might behave like $prog_i$.

with probability at least p'; we assume $p + p' \leq 1$.

By $_{\geq p}\oplus$ we mean $_p\oplus_0$, and so on. (See also (B.3) on p. 328.)

Demonic choice — We write demonic choice between assignments to the same variable x as

$$x{:}\in \{expr_1, expr_2, \cdots\} \,,$$
$$\text{or} \quad x{:}= expr_1 \sqcap expr_2 \sqcap \cdots \,, \tag{1.17}$$

in each case abbreviating $x{:}= expr_1 \sqcap x{:}= expr_2 \sqcap \cdots$. More generally we can write $x{:}\in expr$ or $x{:}\notin expr$ if $expr$ is set-valued, provided the implied choice is finite.[32]

Iteration — The construct (**mu** $xxx \cdot C$) behaves as prescribed by the program context C except that it invokes itself recursively whenever it reaches a point where the program name xxx appears in C. Then, in the usual way, iteration is a special case of recursion:

$$\begin{aligned} &\textbf{do } pred \rightarrow body \textbf{ od} \\ := \quad &(\textbf{mu } xxx \cdot (body\,;xxx) \textbf{ if } pred \textbf{ else skip}) \,. \quad {}^{33} \end{aligned} \tag{1.18}$$

1.5.3 Example of syntax: the "Monty Hall" game

We illustrate the syntax of our language with the example program of Fig. 1.5.1. There are three curtains, labelled A, B and C, and a prize is hidden nondeterministically behind one of them, say pc. A contestant hopes to win the prize by guessing where it is hidden: he chooses randomly to

---

[32]None of our examples requires a choice from the empty set. We see later that the finiteness requirement is so that our programs will be continuous (Footnote 60 on p. 71); and in some cases — for example, the third and fourth statements of the program shown in Fig. 1.5.1 — we rely on type information for that finiteness.

[33]An equivalent but simpler formulation is given by the least fixed-point definition

$$wp.(\textbf{do } pred \rightarrow body \textbf{ od}).R \quad := \quad (\mu Q \cdot wp.body.Q \textbf{ if } pred \textbf{ else } R) \,, \tag{1.19}$$

which matches Dijkstra's original formulation more closely [Dij76]. But there is some technical work required to get between the two, as we explain later at (7.12). The expression on the right can be read

the *least* pre-expectation Q such that

$$Q \quad \equiv \quad wp.body.Q \textbf{ if } pred \textbf{ else } R \,,$$

and is called a FIXED POINT because placing Q in the expression does not alter its value — this is the mathematical equivalent of "and the same again" when the loop returns to its starting point for potentially more iterations.

The "least," for us, means the *lowest* expectation — that reflects the view, appropriate for elementary sequential programming, that unending iteration should have little worth (in fact, zero). For standard programming, the order is false $\leq$ true so that taking the least fixed-point means adopting the view that an infinite loop does not establish any postcondition (*i.e.*, has precondition false).

A more discriminating treatment of unending computations is given in Part III.

$pc{:}\in \{A,B,C\}$;	Prize hidden behind curtain.
$cc{:}= (A \mathbin{@} \frac{1}{3} \mid B \mathbin{@} \frac{1}{3} \mid C \mathbin{@} \frac{1}{3})$;	Contestant chooses randomly.
$ac{:}\notin \{pc,cc\}$;	Another curtain opened; it's empty.
$(cc{:}\notin \{cc,ac\})$ **if** *clever* **else skip**	Changes his mind — or not?

The three "curtain" variables ac, cc, pc are of type $\{A,B,C\}$.
Written in full, the first three statements would be

$$pc{:}= A \sqcap pc{:}= B \sqcap pc{:}= C;$$
$$cc{:}= A \underset{\frac{1}{3}}{\oplus} (cc{:}= B \underset{\frac{1}{2}}{\oplus} cc{:}= C);$$
$$ac{:}\in \{A,B,C\} - \{pc,cc\}\,.$$

The fourth statement is written using $\notin$ just for convenience — in fact it executes deterministically, since cc and ac are guaranteed to be different at that point.

Figure 1.5.1. THE "MONTY HALL" PROGRAM

---

point to curtain cc. The host then tries to get the contestant to change his choice, showing that the prize is *not* behind some other curtain ac — which means that either the contestant has chosen it already or it is behind the other closed curtain.

Should the contestant change his mind?

1.5.4 *Intuitive interpretation of pGCL expectations*

In its full generality, an expectation is a function describing how much each program state is "worth."

The special case of an embedded predicate [*pred*] assigns to each state a worth of zero or of one: states satisfying *pred* are worth one, and states not satisfying *pred* are worth zero. The more general expectations arise when one estimates, in the *initial* state of a probabilistic program, what the worth of its *final* state will be. That estimate, the "expected worth" of the final state, is obtained by summing over all final states

> the worth of the final state multiplied by the probability the program "will go there" from the initial state.

Naturally the "will go there" probabilities depend on "from where," and so that expected worth is a function of the initial state.

When the worth of final states is given by [*post*], the expected worth of the initial state turns out to be just the probability that the program will reach *post*. That is because expected worth of initial state

$\equiv$ (probability *prog* reaches *post*)
          *   (worth of states satisfying *post*)

    $+$ (probability *prog* does not reach *post*)
          *   (worth of states not satisfying *post*)

$\equiv$ (probability *prog* reaches *post*) $* 1$
    $+$ (probability *prog* does not reach *post*) $* 0$

$\equiv$ probability *prog* reaches *post* ;

note we have relied on the fact that all states satisfying *post* have worth one.

More general analyses of programs *prog* in practice lead to conclusions of the form

$$p \quad \equiv \quad wp.prog.[post]$$

for some p and *post* which, given the above, we can interpret in two equivalent ways:

- the expected worth of $[post]$ of the final state is at least the value of p in the initial state; or

- the probability that *prog* will establish *post* is at least p.[34]

Each interpretation is useful, and in the following example we can see them acting together: we ask for the probability that two fair coins when flipped will show the same face, and calculate

$$wp. \left(\begin{array}{l} x := H \; {}_{\frac{1}{2}}\oplus x := T \quad ; \\ y := H \; {}_{\frac{1}{2}}\oplus y := T \end{array} \right).[x = y]$$

$\equiv$ ${}_{\frac{1}{2}}\oplus$, $:=$ and sequential composition [35]
$\quad wp.(x := H \; {}_{\frac{1}{2}}\oplus x := T).([x = H]/2 + [x = T]/2)$

$\equiv$ $(1/2)([H = H]/2 + [H = T]/2)$
    $+$ $(1/2)([T = H]/2 + [T = T]/2)$ ${}_{\frac{1}{2}}\oplus$ and $:=$

---

[34]We must say "at least" in general, because possible demonic choice in *prog* means that the pre-expectation is only a lower bound for the actual expected value the program could deliver; and some analyses give only the weaker $p \Rrightarrow wp.prog.[post]$ in any case. See also Footnote 14 on p. 89.

[35]See Fig. 1.5.3 for this definition.

$\equiv$ $(1/2)(1/2 + 0/2) + (1/2)(0/2 + 1/2)$ definition $[\cdot]$

$\equiv$ $1/2$.   arithmetic

We can then use the second interpretation above to conclude that the faces are the same with probability $1/2$.[36]

But part of the above calculation involves the more general expression

$$wp.(x := H \; {\textstyle\frac{1}{2}} \oplus x := T).([x = H]/2 + [x = T]/2) , \qquad (1.20)$$

and what does that mean on its own? It must be given the first interpretation, that is as an expected worth, since "will *establish* $[x = H]/2 + [x = T]/2$" makes no sense. Thus it means

> the expected value of the expression $[x = H]/2 + [x = T]/2$ after executing the program $x := H \; {\textstyle\frac{1}{2}} \oplus x := T$,

which the calculation goes on to show is in fact $1/2$. But for our overall conclusions we do not need to think about the intermediate expressions — they are only the "glue" that holds the overall reasoning together.[37]

1.5.5 *Semantics*

The probabilistic semantics is derived from generalising the standard semantics in the way suggested in Sec. 1.3. Let the state space be S.

Definition 1.5.2 EXPECTATION SPACE   The space of expectations over S is defined

$$\mathbb{E}S \quad := \quad (S \to \mathbb{R}_{\geq}, \Rrightarrow) ,$$

where the entailment relation $\Rrightarrow$, as we have seen, is inherited pointwise from the normal $\leq$ ordering in $\mathbb{R}_{\geq}$. The expectation-transformer model for programs is

$$\mathbb{T}S \quad := \quad (\mathbb{E}S \leftarrow \mathbb{E}S, \sqsubseteq) ,$$

where we write the functional arrow backward just to emphasise that such transformers map final post-expectations to initial pre-expectations, and where the *refinement* order $\sqsubseteq$ is derived pointwise from entailment $\Rrightarrow$ on $\mathbb{E}S$.   □

---

[36](Recall Footnote 34.) If we do know, by other means say, that the program is deterministic (though still probabilistic), then we can say the pre-expectation is exact.

[37]See p. 271 for an example of this same analogy, but in the context of temporal logic.

Although both $\mathbb{E}S$ and $\mathbb{T}S$ are lattices, neither is a complete partial order,[38] because $\mathbb{R}_\geq$ itself is not. (It lacks an adjoined ∞ element.) In addition, when S is infinite (see *e.g.* Sec. 8.2 of Part II) we must impose the condition on elements of $\mathbb{E}S$ that each of them be *bounded above* by some non-negative real.[39]

In Fig. 1.5.3 we give a probabilistic semantics to the constructs of our language. It has the important feature that the standard programming constructs behave as usual, and are described just as concisely.

Note that our semantics states how *wp.prog* in each case transforms an *expression* in the program variables: that is, we give a procedure for calculating the greatest pre-expectation by purely syntactic manipulation. An alternative view is to see the post-expectations as mathematical *functions* of type $\mathbb{E}S$, and the expressions *wp.prog* are then of type $\mathbb{T}S$.

The expression-based view is more convenient in an introduction, and for the treatment of specific programs; the function-based view is more convenient (and, for recursion, necessary) for general properties of expectation transformers. In this chapter and the rest of Part I we retain the

---

[38]A PARTIAL ORDER differs from the familiar "total" orders like "$\leq$" in that two elements can be "incomparable"; the most common example is subset $\subseteq$ between sets, which satisfies REFLEXIVITY (a set is a subset of itself), ANTI-SYMMETRY (two sets cannot be subsets of each other without being the same set) and TRANSITIVITY (one set within a second within a third is a subset of the third directly as well). But it is not true that for any two sets one is necessarily a subset of the other.

A LATTICE is a non-empty partially ordered set where for all x, y in the set there is a GREATEST LOWER BOUND $x \sqcap y$ and and a LEAST UPPER BOUND $x \sqcup y$. This holds *e.g.* for the lattice of sets, as above; but the collection of *non-empty* sets is not a lattice, because $x \cap y$ (which is how $x \sqcap y$ is written for sets) is not necessarily non-empty even if x and y are.

A partial order $\sqsubseteq$ is CHAIN- or DIRECTED COMPLETE — then called a CPO — when it contains all limits of chains or directed sets respectively, where a CHAIN is a set totally ordered by $\sqsubseteq$ and a set is $\sqsubseteq$-DIRECTED if for any x, y in the set there is a z also in the set such that $x, y \sqsubseteq z$. (Since a chain is directed, directed completeness implies chain completeness; in fact with the Axiom of Choice, chain- and directed completeness are equivalent.)

All of these details can be found in standard texts [DP90].

[39]There is a difference between requiring that there be an upper bound for all expectations (we do not) and requiring that each expectation separately have an upper bound (we do).

In the first case, we would be saying that there is some M such that every expectation α in $\mathbb{E}S$ satisfied $\alpha \Rightarrow \underline{M}$. That would be convenient because it would make both $\mathbb{E}S$ and $\mathbb{T}S$ complete partial orders, trivially; and that would *e.g.* allow us to use a standard treatment of fixed points.

But we adopt the second case where, for each expectation α separately, there is some M_α such that $\alpha \Rightarrow \underline{M_\alpha}$; and, as α varies, these M_α's can increase without bound. That is why $\mathbb{E}S$ is not complete and is, therefore, why we will need a slightly special argument when dealing with fixed points.

$$
\begin{aligned}
wp.\mathbf{abort}.postE &:= 0 \\
wp.\mathbf{skip}.postE &:= postE \\
wp.(x := expr).postE &:= postE \;\langle x \mapsto expr \rangle \\
wp.(prog\,;prog').postE &:= wp.prog.(wp.prog'.postE) \\
wp.(prog \sqcap prog').postE &:= wp.prog.postE \;\; \mathsf{min} \;\; wp.prog'.postE \\
wp.(prog\,_p\oplus prog').postE &:= p * wp.prog.postE \;\; + \;\; \overline{p} * wp.prog'.postE
\end{aligned}
$$

Recall that $\overline{p}$ is the complement of p.

The expression on the right gives the *greatest pre-expectation* of *postE* with respect to each *pGCL* construct, where *postE* is an expression of type $\mathbb{E}S$ over the variables in state space S. (For historical reasons we continue to write *wp* instead of *gp*.)

In the case of recursion, however, we cannot give a purely syntactic definition. Instead we say that

$$
(\mathbf{mu}\; xxx \cdot C) \quad := \quad \text{least fixed-point of the function } cntx\colon \mathbb{T}S \to \mathbb{T}S
$$
$$
\text{defined so that } cntx.(wp.xxx) = wp.C. \;^{40}
$$

Figure 1.5.3. Probabilistic *wp*-semantics of *pGCL*

---

expression-based view as far as possible; but in Part II we use the more mathematical notation. (See for example Sec. 5.3.)

The worst program **abort** cannot be guaranteed to terminate in any proper state and therefore maps every post-expectation to 0. The immediately terminating program **skip** does not change anything, therefore the expected value of post-expectation *postE* after execution of **skip** is just its actual value before. The pre-expectation of the assignment $x := expr$ is the postcondition with the expression *expr* substituted for x. Sequential composition is functional composition. The semantics of demonic choice $\sqcap$ reflects the dual metaphors for it: as abstraction, we must take the minimum because we are giving a guarantee over all possible implementations; as a demon's behaviour, we assume he acts to make our expected winnings as small as possible.

The pre-expectation of probabilistic choice is the weighted average of the pre-expectations of its branches. Since any such average is no less than the minimum it follows immediately that probabilistic choice refines demonic

---

[40]Because $\mathbb{T}S$ is not complete, to ensure existence of the fixed point we insist that the transformer-to-transformer function *cntx* be "feasibility-preserving," *i.e.* that if applied to a feasible transformer it returns a feasible transformer again. "Feasibility" of transformers is one of the "healthiness conditions" we will encounter in Sec. 1.6. For convenience, we usually assume that *cntx* is continuous as well.

See Lem. 5.6.8 on p. 148.

choice, which corresponds to our intuition. In fact we consider probabilistic choice to be a *deterministic* programming construct; that is we say that a program is deterministic if it is free of demonic nondeterminism unless it aborts.[41]

Finally, recursive programs have least-fixed-point semantics as usual.

1.5.6 Example of semantics: Monty Hall again

We illustrate the semantics by returning to the program of Fig. 1.5.1. Consider the post-expectation $[pc = cc]$, which takes value one just in those final states in which the candidate has correctly chosen the prize. Working backwards through the program's four statements, we have first (by standard *wp* calculations) that

$$wp. \ ((cc: \notin \{cc, ac\}) \ \textbf{if} \ clever \ \textbf{else skip}) \ . [pc = cc]$$
$$\equiv \quad [clever] * [\{ac, cc, pc\} = \{A, B, C\}] \ + \ [\neg clever] * [pc = cc] \ ,$$

because (in case *clever*) the nondeterministic choice is guaranteed to pick *pc* only when it cannot avoid doing so.[42]

Standard reasoning suffices for our next step also:

$$wp. \ (ac: \notin \{pc, cc\}).$$
$$([clever] * [\{ac, cc, pc\} = \{A, B, C\}] + [\neg clever] * [pc = cc])$$

$$\equiv \quad [clever] * [pc \neq cc] + [\neg clever] * [pc = cc] \ .$$

For the *clever* case note that $\{ac, cc, pc\} = \{A, B, C\}$ holds (in the post-expectation) iff all three elements differ, and that the statement itself establishes only two of the required three inequalities — that $ac \neq pc$ and $ac \neq cc$. The weakest precondition supplies the third.

For the ¬*clever* case note that neither *pc* nor *cc* is assigned to by $ac: \notin \{pc, cc\}$, so that $pc = cc$ holds afterwards iff it held before.

The next statement is probabilistic, and so produces a probabilistic pre-expectation involving the factors 1/3 given explicitly in the program; we have

$$wp. \ (cc := (A @ \tfrac{1}{3} \mid B @ \tfrac{1}{3} \mid C @ \tfrac{1}{3})).$$
$$([clever] * [pc \neq cc] + [\neg clever] * [pc = cc])$$

$$\equiv \qquad [clever]/3 \ * \ ([pc \neq A] + [pc \neq B] + [pc \neq C])$$
$$+ \ [\neg clever]/3 \ * \ ([pc = A] + [pc = B] + [pc = C])$$

---

[41]Some writers call that PRE-DETERMINISM: "deterministic if terminating."

[42]In Fig. 1.5.1 we said that this fourth statement "executes deterministically"; yet here we have called it nondeterministic.

On its own, it is nondeterministic; but in the context of the program its nondeterminism is limited to making a choice from a singleton set, as our subsequent calculations will show.

$$\equiv \qquad ([clever]/3) * 2 \; + \; ([\neg clever]/3) * 1 \qquad \text{type of } pc \text{ is } \{A, B, C\} \;^{43}$$
$$\equiv \qquad 2 [clever]/3 + [\neg clever]/3 \; .$$

Then for the first statement $pc \colon \in \{A, B, C\}$ we only note that pc does not appear in the final condition above, thus leaving it unchanged under *wp*: with simplification it becomes

$$(1 + [clever])/3 \; ,$$

which is thus the pre-expectation for the whole program.

Since the post-expectation $[pc = cc]$ is standard (it is the characteristic function of the set of states in which $pc = cc$), we are able to interpret the pre-expectation directly as the probability that $pc = cc$ will be satisfied on termination: we conclude that the contestant has $2/3$ probability of finding the prize if he is clever, and only $1/3$ if he is not.

1.6   Healthiness and algebra for *pGCL*

Recall that all standard *GCL* constructs satisfy the important property of conjunctivity[44] — that is, for any *GCL* command *prog* and post-conditions *post, post'* we have

$$wp.prog.(post \wedge post') \quad = \quad wp.prog.post \wedge wp.prog.post' \; .$$

That "healthiness condition" [Dij76] is used to prove many general properties of programs.

In *pGCL* the healthiness condition becomes "sublinearity," a generalisation of conjunctivity:[45]

Definition 1.6.1 SUBLINEARITY OF *pGCL*   Let c_0, c_1, c_2 be non-negative reals, and $postE_1, postE_2$ expectations; then all *pGCL* constructs *prog* satisfy

$$wp.prog.(c_1 * postE_1 + c_2 * postE_2 \ominus c_0)$$
$$\Leftarrow \qquad c_1 * wp.prog.postE_1 \; + \; c_2 * wp.prog.postE_2 \; \ominus \; c_0 \; ,$$

which property of *prog* is called *sublinearity*. Truncated subtraction $\ominus$ is defined

$$x \ominus y \quad := \quad (x - y) \, \mathsf{max} \, 0 \; ,$$

---

[43] Footnote 50 on p. 33 explains how typing might be propagated this way.

[44] They satisfy monotonicity too, which is implied by conjunctivity.

[45] Having discovered a probabilistic analogue of conjunctivity, we naturally ask for an analogue of disjunctivity. That turns out to be "super-linearity" — which when combined with sublinearity gives (just) linearity, and is characteristic of *deterministic* probabilistic programs, just as disjunctivity (with conjunctivity) characterises deterministic standard programs. See Sec. 8.3.

the maximum of the normal difference and zero. It has syntactic precedence lower than $+$.   □

Although it has a strange appearance, from sublinearity we can extract a number of very useful consequences, as we now show. We begin with monotonicity, feasibility and scaling.[46]

Definition 1.6.2 HEALTHINESS CONDITIONS

- *monotonicity:* increasing a post-expectation can only increase the pre-expectation. Suppose $postE \Rightarrow postE'$ for two expectations $postE, postE'$; then

$$wp.prog.postE'$$
$$\equiv \quad wp.prog.(postE + (postE' - postE))$$

$$\Leftarrow \qquad\qquad\qquad\qquad postE' - postE \Leftarrow 0, \text{ hence well defined;}$$
$$\qquad\qquad\qquad\qquad \text{sublinearity with } c_0, c_1, c_2 := 0, 1, 1$$
$$wp.prog.postE + wp.prog.(postE' - postE)$$

$$\Leftarrow \quad wp.prog.postE . \qquad\qquad\qquad 0 \Rightarrow wp.prog.(postE' - postE)$$

- *feasibility:* pre-expectations cannot be "too large." First note that

$$wp.prog.0$$
$$\equiv \quad wp.prog.(2 * 0)$$
$$\Leftarrow \quad 2 * wp.prog.0 , \qquad\qquad \text{sublinearity with } c_0, c_1, c_2 := 0, 2, 0$$

so that $wp.prog.0$ must be zero.
Now write max $postE$ for the maximum of $postE$ over all its variables' values; then

$$0$$
$$\equiv \quad wp.prog.0 \qquad\qquad\qquad\qquad\qquad\qquad \text{feasibility above}$$
$$\equiv \quad wp.prog.(postE \ominus \text{max } postE) \qquad\qquad postE \ominus \text{max } postE \equiv 0$$
$$\Leftarrow \quad wp.prog.postE \ominus \text{max } postE . \qquad c_0, c_1, c_2 := \text{max } postE, 1, 0$$

But from $0 \Leftarrow wp.prog.postE \ominus$ max $postE$ we have trivially that

$$wp.prog.postE \quad \Rightarrow \quad \text{max } postE , \qquad\qquad (1.21)$$

which we identify as the *feasibility* condition for $pGCL$.[47]

- *scaling:* multiplication by a non-negative constant distributes through commands. Note first that $wp.prog.(c * postE) \Leftarrow c * wp.prog.postE$ directly from sublinearity.

---

[46]These properties are collected together in Sec. 5.6, and restated in Part II as Defs. 5.6.3–5.6.5.

[47]Note how the general (1.21) implies the STRICTNESS condition $wp.prog.0 \equiv 0$, a direct numeric embedding of Dijkstra's *Law of the Excluded Miracle*.

For $\Rightarrow$ we have two cases: when c is zero, trivially from feasibility

$$wp.prog.(0 * postE) \quad\equiv\quad wp.prog.0 \quad\equiv\quad 0 \quad\equiv\quad 0 * wp.prog.postE \ ;$$

and for the other case $c \neq 0$ we reason

$$
\begin{array}{lll}
& wp.prog.(c * postE) & \\
\equiv & c(1/c) * wp.prog.(c * postE) & c \neq 0 \\
\Rightarrow & c * wp.prog.((1/c)c * postE)) & \text{sublinearity using } 1/c \\
\equiv & c * wp.prog.postE \ , &
\end{array}
$$

thus establishing $wp.prog.(c * postE) \equiv c * wp.prog.postE$ generally. (See p. 53 for an example of scaling's use.)

$\square$

The remaining property we examine is so-called "probabilistic conjunctivity." Since standard conjunction "$\wedge$" is not defined over numbers, we have many choices for a probabilistic analogue "&" of it, requiring only that

$$
\begin{array}{rcl}
0 \,\&\, 0 & = & 0 \\
0 \,\&\, 1 & = & 0 \\
1 \,\&\, 0 & = & 0 \\
1 \,\&\, 1 & = & 1
\end{array}
\tag{1.22}
$$

for consistency with embedded Booleans.

Obvious possibilities for & are multiplication $*$ and minimum min, and each of those has its uses; but neither satisfies anything like a generalisation of conjunctivity. Return for example to the program of Fig. 1.5.1, and consider its second statement

$$cc := (A \,@\, \tfrac{1}{3} \mid B \,@\, \tfrac{1}{3} \mid C \,@\, \tfrac{1}{3}) \ .$$

Writing *prog* for the above, with postcondition $[cc \neq C]$ min $[cc \neq A]$ we find

$$
\begin{array}{ll}
& wp.prog.(\, [cc \neq C] \text{ min } [cc \neq A]\,) \\
\equiv & wp.prog.[cc \neq C \wedge cc \neq A] \\
\equiv & wp.prog.[cc = B] \\
\equiv & 1/3 \\
\neq & 2/3 \text{ min } 2/3 \\
\equiv & wp.prog.[cc \neq C] \text{ min } wp.prog.[cc \neq A] \ .
\end{array}
$$

Thus probabilistic programs do *not* distribute min in general, and we must find something else. Instead we define

$$exp \,\&\, exp' \quad := \quad exp + exp' \ominus 1 \ , \tag{1.23}$$

whose right-hand side is inspired by sublinearity when $c_0, c_1, c_2 := 1, 1, 1$. The operator is commutative; and if we restrict expectations to $[0, 1]$ it is associative as well. Note however that it is not idempotent.[48]

We now state a (sub-)distribution property for &, a direct consequence of sublinearity.

sub-conjunctivity: the operator & sub-distributes through expectation transformers. From sublinearity with $c_0, c_1, c_2 := 1, 1, 1$ we have

$$wp.prog.(postE \,\&\, postE') \quad \Lleftarrow \quad wp.prog.postE \,\&\, wp.prog.postE'$$

for all $prog$.

(Unfortunately there does not seem to be a full ($\equiv$) conjunctivity property for expectation transformers.)

Beyond sub-conjunctivity, we say that & generalises conjunction for several other reasons as well. The first is of course that it satisfies the standard properties (1.22).

The second reason is that sub-conjunctivity (a consequence of sublinearity) implies "full" conjunctivity for standard programs. Standard programs, containing no probabilistic choices, take standard $[post]$-style post-expectations to standard pre-expectations: they are the embedding of GCL in $pGCL$, and for standard $prog$ we now show that

$$wp.prog.([post] \,\&\, [post'])$$
$$\equiv \quad wp.prog.[post] \,\&\, wp.prog.[post'] \,. \tag{1.24}$$

First note that "$\Lleftarrow$" comes directly from sub-conjunctivity above, taking $postE, postE'$ to be $[post], [post']$.

For "$\Rrightarrow$" we appeal to monotonicity, because $[post] \,\&\, [post'] \Rrightarrow [post]$ whence $wp.prog.([post] \,\&\, [post']) \Rrightarrow wp.prog.[post]$, and similarly for $post'$. Putting those together gives

$$wp.prog.([post] \,\&\, [post']) \quad \Rrightarrow \quad wp.prog.[post] \text{ min } wp.prog.[post'] \,,$$

by elementary arithmetic properties of $\Rrightarrow$. But on standard expectations — which $wp.prog.[post]$ and $wp.prog.[post']$ are, because $prog$ is standard — the operators min and & agree.

A last attribute linking & to $\wedge$ comes straight from elementary probability theory. Let X and Y be two events, not necessarily independent: then

> if the probability of X is at least p, and the probability of Y is at least q, the most that can be said in general about the joint event $X \cap Y$ is that it has probability at least $p \,\&\, q$.

---

[48]A binary operator $\odot$ is IDEMPOTENT just when $x \odot x = x$ for all x.

To see this, we begin by recalling that for any events X, Y and any probability distribution Pr we have[49]

$$\text{Pr}.(X \cap Y)$$
$$= \quad \text{Pr}.X + \text{Pr}.Y - \text{Pr}.(X \cup Y)$$

$$\geq \qquad\qquad\qquad \text{because Pr}.(X \cup Y) \leq 1 \text{ and Pr}.(X \cap Y) \geq 0$$
$$(\text{Pr}.X + \text{Pr}.Y - 1) \sqcup 0 \ .$$

We are not dealing with exact probabilities however: when demonic non-determinism is present we have only lower bounds. Thus we address the question

> Given only $\text{Pr}.X \geq p$ and $\text{Pr}.Y \geq q$, what is the most precise lower bound for $\text{Pr}.(X \cap Y)$ in terms of p and q?

From the reasoning above we obtain

$$(p + q - 1) \sqcup 0 \tag{1.25}$$

immediately as a lower bound. But to see that it is the *greatest* lower bound we must show that for any X, Y, p, q there is a probability distribution Pr such that the bound is attained; and that is illustrated in Fig. 1.6.3, where an explicit distribution is given in which $\text{Pr}.X = p$, $\text{Pr}.Y = q$ and $\text{Pr}.(X \cap Y)$ is as low as possible, reaching $(p + q - 1) \sqcup 0$ exactly.

Returning to our example, but using &, we now have equality:

$$wp.prog.([cc \neq C] \ \& \ [cc \neq A])$$
$$\equiv \quad wp.prog.[cc = B]$$
$$\equiv \quad 1/3$$
$$\equiv \quad 2/3 \ \& \ 2/3$$
$$\equiv \quad wp.prog.[cc \neq C] \ \& \ wp.prog.[cc \neq A] \ .$$

The & operator also plays a crucial role in the proof (Chap. 7) of our probabilistic loop rule, presented in Chap. 2 and used in the examples to come.

1.7   Healthiness example: modular reasoning

As an example of the use of healthiness conditions, we formulate and prove a simple but very powerful property of *pGCL* programs, important for "modular" reasoning about them.

By *modular* reasoning in this case we mean determining, first, that a program *prog* of interest has some standard property; then for subsequent (possibly probabilistic) reasoning we assume that property. This makes

---

[49]The first step is the *modularity law* for probabilities.

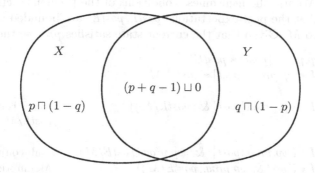

$$\mathrm{Pr}.X = p \sqcap (1-q) + (p+q-1) \sqcup 0 = p$$
$$\mathrm{Pr}.Y = q \sqcap (1-p) + (p+q-1) \sqcup 0 = q$$
$$\mathrm{Pr}.(X \cap Y) = (p+q-1) \sqcup 0 = p \,\&\, q$$

The lower bound $p \,\&\, q$ is the best possible.

Figure 1.6.3. PROBABILISTIC CONJUNCTION & DEPICTED

---

the reasoning modular in the sense that we do not have to prove all the properties at once.[50]

We formulate the principle as a lemma.

Lemma 1.7.1 MODULAR REASONING   Suppose for some program *prog* and predicates *pre* and *post* we have

$$[pre] \;\Rightarrow\; wp.prog.[post] \,, \qquad\qquad (1.26)$$

which is just the embedded form of a standard Hoare-triple specification. Then in any state satisfying *pre* we have for any bounded post-expectations *postE, postE'* that

$$wp.prog.postE \;=\; wp.prog.postE' \,, \qquad [51]$$

provided *post* implies that *postE* and *postE'* are equal.

That is, with (1.26) we can assume the truth of *post* when reasoning about the post-expectation, provided *pre* holds in the initial state.

---

[50]A typical use of this appeals to standard reasoning, in a "first pass," to establish that some (Boolean) property — such as a variable's typing — is invariant in a program; then, in the "second pass" during which probabilistic reasoning might be carried out, we can assume that invariant everywhere without comment. Recall Footnote 43 on p. 28; see also the treatment of Fig. 7.7.11 on p. 211 to come.

[51]We write "=" rather than "≡" because the equality holds only in some states (those satisfying *pre*), as indicated in the text above. Thus writing "≡, ⇒, ⇐" as we do elsewhere is just an alternative for the text "in all states".

Proof   We use the healthiness conditions of the previous section, and we assume that the post-expectations $postE, postE'$ are bounded above by some nonzero M. Given that the current state satisfies pre, we then have

$$wp.prog.([post] * postE)$$
$$= \quad M * wp.prog.([post] * postE/M) \qquad\qquad\qquad \text{scaling}$$

$$= \quad M * wp.prog.([post] \,\&\, (postE/M)) \qquad\qquad [post] \text{ is standard;}$$
$$postE/M \Rrightarrow 1$$

$$\geq \quad M * (wp.prog.[post] \,\&\, wp.prog.(postE/M)) \qquad \text{sub-conjunctivity}$$
$$\geq \quad M * ([pre] \,\&\, wp.prog.(postE/M)) \qquad\qquad \text{Assumption (1.26)}$$
$$= \quad M * (1 \,\&\, wp.prog.(postE/M)) \qquad\qquad pre \text{ holds in current state}$$
$$= \quad M * wp.prog.(postE/M) \qquad\qquad\qquad\qquad \text{arithmetic}$$
$$= \quad wp.prog.postE \;. \qquad\qquad\qquad\qquad\qquad\qquad \text{scaling}$$

The opposite inequality is immediate (in all states) from the monotonicity healthiness property, since $[post] * postE \Rrightarrow postE$. Thus, still assuming pre in the current state, we conclude with

$$wp.prog.postE$$
$$= \quad wp.prog.([post] * postE) \qquad\qquad\qquad\qquad\qquad \text{above}$$
$$= \quad wp.prog.([post] * postE') \qquad\qquad \text{assumption about } postE, postE'$$
$$= \quad wp.prog.postE' \;. \qquad\qquad\qquad \text{as above, but for } postE'$$

$$\square$$

This kind of reasoning is nothing new for standard programs, and indeed is usually taken for granted (although its formal justification appeals to conjunctivity). It is important that it is available in $pGCL$ as well.[52]

1.8   Interaction of probabilistic- and demonic choice

We conclude with some illustrations of the interaction of demonic and probabilistic choice. Consider two variables x, y, one chosen demonically and the other probabilistically. Suppose first that x is chosen demonically and y probabilistically, and take post-expectation $[x = y]$. Then

---

[52]Lem. 1.7.1 holds even when $postE, postE'$ are unbounded, provided of course that $wp.prog$ is defined for them; the proof of that can be given by direct reference to the definition of wp over the model, as set out in Chap. 5.

We will need that extension for our occasional excursions beyond the "safe" bounded world we have formally dealt with in the logic ($e.g.$ Sections 2.11 and 3.3).

$$wp.(\,(x:=1\sqcap x:=2);\ (y:=1\,{}_{\frac12}\!\oplus y:=2)\,).[x=y]$$
$$\equiv\quad wp.(x:=1\sqcap x:=2).([x=1]/2+[x=2]/2)$$
$$\equiv\quad ([1=1]/2+[1=2]/2)\ \min\ ([2=1]/2+[2=2]/2)$$
$$\equiv\quad (1/2+0/2)\ \min\ (0/2+1/2)$$
$$\equiv\quad 1/2\ ,$$

from which we see that program establishes $x=y$ with probability at least $1/2$: no matter which value is assigned to x, with probability $1/2$ the second command will assign the same to y.

Now suppose instead that it is the second choice that is demonic. Then we have

$$wp.(\,(x:=1\,{}_{\frac12}\!\oplus x:=2);\ (y:=1\sqcap y:=2)\,).[x=y]$$
$$\equiv\quad wp.(x:=1\,{}_{\frac12}\!\oplus x:=2).([x=1]\ \min\ [x=2])$$
$$\equiv\quad ([1=1]\ \min\ [1=2])/2\ +\ ([2=1]\ \min\ [2=2])/2$$
$$\equiv\quad (1\ \min\ 0)/2\ +\ (0\ \min\ 1)/2$$
$$\equiv\quad 0\ ,$$

reflecting that no matter what value is assigned probabilistically to x, the demon could choose subsequently to assign a different value to y.

Thus it is clear that the execution order of occurrence of the two choices plays a critical role in their interaction, and in particular that the demon in the first case cannot make the assignment "clairvoyantly" to x in order to avoid the value that later will be assigned to y.

1.9 Summary

Being able to reason formally about probabilistic programs does not of course remove *per se* the complexity of the mathematics on which they rely: we do not now expect to find astonishingly simple correctness proofs for all the large collection of randomised algorithms that have been developed over the decades [MR95]. However it should be possible in principle to locate and determine reliably what are the probabilistic/mathematical facts the construction of a randomised algorithm needs to exploit... which is of course just what standard predicate transformers do for conventional algorithms.

In the remainder of Part I we concentrate on proof rules that can be derived for *pGCL* — principally for loops — and on examples.

The theory of expectation transformers with nondeterminism is given in Part II, where in particular the role of sublinearity is identified and proved: it characterises a subspace of the predicate transformers that has an equivalent operational semantics of relations between initial and final probabilistic distributions over the state space — a formalisation of the

gambling game of Sec. 1.3. All the programming constructs of the probabilistic language of guarded commands belong to that subspace, which
means that the programmer who uses the language can elect to reason
about it either axiomatically or operationally.

Chapter notes

In the mid-1970's, Rabin demonstrated how randomisation could be used to solve
a variety of programming problems [Rab76]; since then, the range of applications
has increased considerably [MR95], and indeed we analyse several of them as case
studies in later chapters. In the meantime — fuelled by randomisation's impres-
sive applicability — the search for an effective logic of probabilistic programs
became an important research topic around the beginning of the 1980's, and
remained so until the mid-1990's. Ironically, the major technical difficulty was
due, in the main, to one of standard programming's major successes: *demonic
nondeterminism*, the basis for abstraction. It was a challenging problem to decide what to do about it, and how it should interact with the new *probabilistic
nondeterminism*.

The first probabilistic logics did not treat demonic nondeterminism at all —
Feldman and Harel [FH84] for instance proved soundness and completeness for
a probabilistic *PDL* which was (in our terms) purely deterministic. The logical
language allowed statements about programs to be made at the level of probability *distributions* and, as we discuss in Sec. A.2, that proves to be an impediment
to the natural introduction of a demon. A Hoare-style logic based on similar
principles has also been explored by den Hartog and de Vink [dHdV02].

The crucial step of a *quantitative* logic of expectations was taken by Kozen
[Koz85]. Subsequently Jones [Jon90], with Plotkin and using the *evaluations* from
earlier work of Saheb-Djahromi [SD80] that were based directly on topologies
rather than on σ- or Borel algebras, worked on more general probabilistic powerdomains; as an example of her technique she specialised it to the Kozen-style
logic for deterministic programs, resulting in the *sub-probability measures* that
provide a neat way to quantify nontermination.[53]

In 1997 He *et al.* [HSM97] finally proposed the *operational model* containing
all the ingredients for a full treatment of abstraction and program refinement in the context of probability — and that model paved the way for the
"demonic/probabilistic" program logic based on expectation transformers. Subsequently Ying [Yin03] has worked towards a probabilistic *refinement calculus* in
the style of Back [BvW98].

---

[53]The notion of sub-probability measures to characterise termination was present much
earlier, for example in the work of Feldman and Harel [FH84].

2
Probabilistic loops:
Invariants and variants

2.1   Introduction: loops via recursion

We saw in Chap. 1 that iteration is a special case of recursion. But the weakest pre-expectation $wp.prog.postE$ cannot be given a purely syntactic definition for general recursive $prog$ — the definition given earlier (Fig. 1.5.3) is semantic, a least fixed-point over expectation transformers. It does give us an algebraic property of recursive programs, $viz.$

$$(\mathbf{mu}\, X \cdot C) \quad = \quad C \, \langle X \mapsto (\mathbf{mu}\, X \cdot C) \rangle \, , \qquad\qquad (2.1)$$

but that is an *equation* rather than a definition.[1] See (2.6) below however for an example of its use in spite of that.

Iteration's special form (1.18) is the case of recursion where the context C is [2]

$$(prog; X) \ \mathbf{if} \ pred \ \mathbf{else} \ \mathbf{skip} \, ,$$

from which with (2.1) we have immediately the property that

$$\mathbf{do}\ pred \rightarrow prog\ \mathbf{od} \quad = \quad (\, prog; \mathbf{do}\ pred \rightarrow prog\ \mathbf{od}\,) \qquad (2.2)$$
$$\mathbf{if}\ pred\ \mathbf{else}$$
$$\mathbf{skip} \, .$$

Applying wp to both sides, with respect to an arbitrary post-expectation $postE$ we see that

if $\qquad\qquad preE \ = \ wp.(\mathbf{do}\ pred \rightarrow prog\ \mathbf{od}).postE$

then also $\qquad preE \ = \ wp.prog.preE \ \mathbf{if}\ pred\ \mathbf{else}\ postE \, ,$ $\qquad (2.3)$

of which the latter at least is now an equation in expectations (in $preE$) rather than in programs (contrast (2.1)). However, because $preE$ occurs on both sides, we still do not have a definition.[3]

Because of the issues above, and for historical reasons [Flo67, Hoa69], for standard programs rather than ask for the *weakest* precondition for a loop we are content with a precondition that is sufficiently strong (*i.e.* might be stronger than required, but still implies the weakest one).

---

[1] To turn this into a definition we would add that $(\mathbf{mu}\, X \cdot C)$ is the LEAST PROGRAM IN THE REFINEMENT ORDER ($\sqsubseteq$) that has Property (2.1).

[2] It is TAIL RECURSION, which is what allows the following treatment.

[3] Recalling Footnote 1 just above, we see that a definition of $preE$ would then be that it is the least such $preE$ in the probabilistic implication order $\Rrightarrow$; but "taking the least" is not a syntactic operation. Dijkstra finessed this in his presentation of wp for standard loops by using the usual iterative formulation for a least fixed-point [Tar55], the iterates being his predicates H_k [Dij76, p. 35], and then taking the limit with a quantification $(\exists k \cdot H_k)$. But that quantification, being over predicates, is second-order and thus lies outside the first-order logic of wp used here (see however *e.g.* Back and von Wright, and Ward [War89, BvW93]).

We call this "finessing," rather than *e.g.* "cheating," because with it Dijkstra successfully confined the second-order reasoning to just the place where it was required.

For probabilistic programs, by analogy, we look not necessarily for the *greatest* pre-expectation but merely for one that is sufficiently small.

Whether standard or probabilistic, the general techniques for loops involve "invariants" and "variants," and we now consider their probabilistic versions.

In the rest of Part I we use single capital letters $A, B, \cdots, P, Q, \cdots$ for expressions over the state, that is Boolean-valued predicates, real-valued expectations *etc.* — rather than words $preE, postE$ as earlier — unless the occasion demands extra clarity. As an exception we will write $Inv, Term$ for standard invariants and termination conditions, reserving I, T for their probabilistic counterparts.

2.2 Probabilistic invariants

In a standard loop, the invariant is proved to hold at the beginning of every iteration: as a result it describes a set of states from which the loop body cannot escape and so termination — if it occurs — must also lie within that set. The proof obligation for a standard loop is

$$G \wedge Inv \quad \Rrightarrow \quad wp.body.Inv\ , \quad ^{4}$$

where G is the loop guard and Inv is the predicate describing the invariant set of states.

For a probabilistic loop we have a post-expectation rather than a post-condition; but otherwise the situation is much the same. We therefore say that an expectation I is an invariant of a loop under these conditions:

Definition 2.2.1 PROBABILISTIC INVARIANT   Consider a program *loop* defined **do** $G \to body$ **od**, with predicate G its *guard*. We say that expectation I is an *invariant* of *loop* just when

$$[G] * I \quad \Rrightarrow \quad wp.body.I\ .$$

□

The use of multiplication in the idiom $[G] * I$ is just a convenient way of writing I **if** G **else** 0, with the effect in this case of restricting the probabilistic implication $\Rrightarrow$ to require proof only in states satisfying G:

---

[4] We now dispense with the distinction between Boolean- and $\{0, 1\}$ types for standard predicates, and overload the $\Rrightarrow$-style and wp operators in order to reduce the occurrence of embedding-brackets [·]. In particular we use $\Rrightarrow$ for "everywhere implies" between predicates as well as "is everywhere no more than" between expectations, to achieve a consistency of notation.

for if G does not hold, then the left-hand side is zero, and so "$\Rrightarrow$-implies" anything.[5]

As with standard programming, finding invariants is a skill — perhaps one of the main skills, for intricate algorithms — and heuristics for it are important. One such is the following: if the post-expectation of the loop is standard,[6] some $[post]$ say, then as an aid to the intuition we can look for an expression that gives a lower bound on the probability that we will establish *post* by (continuing to) execute the loop body.[7] Often that expression will have the form

$$p * [pred] \qquad (2.4)$$

with p a probability and *pred* a predicate, both expressions over the state.[8] The standard part *pred* might be found by conventional techniques, with the probability p estimated by a separate "worst-case" analysis.

From the definition of the embedding $[pred]$ we know that one interpretation of (2.4) is

> with probability p if *pred* holds, and with probability zero otherwise,

and in many cases it will serve directly as an invariant for the loop — whence our interpretation becomes

> if *pred* holds then with probability at least p the loop executed from this state will satisfy *post* on termination; but

> if *pred* does not hold, so that (2.4) is zero, then we have no information about the probability of achieving *post*, since all probabilities are at least zero anyway.

We see several examples of such invariants in Sec. 2.5.

2.3   Probabilistic termination

The *probability* that a program will terminate generalises the usual definition of termination: recalling that $[\text{true}] \equiv 1$ we see that a program's greatest guaranteed probability of termination is just

$$wp.prog.1 \ . \qquad (2.5)$$

---

[5]Recall that expectations are non-negative.

[6]We say an expectation is STANDARD if it can be written $[pred]$ for some predicate *pred*, or equivalently if it is $\{0, 1\}$-valued.

[7]We give examples of this technique in Sec. 2.5, and return to it in Sec. 7.7.8 for loops "**do** $1/2 \to \cdots$" with probabilistic guards.

[8]We extend our naming convention to allow lower-case letters like "p" for probability-valued expressions over the state.

That raises the issue of whether there is a difference between

> certain $wp.(\cdots).\text{true}$ for a standard program,
> termination

and what we could call

> "almost-certain" $wp.(\cdots).[\text{true}]$ for a probabilistic one.
> termination

In fact there is a difference: we say that a computation — regarded as a branching tree of steps — terminates *absolutely* if *every* path in the tree leads to termination, *i.e.* is finite; if there are any infinite paths at all, then termination is not absolute.

Even if the computation is probabilistic, the same applies; but we can be more discriminating if we wish. Each step on a path will have an associated probability (often probability one, for standard steps in the computation) and — speaking informally[9] — the probabilities on those individual steps, multiplied together, determine probabilities for paths and, when added up, for sets of paths as a whole. If the collective probability of the infinite paths in a computation is zero, then we say it terminates *almost-certainly*, and that nontermination is *almost impossible*.

Our semantics, and logic, do not distinguish absolute- and almost-certain termination — in other words, we take the position that the benefit of a simpler logic outweighs any disadvantage of ignoring almost-impossible events.

As a simple example of recursive termination, suppose *prog* is the program defined[10]

$$prog \quad := \quad prog \,_p{\oplus}\, \textbf{skip} \;, \tag{2.6}$$

in which we assume that p is some fixed probability not equal to one: on each recursive call, *prog* has probability $\overline{p}$ of termination, continuing otherwise with further recursion. Elementary probability theory shows that *prog* terminates with probability one (after an expected $p/\overline{p}$ recursive calls). In the expectation logic we calculate

> $wp.prog.1$

---

[9]A rigorous treatment of probability distributions over infinite sets involves σ-algebras [GS92]; but in this discussion we do not need those details. See Footnote 7 on p. 297 for an example.

[10]That is, *prog* is the program $(\textbf{mu}\, X \bullet X \,_p{\oplus}\, \textbf{skip})$; we have merely written out its property (2.1), and have assumed we mean the $\sqsubseteq$-least such program.

In Sec. 7.7 we see that this kind of program can be written

$$\textbf{do}\, p \to \textbf{skip}\, \textbf{od} \;.$$

(Compare *e.g.* (7.24) on p. 203.) We explain there how an extended form of variant-argument can show termination of such loops.

$$\equiv \qquad p * wp.prog.1 \ + \ \overline{p} * wp.\textbf{skip}.1 \qquad\qquad (2.5); \ wp \text{ for } {}_p\oplus$$
$$\equiv \qquad p * wp.prog.1 \ + \ \overline{p} \ , \qquad\qquad\qquad wp \text{ for } \textbf{skip}$$

so that $\overline{p} * wp.prog.1 \equiv \overline{p}$. Since p is not one, we can divide by $\overline{p}$ to see that indeed $wp.prog.1 \equiv 1$, agreeing with the elementary theory.

However proved, it is true that *prog* terminates with probability one — *i.e.* it terminates almost certainly. But it does not terminate absolutely: regarded as a tree of possible recursive calls, the computation contains an infinite path *recurse, recurse, recurse...* That path's overall probability, however, is no more than the probability p^n of taking its first n steps, for any n no matter how large — *i.e.* for $p < 1$ it is zero for the whole path. Our logic ignores it.

We return to probabilistic termination in Sec. 2.6.

2.4 Invariance and termination together: the loop rule

The conventional technique for standard loops is to prove "partial correctness" and termination separately, and then to put them together to establish "total correctness."

Suppose *loop* is standard: if *Inv* is an invariant for it, then we know *Inv* initially is sufficient to achieve $Inv \wedge \overline{G}$ finally, provided the loop terminates. That is *partial correctness*. A separate proof is required to establish the initial states from which termination is guaranteed; from those initial states which additionally satisfy *Inv* the loop *will* establish $Inv \wedge \overline{G}$. That is *total correctness*.

Together, we thus have that if *Inv* is the invariant and *Term* describes states from which termination is guaranteed then

$$Inv \ \wedge \ Term \quad \Rrightarrow \quad wp.loop.(Inv \wedge \overline{G})$$

for a standard loop.

For probabilistic loops we pursue a similar strategy, of joining partial correctness and termination, although of course we cannot use $\wedge$ for the purpose since we are no longer dealing with Booleans. The general methods are the subject of Chap. 7; here we set out three useful specialisations.[11]

---

[11]We will see that the proof of the general methods is relatively complicated — which is a surprise given the simplicity of Def. 2.2.1 and the fact that it corresponds to a simple recursive argument in elementary probability theory. See Footnote 66 on p. 73 for reassurance that the extra complexity we have deferred until Chap. 7 is probably unavoidable.

Lemma 2.4.1 TOTAL CORRECTNESS FOR PROBABILISTIC LOOPS
Let expectation T be the termination probability of *loop*, so that

$$T \quad := \quad wp.(\textbf{do } G \to prog \textbf{ od}).1 \ ,$$

and let I be a probabilistic invariant for it (Def. 2.2.1). We consider three cases:

1. If $I \equiv [Inv]$ for some standard *Inv*, define $preE := T * [Inv]$.

2. If $[Term] \Rrightarrow T$ for some standard *Term*, so that *Term* contains only states where termination is almost certain, define $preE := I * [Term]$. (Thus if T is itself standard, we can again define $preE := I * T$.)

3. If $I \Rrightarrow T$, then define $preE := I$.[12]

Then in each case we have that $preE$ is a sufficient pre-expectation for the loop to terminate while maintaining the invariant: [13]

$$preE \quad \Rrightarrow \quad wp.loop.([\overline{G}] * I) \ .$$

Proof   Given in Sec. 7.4 of Part II. (For an alternative but "model-based" argument, see also Sec. 2.12 in this chapter.)   □

Notice that Lem. 2.4.1 subsumes the standard loop rule, since if I and T are both standard then both Cases 1 and 2 will apply — whence the pre-expectation is just $[Inv] * [Term] \equiv [Inv \wedge Term]$ either way, as usual.

Case 1 occurs when a loop's calculations are captured by a standard invariant, but its termination is not guaranteed. Typically "Las-Vegas" algorithms have this property, that they are correct if they terminate but the termination occurs only with some probability [BB96]. (See Sec. 2.5.1.)

---

[12]See however Sec. 2.6 in this chapter for a generalisation of this case.

[13]A "sufficient precondition" is a predicate which is sufficiently strong to imply the weakest precondition, and the "sufficient" is usually implicit — so that we simply say "a precondition." Thus by analogy here we mean a (pre-)expectation which is sufficiently low to imply ($\Rrightarrow$) the *greatest* pre-expectation.

As Hehner points out, however, in its normal English sense "a precondition" is *not* implicitly sufficient — rather it is "necessary" (*i.e.* is *implied by* rather than implying the weakest-). We are stuck with the Computer-Science sense which, as we now see, is the opposite.

Case 2 occurs when a loop's termination is guaranteed, but its calculations are correct only with some probability. This is characteristic of "Monte-Carlo" algorithms [*op. cit.*].[14] (See Sec. 2.5.2.)

Case 3 applies when neither partial correctness nor termination are assured (Sec. 2.5.3).

2.5   Three examples of probabilistic loops

In the examples of this section we illustrate the three cases of Lem. 2.4.1. The formal calculations justifying the invariant or termination conditions themselves are only sketched, so that the interaction of invariance and termination can be highlighted.

2.5.1   The martingale

We begin with an example of Case 1, where correctness is certain if the loop terminates, but termination occurs only with some probability.

Imagine a gambler who is placing 50/50 bets: that is he expects to win each bet with probability 1/2 and receives twice his bet back if he does. He starts with a bet of 1 unit; if he wins, he has increased his capital by 1. If he loses, however, he doubles his bet to 2 and tries again; winning this time also increases his original capital by 1, since he wins 2 but lost 1 before. If he again loses, he continues with 4 units, and the same applies (might win 4, but lost $1 + 2$ before). Eventually he must win, he reasons, in which case he will receive 1 unit more than the total he has already lost; and he will have increased his capital by exactly 1.

Thus by following this strategy, he believes, he is guaranteed eventually to win 1 unit if he continues long enough.

In fact the gambler's reasoning is only "partially correct," and the flaw is that he has not assured (successful) termination: there is a chance that he will suffer a losing streak so long that he eventually cannot double his bet any more, because he has exhausted his entire capital. Though that possibility may be remote, his loss if it occurs could be very large.

---

[14]Since we do not distinguish absolute- from almost-certain termination, we must admit that our requirement "termination is guaranteed" does allow nontermination — but only if it is almost impossible. Most descriptions of Monte-Carlo algorithms however assume *absolute* termination, which we would express by using standard *wp* over a version of the program text in which all occurrences of probabilistic choice $_p\oplus$ had been replaced by demonic choice $\sqcap$. (Since with our quantitative *wp* we have *identified* the two forms of termination, we cannot expect to give a semantic characterisation of absolute- as opposed to almost-certain termination in our model; that is why we refer to the program text. See also Footnote 25 on p. 235.)

$$
\begin{aligned}
init &\rightarrow & c,b:&= C,1; \\
loop &\rightarrow & \mathbf{do}\ & b \neq 0 \rightarrow \\
& & & \mathbf{if}\ b \leq c\ \mathbf{then} \\
& & & \quad c:= c-b; \\
& & & \quad c,b:= c+2b,0\ {}_{1/2}\oplus\ b:=2b \\
& & & \mathbf{fi} \\
& & \mathbf{od} &
\end{aligned}
$$

The program *prog* is the whole of the above.

The gambler's capital c is initially C, and his intended bet b is initially one.

On each iteration, if his intended bet does not exceed his capital, he is allowed to place it and he has a 1/2 chance of winning.

If he wins, he receives twice his bet in return and sets his intended bet to 0 to indicate he is finished; if he loses, he receives nothing and doubles his intended bet — hoping to win next time.

If he loses sufficiently often in succession, his intended bet b will eventually be more than he can afford — more than his remaining capital c — and he will then be "trapped" forever within the iteration.

Figure 2.5.1. THE MARTINGALE

We model the martingale as in Fig. 2.5.1. It is a Las-Vegas algorithm because its termination is uncertain — the way the program is written, if the gambler runs out of money he never leaves the casino. If termination does occur, then the correct postcondition is established, that $c = C + 1$, and he goes home with a win of exactly one unit.

It is easy arithmetic to show that $c + b = C + 1$ is a standard invariant of *loop* — treat probabilistic choice ${}_{1/2}\oplus$ as demonic choice $\sqcap$ and apply standard reasoning: any invariant of that demonic abstraction will be an invariant of its probabilistic refinement, the loop as given.

In general, the fact that probabilistic choice refines demonic choice is the mathematical relationship that allows us to apply standard reasoning to probabilistic programs, when we need only standard facts about them.

For termination, we consider an initial state in which $1 \leq b \leq c$, meaning that the gambler is still betting (he has not retired by setting b to 0) and that he has sufficient capital to make his next bet (because $c \geq b$). For the loop *not* to terminate, from such a state, the gambler must lose N times in succession from then on, for some N such that the total capital $b+2b+\cdots 2^N b$ required to reach and make the $(N+1)^{\text{st}}$ bet strictly exceeds his current capital c — we therefore define $N_{b,c}$ to be the least N such that $(2^{N+1} - 1)b > c$.

Thus the probability of nontermination of *loop* is *no more* than $1/2^{N_{b,c}}$ if $1 \leq b \leq c$, and so its converse, the probability of termination under those conditions, must be *no less* than $1 - 1/2^{N_{b,c}}$, which we write $T_{b,c}$; that is, whatever the exact termination probability T might be in general, we have at least that[15]

$$T_{b,c} * [1 \leq b \leq c] \quad \Rrightarrow \quad T \ .$$

Thus our *preE* from Case 1 must satisfy

$$
\begin{aligned}
&\quad\ preE \\
\equiv &\quad\ I * T \\
\Lleftarrow &\quad\ [c + b = C + 1] * T_{b,c} * [1 \leq b \leq c] &\text{above} \\
\equiv &\quad\ T_{b,c} * [c + b = C + 1 \ \wedge\ 1 \leq b \leq c] \ . &\text{combine first and third terms}
\end{aligned}
$$

Referring again to Case 1, we see that the post-expectation of the loop is then

$$[\overline{G}] * I \quad \equiv \quad [b = 0] * [c + b = C + 1] \quad \Rrightarrow \quad [c = C + 1] \ ,$$

indeed establishing that the gambler's capital will increase by exactly one. We finish by calculating a pre-expectation for the whole program, *viz.*

$$
\begin{aligned}
&\quad\ wp.prog.[c = C + 1] \\
\equiv &\quad\ wp.(init;\ loop).[c = C + 1] \\
\Lleftarrow &\quad\ wp.init.preE &\text{meaning of } preE \text{ here} \\
\Lleftarrow &\quad\ (T_{b,c} * [c + b = C + 1 \ \wedge\ 1 \leq b \leq c]) \langle c, b \mapsto C, 1 \rangle &\text{above} \\
\equiv &\quad\ T_{1,C} * [C + 1 = C + 1 \ \wedge\ 1 \leq 1 \leq C] \\
\equiv &\quad\ T_{1,C} * [C \geq 1] \ ,
\end{aligned}
$$

so showing that *prog* establishes $c = C + 1$ with probability at least $T_{1,C}$ provided he has some capital to start with.

Thus if we set $P_C := T_{1,C}$, we see that the gambler has a large chance P_C of winning a little (one unit), but a small chance $\overline{P_C}$ of losing a lot — more than half his initial capital.[16]

In Sec. 2.10.1 we return to the martingale, to determine his expected winnings either way.

2.5.2 Probabilistic amplification

Our second example relates to Case 2, where termination is certain but correctness is probabilistic.

---

[15]Note again the technique of writing "$*[pred]$" on the left of $\Rrightarrow$ to restrict attention to only those states.

[16]If he starts with \$1,000, then he has chance $P_{1000} \geq 99.8\%$ of winning his \$1. But with the remaining probability 0.2% he will lose \$511.

$$
\begin{aligned}
init \quad &\rightarrow && a, n\!: = \; \mathsf{true}, N; \\
loop \quad &\rightarrow && \mathbf{do}\ n \neq 0 \ \wedge \ a \rightarrow \\
& && \quad a\!: = \; Q \;_{p}\!\oplus\, \mathsf{true}; \qquad\quad \leftarrow (2.7) \\
& && \quad n\!: = \; n - 1 \\
& && \mathbf{od}
\end{aligned}
$$

The program *prog* is the whole of the above; it attempts to answer in *a* the question posed in *Q*.

Given a natural number N, the refutation procedure (2.7) is applied N times, with overall refutation occurring if any of those N separate applications refutes.

Figure 2.5.2. PROBABILISTIC AMPLIFICATION

Suppose there is a Boolean question Q for which there is a probability-p refutation procedure

$$
a\!: = \; Q \;_{p}\!\oplus\, \mathsf{true}\ , \tag{2.7}
$$

which certainly delivers some answer a, but does not guarantee that the answer is correct. It is thus a Monte-Carlo algorithm.[17],[18]

For example, if Q were the question "is N prime," then the Miller-Rabin test for primality [MR95] is an example of the refutation procedure above with $p := 1/2$: if N is prime (so that Q is true), then the Miller-Rabin test is guaranteed to report "indeed N is prime" (setting answer a to true). If however N is not prime, then with probability $1/2$ the test will state correctly "N is not prime" (setting a to false)[19] — but with the remaining probability $1 - 1/2$ the test fails to refute N's primality (and leaves a set "incorrectly" to true even though Q is false).

With a failure probability of $1/2$, the Miller-Rabin test used on its own could not be called a "reliable" prime-checker; but if used independently say 10 times in a row, it would fail to give the correct answer with probability less than 1 in 1000 — since for the answer to be wrong overall every one of the individual tests would have had to fail. That technique is called *probabilistic amplification,* and is illustrated in Fig. 2.5.2.

---

[17] A more general description of a Monte-Carlo algorithm would be

$$
a\!: = \; Q \;_{p}\!\oplus\, (a\!: = \; \mathsf{true} \sqcap a\!: = \; \mathsf{false})\ ,
$$

where the answer can be arbitrarily wrong with some probability $\overline{p}$. We must use a more specific case, here, for probabilistic amplification.

[18] Monte-Carlo algorithms give rise to the complexity class *bounded-error probabilistic polynomial-time* $\mathcal{BPP}$, and Las-Vegas algorithms suggest the class *zero-error probabilistic polynomial-time* $\mathcal{ZPP}$ [BB96, p. 463].

[19] In fact in the Miller-Rabin test the probability is *at least* $1/2$, which we discuss further below (p. 50).

Let *prog*, *init* and *loop* be as in Fig. 2.5.2: given arbitrary question Q, we are interested in the probability of a correct answer a at the end — that is, our post-expectation is $[a = Q]$. Using our heuristic for invariants (p. 40) we ask "given arbitrary values for a, n, Q, can we estimate the probability of achieving $a = Q$ finally if the loop were executed from there?" We reason informally as follows:

- If a does not hold, then termination is immediate and achieving $a = Q$ depends only on Q's current value: the probability is therefore $[\overline{Q}]$ in this case.[20]

- If a and Q both hold, then achieving $a = Q$ is certain because the test (2.7) never refutes a true Q.[21]

- If a holds but Q does not, then with probability $1 - \overline{p}^n$ at least one of the n remaining tests will correctly refute Q and set a to false.

With some experimentation we formulate the above observations as an expression[22]

$$[a] \text{ if } Q \text{ else } (1 - \overline{p}^n [a]) \ ,$$

which we now show is indeed an invariant for *loop*.

As a notational convenience in the calculation, we start with the overall post-expectation and reason backwards towards the pre-expectation, indicating between expectations when *wp* is applied to give the lower (the right-hand side of a reasoning step) from the upper (the left-hand side).[23]

Working from Fig. 2.5.2, we have

---

[20]Note that we do *not* reason here that "a can be false only if Q is" (and hence that the probability of achieving $a = Q$ in this case is actually one), since using that extra fact would depend on the assumption that the current state has been reached by executing the loop from its initialisation. (See also Sec. 7.7.8 below.)

Making a mistake of that kind in a heuristic is not "dangerous" of course: it just wastes time, since later when we check the putative invariant we will reject it.

[21]One of the reasons this is informal here is that we are ignoring termination for the moment; as remarked in Footnote 20 preceding, rigour comes when we prove, below, that our "guessed" invariant really is one.

[22]That is, we try various formulations and see which leads to the simplest calculations to follow.

[23]Admittedly this seems strange at first, since it appears we are writing $\equiv$ between expressions that are not equal — but in practice its convenience outweighs any initial uneasiness [vGB98]. As a warning we will include a small dot to the left of the relational symbol (thus "$\cdot \equiv$" but sometimes "$\cdot \Leftarrow$") to indicate when we are using this notational device. Thus when we write

$$\begin{array}{ll} & post \\ \cdot \Leftarrow & pre \qquad\qquad\qquad\qquad\qquad\qquad \text{applying } wp.prog \end{array}$$

we mean that we have established "$pre \Rrightarrow wp.prog.post$" by writing down *post*, then *prog* and finally calculating *pre* from those. The advantage will be clear when we chain several such steps together.

$$[a] \text{ if } Q \text{ else } (1 - \overline{p}^n [a])$$

$\cdot \equiv$ $[a] \text{ if } Q \text{ else } (1 - \overline{p}^{n-1} [a])$ applying $wp.(n := n - 1)$

$\cdot \equiv$ applying $wp.(a := Q \ {}_{p}{\oplus} \text{ true})$

$$\begin{aligned} &p &*& \quad [Q] \text{ if } Q \text{ else } (1 - \overline{p}^{n-1} [Q]) \\ + &\overline{p} &*& \quad [\text{true}] \text{ if } Q \text{ else } (1 - \overline{p}^{n-1} [\text{true}]) \end{aligned}$$

$\equiv$ $\begin{aligned} &p &*& \quad 1 \text{ if } Q \text{ else } (1 - \overline{p}^{n-1} * 0) \\ + &\overline{p} &*& \quad 1 \text{ if } Q \text{ else } (1 - \overline{p}^{n-1} * 1) \end{aligned}$ arithmetic

$\equiv$ $p \quad + \quad \overline{p} * (1 \text{ if } Q \text{ else } (1 - \overline{p}^{n-1}))$ arithmetic

$\equiv$ $1 \text{ if } Q \text{ else } (1 - \overline{p}^n)$ arithmetic

$\Leftarrow$ guard on left allows us to assume $[a] = 1$ on right

$$[n \neq 0 \wedge a] \quad * \quad [a] \text{ if } Q \text{ else } (1 - \overline{p}^n [a]) ,$$

as required by Def. 2.2.1.

For termination we see by inspection that it is certain whenever $n \geq 0$, so that in Case 2 we can take $Term := 0 \leq n$ and our $preE$ becomes

$$[0 \leq n] \quad * \quad [a] \text{ if } Q \text{ else } (1 - \overline{p}^n [a]) .$$

And for the post-expectation we reason

$\begin{aligned} &[n = 0 \vee \overline{a}] \\ * \quad &[a] \text{ if } Q \text{ else } (1 - \overline{p}^n [a]) \end{aligned}$ negated-guard $*$ invariant

$\equiv$ $\begin{aligned} &[n = 0 \vee \overline{a}] \\ * \quad &[a] \text{ if } Q \text{ else } [\overline{a}] \end{aligned}$ both $n = 0$ and $\overline{a}$ imply $1 - \overline{p}^n [a] \equiv [\overline{a}]$

$\Rightarrow$ $[a] \text{ if } Q \text{ else } [\overline{a}]$ drop negated guard

$\equiv$ $[a = Q] ,$

as required.

Finally, for the overall pre-expectation we calculate

$\quad wp.prog.[a = Q]$

$\Leftarrow$ $wp.init.preE$

$\equiv$ $([0 \leq n] \quad * \quad [a] \text{ if } Q \text{ else } (1 - \overline{p}^n [a])) \langle a, n \mapsto \text{true}, N \rangle$

$\equiv$ $[0 \leq N] \quad * \quad 1 \text{ if } Q \text{ else } (1 - \overline{p}^N) ,$

$\Leftarrow$ $[0 \leq N] \quad * \quad (1 - \overline{p}^N) ,$

which is to say that, provided $0 \leq N$, the program has an error probability of no more than $\overline{p}^N$ and moreover (second-last line) is in fact always correct if Q is true.

But what if the refutation probability p is not exact?

Because our overall pre-expectation $1 - \overline{p}^N$ is an increasing function of p, it is tempting to conclude that we can replace $_p\oplus$ by $_{\geq p}\oplus$ in Fig. 2.5.2 without affecting our result. For example, the procedure for the Miller-Rabin primality test refutes with probability *at least* rather than exactly $1/2$, but we still expect its amplified error probability to be no more than $1/2^N$. Although that conclusion does hold for Fig. 2.5.2 (and thus for Miller-Rabin), such reasoning is not valid in general — as we now show by example.

By inspection, the program

$$x := \quad \text{true } _{3/4}\oplus \text{ false};$$
$$y := \quad \text{true } _{p}\oplus \text{ false}$$

establishes postcondition $x = y$ with probability at least $p/2 + 1/4$; and we can confirm that with the calculation

$$
\begin{array}{lll}
& [x = y] & \\
\cdot \equiv & p\,[x] + \overline{p}\,[\overline{x}] & \text{applying } wp.(y := \text{ true }_p\oplus \text{ false}) \\
\equiv & \overline{p} + (p - \overline{p})\,[x] & \\
\cdot \equiv & 3/4 * (\overline{p} + p - \overline{p}) + 1/4 * \overline{p} & \text{applying } wp.(x := \text{ true }_{3/4}\oplus \text{ false}) \\
\equiv & p/2 + 1/4 \;. &
\end{array}
$$

Now $p/2 + 1/4$ is an increasing function of p, but it is *not* true that the related program

$$x := \quad \text{true } _{3/4}\oplus \text{ false};$$
$$y := \quad \text{true } _{\geq p}\oplus \text{ false} \tag{2.8}$$

establishes $[x = y]$ with *at least* that probability: repeating the calculation, we find instead

$$
\begin{array}{lll}
& [x = y] & \\
\cdot \equiv & (p\,[x] + \overline{p}\,[\overline{x}]) \quad \min \quad [x] & \text{applying } wp.(y := \text{ true }_{\geq p}\oplus \text{ false}) \\
\equiv & p\,[x] & \text{arithmetic: cases on Boolean } x \\
\cdot \equiv & 3p/4 \;, & \text{applying } wp.(x := \text{ true }_{3/4}\oplus \text{ false})
\end{array}
$$

which is not necessarily at least $p/2 + 1/4$.

This effect is due to our interpretation of $_{\geq p}\oplus$, where the choice of which probability $p \leq p' \leq 1$ to use each time $_{\geq p}\oplus$ is executed may depend on many factors, and is not determined "just once, at the beginning." For example, in the Miller-Rabin test the actual probability of a correct answer will depend on internal features of the implementation, which we must abstract as $\sqcap$ since we cannot know what they are: in general, different values of p' might be used on different occasions, though each time still in the required interval $[p, 1]$.

Behaviour like that is an example of demonic choice, modelling the fact that we don't know which p' will be chosen. In the program above, one possibility — the worst, if we want to achieve $x = y$ finally — is to take $p' := p$ **if** x **else** 1, thus depending explicitly on x. When the first statement sets x to true, the second selects false for y with the maximum probability $\overline{p}$ allowed; but when x is set to false, the second statement sets y to true with probability one.

In fact the program (2.8) written in more primitive terms (recall p. 21) would be

$$
\begin{aligned}
&x{:}= \text{ true } {}_{3/4}\oplus \text{ false}; \\
&y{:}= \text{ true } {}_{p}\oplus \text{ false} \quad \sqcap \quad y{:}= \text{ true} ,
\end{aligned}
\tag{2.9}
$$

where the demonic choice could be implemented (*lhs* **if** x **else** *rhs*) in order to make the chance of establishing $x = y$ as low as possible, as we showed above.

We are seeing an example of the sometimes subtle interaction between probabilistic and demonic choice: in (2.9) the "demon" at $\sqcap$ is "aware" of the outcome of the earlier probabilistic assignment to x, and "uses" that information to avoid our postcondition.

Our earlier, but informal, reasoning — based on $1 - \overline{p}^N$ being increasing in p — delivers a safe conclusion only because in that program the demon's best strategy (worst, for us) is to choose the lowest probability p allowed, every time, no matter what the values of the state variables. If we altered the program, replacing ${}_{p}\oplus$ by ${}_{\geq p}\oplus$, and restarted the calculation at that point, we would verify that by calculating

$$
\begin{aligned}
\cdot \equiv \quad & \qquad\qquad\qquad\qquad\qquad \text{applying } wp.(a{:}= Q {}_{\geq p}\oplus \text{ true}) \\
& p \quad * \quad [Q] \text{ if } Q \text{ else } (1 - \overline{p}^{n-1}[Q]) \\
& + \quad \overline{p} \quad * \quad [\text{true}] \text{ if } Q \text{ else } (1 - \overline{p}^{n-1}[\text{true}])
\end{aligned}
$$

$$
\min \quad [Q] \text{ if } Q \text{ else } (1 - \overline{p}^{n-1}[Q])
$$

$$
\begin{aligned}
\equiv \quad & \qquad\qquad\qquad \text{for the above min we always have } lhs \Rightarrow rhs \\
& p \quad * \quad [Q] \text{ if } Q \text{ else } (1 - \overline{p}^{n-1}[Q]) \\
& + \quad \overline{p} \quad * \quad [\text{true}] \text{ if } Q \text{ else } (1 - \overline{p}^{n-1}[\text{true}]) ,
\end{aligned}
$$

after which the reasoning would proceed as before. The justification "*lhs* $\Rightarrow$ *rhs*," that in every state *lhs* $\leq$ *rhs*, is the arithmetic that shows the demon — no matter what the state contains — will in this program select the left-hand branch on each occasion.

See also the discussion of Program (B.1) on p. 326.

2.5.3 Faulty factorial

Our final example illustrates Case 3, where both the termination condition and the invariant are probabilistic. Given a natural number N, the program is to (attempt to) set f to $N!$ in spite of a probabilistically faulty subtraction.

The program is shown in Fig. 2.5.3, and is the conventional factorial algorithm except that the decrement of n sometimes increments instead.

$$
\begin{aligned}
&init &\rightarrow\quad &n, f := N, 1; \\
&loop &\rightarrow\quad &\textbf{do } n \neq 0 \rightarrow \\
&&&\quad f := f * n; \\
&&&\quad n := n - 1 \;_p\!\oplus\; n := n + 1 \\
&&&\textbf{od}
\end{aligned}
$$

The program *prog* is the whole of the above.

The decrementing of n fails probabilistically, sometimes incrementing instead.

Figure 2.5.3. FAULTY FACTORIAL

When p is one, making the program standard (and decrementing of n certain), the invariant $n \geq 0 \wedge N! = f*n!$ suffices in the usual way to show that $wp.prog.[f = N!] \Lleftarrow [N \geq 0]$. In general, however, that postcondition is achieved only if the decrement alternative is chosen on each of the N executions of the loop body, thus with probability p^N. In the probabilistic case therefore we define invariant $I := p^n * [n \geq 0 \wedge N! = f * n!]$, showing its preservation with this calculation:

$$p^n * [n \geq 0 \wedge N! = f * n!]$$

$\cdot \equiv$ applying $wp.(n := n - 1 \;_p\!\oplus\; n := n + 1)$

$$
\begin{aligned}
&p(p^{n-1}) * [n - 1 \geq 0 \wedge N! = f * (n - 1)!] \\
+\; &\overline{p}(p^{n+1}) * [n + 1 \geq 0 \wedge N! = f * (n + 1)!]
\end{aligned}
$$

$\Lleftarrow$ $p^n * [n > 0 \wedge N! = f * (n - 1)!]$ dropping the right additive term

$\cdot \equiv$ $p^n * [n > 0 \wedge N! = f * n * (n - 1)!]$ applying $wp.(f := f * n)$

$\equiv$ $[n \neq 0] * p^n [n \geq 0 \wedge N! = f * n!]$,   property of factorial

as required by Def. 2.2.1.

The exact termination condition depends on p. Standard *random walk* results [GW86] show that for $n \geq 0$ the loop terminates certainly when $p \geq 1/2$, but with probability only $(p/\overline{p})^n$ otherwise. In any case, however, we have $T \Lleftarrow p^n * [n \geq 0]$, since that is the probability of no subtraction errors occurring at all. Thus

$$I \equiv p^n * [n \geq 0 \wedge N! = f * n!] \Rrightarrow p^n * [n \geq 0] \Rrightarrow T,$$

so that Case 3 of Lem. 2.4.1 applies and $preE := I$. The post-expectation achieved is $[n = 0] * p^n * [n \geq 0 \wedge N! = f * n!] \Rrightarrow [N! = f]$.

We conclude with

$$wp.prog.[N! = f]$$

$\Lleftarrow$ $wp.init.preE$

$\equiv$ $(p^n * [n \geq 0 \wedge N! = f * n!]) \langle n, f \mapsto N, 1 \rangle$

$\equiv$ $p^N * [N \geq 0 \wedge N! = 1 * N!]$

$\equiv$ $p^N * [N \geq 0]$,

i.e. that the factorial is correctly calculated with probability at least p^N provided $N \geq 0$.

2.6   The Zero-One Law for termination

In this section we look more carefully at probabilistic termination on its own, showing that Case 3 of Lem. 2.4.1 implies a "Zero-One law" which, in turn, will justify our approach to probabilistic variants in Sec. 2.7.

Let *Inv* be some standard invariant for our usual (probabilistic) loop, so that

$$G \wedge Inv \quad \Rrightarrow \quad wp.body.Inv \,, \tag{2.10}$$

and suppose we have additionally that

$$\varepsilon \, [Inv] \quad \Rrightarrow \quad T \quad ^{24} \tag{2.11}$$

for some fixed real $\varepsilon > 0$, where as usual $T := wp.loop.1$. Informally (2.11) says that from every state satisfying *Inv* the probability of loop termination is at least ε.

We now show that under these conditions the probability of termination from *Inv* is not just ε — *in fact it is one.*

Define expectation $I' := \varepsilon \, [Inv]$; we then have

$$
\begin{array}{ll}
& [G] * I' \\
\equiv & [G] * \varepsilon \, [Inv] \\
\equiv & \varepsilon \, [G \wedge Inv] \\
\Rrightarrow & \varepsilon * (wp.body.[Inv]) \quad ^{25} & \text{from (2.10)} \\
\equiv & wp.body.(\varepsilon \, [Inv]) & \text{scaling for } body, \text{ Lem. 1.6.2} \\
\equiv & wp.body.I' \,,
\end{array}
$$

so that I' is also an invariant of *loop*. But from (2.11) we have $I' \Rrightarrow T$; and so we continue

$$
\begin{array}{ll}
& [Inv] \\
\equiv & I'/\varepsilon \\
\Rrightarrow & (wp.loop.(\,[\overline{G}] * I'))/\varepsilon & \text{Lem. 2.4.1 with invariant } I' \\
\equiv & wp.loop.(\,[\overline{G}] * I'/\varepsilon) & \text{scaling for } loop \\
\equiv & wp.loop.[\overline{G}] \wedge Inv \,,
\end{array}
$$

---

[24] As in normal algebra, we sometimes leave out the explicit multiplication sign...

[25] ...but here we use the multiplication sign explicitly since otherwise $\varepsilon(\cdots)$ might look like the traditional function application (for which however we write an explicit dot symbol ".").

showing that from any state satisfying Inv the loop terminates almost certainly in a state satisfying $\overline{G} \wedge Inv$.

In passing, we note that the above reasoning holds even for properly probabilistic invariants I — i.e. not just those of the form $[Inv]$ — and so we have shown that Case 3 of Lem. 2.4.1 can be relaxed to

$$3a. \qquad \text{If } \varepsilon * I \Rrightarrow T \text{ for some fixed } \varepsilon > 0, \\ \text{then define } preE := I. \tag{2.12}$$

(Recall Footnote 12 on p. 43.) That appears to be a very weak condition, since ε is arbitrary; in fact it requires only that in those states where I is nonzero, the termination probability T cannot be arbitrarily small.

If we continue to concentrate on standard invariants Inv, then the general effect we are seeing can be summarised as a *Zero-One law* for probabilistic processes, stated informally as follows.

Lemma 2.6.1 ZERO-ONE LAW FOR PROBABILISTIC PROCESSES
   Let a non-aborting process P act over some state space,[26] and suppose that from every state in a subset of states (described by) Inv the probability of P's eventual escape from Inv is at least ε, for some fixed $\varepsilon > 0$.[27]
   Then P's escape from Inv is almost certain: it escapes with probability one.
   Proof   There are many zero-one laws in probability theory, of which ours here is a special case [GS92, p. 189ff].   □

We can say more succinctly that the infimum over subset Inv of eventual escape probability is either zero or one: it cannot lie properly in between. (That is why Lem. 2.6.1 is called a "Zero-One" law.)
   Note that we do not require that for every state in Inv the probability of *immediate* escape is at least ε — that is a stronger condition, from which the certainty of eventual escape is much more obvious (since it is trivially at least $1 - \overline{\varepsilon}^n$ over n steps).

2.7   Probabilistic variant arguments for almost-certain termination

Termination of standard loops is conventionally shown using "variants" based on the state: they are integer-valued expressions over the state variables, bounded below but still strictly decreased by each iteration of the

---

[26]The non-aborting condition is not usually imposed in the literature, where "aborting" processes do not occur. We mention it to ensure that this lemma is not applied inappropriately to program fragments containing **abort**.
   [27]Note that these conditions are formalised exactly by (2.12) for $I := [Inv]$.

loop. That method is complete (up to expressibility) since informally one can always define a variant

the largest number of iterations possible from the current state,

which satisfies the above conditions trivially if the loop indeed terminates.

For probabilistic programs however the standard variant method is not complete (though clearly it remains sound): for example the program

$$\textbf{do } (n \bmod N) \neq 0 \rightarrow$$
$$n := n + 1 \quad {}_{1/2}\oplus \quad n := n - 1 \qquad (2.13)$$
$$\textbf{od}$$

is almost certain to terminate, yet from the fact that its body can both increment and decrement n it is clear there can be no strictly decreasing variant.

With the results of the previous section however we are able to justify the following variant-based rule for probabilistic termination, sufficient for many practical cases including (2.13).[28]

Lemma 2.7.1 VARIANT RULE FOR LOOPS   Let V be an integer-valued expression in the program variables, defined at least over all states satisfying some predicate *Inv*. Suppose further for our usual *loop* that

1. if the set of states satisfying $G \wedge Inv$ is not finite, then there are fixed integer constants L (low) and H (high) such that

$$G \wedge Inv \quad \Rightarrow \quad L \leq V < H \,, \qquad {}^{29}$$

and

2. the predicate *Inv* is a (standard) invariant for *loop* and

3. for some fixed probability $0 < \varepsilon \leq 1$ and for all integers N we have

$$\varepsilon * [G \wedge Inv \wedge (V{=}N)] \quad \Rightarrow \quad wp.body.[V < N] \,.$$

Then termination is certain from any state in which *Inv* holds: we have $[Inv] \Rightarrow T$, where T is the termination condition of *loop*.

Proof   Informally we argue as follows.

The probability of V's eventual escape from any point in the interval $[L..H)$ cannot be less than ε^{H-L}, since that is the (possibly remote, but still nonzero) probability of the $H - L$ consecutive decrements that would

---

[28]Later, in Chap. 7, we show it complete over finite state spaces: any finite-state loop which terminates with probability one must have a probabilistic variant that shows it does.

[29]Note that either way the existence of such L, H is guaranteed — but in the finite case there is no need to say explicitly what they are. The condition amounts to requiring that V take only finitely many values over $G \wedge Inv$.

cause eventual escape.[30] We then appeal to Sec. 2.6, using ε^{H-L} here for ε there.

See Chap. 7 for a full proof.   □

Lem. 2.7.1 shows termination given an integer-valued variant bounded *above and* below such that on each iteration a strict decrease is guaranteed *with at least fixed probability* $\varepsilon > 0$.[31] Note that the probabilistic variant *is allowed to increase* — but not above H. (We have emphasised the parts that differ from the standard variant rule.)

The termination of Program (2.13) now follows immediately from Lem. 2.7.1 with variant $n \bmod N$ itself, and $L, H := 0, N$ — trivially, every iteration strictly decreases $n \bmod N$ with probability at least $1/2$. [32]

In some circumstances it is convenient to use other forms of variant argument, related to Lem. 2.7.1; one easily proved from it is the more conventional rule in which

> the variant is bounded below (but not necessarily above); it must decrease with fixed probability $\varepsilon > 0$; and it cannot increase.

That rule follows (informally) from Lem. 2.7.1 by noting that since the variant cannot increase its initial value determines the upper bound H required by the lemma, and it shows termination for example of the loop

$$\textbf{do } n > 0 \rightarrow$$
$$n := n - 1 \;\; _{1/2}\oplus \textbf{ skip}$$
$$\textbf{od } ,$$

for which variant n suffices with $L := 0$.

2.8   Termination example: self-stabilisation

As an example of the variant technique, we apply Lem. 2.7.1 to Herman's *probabilistic self-stabilisation* [Her90], a distributed probabilistic algorithm that can be used for leadership election in a ring of synchronously executing processors.[33]

---

[30] For integers a, b, we write $[a..b)$ for the closed-open interval of integers i satisfying $a \leq i < b$.

[31] Since the variant is bounded above as well as below, we might just as well guarantee a strict *increase* with some probability; and in some problems that is more convenient. Clearly the two alternatives are of equal power.

[32] When $n \bmod N = N - 1$, in fact decrease is certain, *i.e.* has probability greater than $1/2$; but "at least" is sufficient. Note that in Program (2.13) the state space is not finite, which is why we must give L, H explicitly.

[33] In Herman's original presentation each processor contains a single bit, and the aim is via pairwise-local and symmetric processing to reach a stable global configuration of those bits (modulo rotation around the ring).   ...

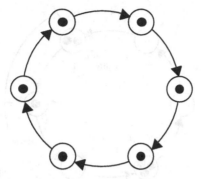

In this example we have $N = 6$ and, initially, one token in each processor.

Figure 2.8.1. HERMAN'S RING

We consider N identical processors connected clockwise in a ring, as illustrated in Fig. 2.8.1. A single processor — a *leader* — is chosen from them in the following way. Initially each processor is given exactly one token; the leader is the first processor to obtain all N of them.

Fix some probability p with $0 < p < 1$. On each step (synchronously) every processor performs the following actions:

- Make a local probabilistic decision either to *pass* (probability p) or to *keep* (probability $\bar{p}$) all its tokens.

- If *passing*, then send *all* its tokens to the next (clockwise) processor; if *keeping*, do nothing.

- Receive tokens passed (if any) from the previous (anticlockwise) processor, adding them to the tokens currently held (if any).

We show easily that with probability one eventually a single processor will have all N tokens. We define

- the invariant *Inv* to be that the total number of tokens is constant (at N),

- the guard G (true while the computation continues) to be that more than one processor holds tokens and

- the variant V to be the shortest length of any ring segment (contiguous sequence of arcs) containing all tokens. (See Fig. 2.8.2.)

---

[33]We transform (data-refine) the algorithm, first by introducing tokens to represent the equality or difference of adjacent processors' bits, and then by allowing the tokens to accumulate rather than requiring them to annihilate each other when they meet.

Our termination argument first appeared in the article on which this chapter is partly based [Mor96].

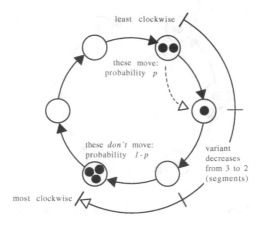

least clockwise

these move:
probability p

variant
decreases
from 3 to 2
(segments)

these *don't* move:
probability $1-p$

most clockwise

The variant is the length of a shortest segment containing all tokens. It decreases with nonzero probability at least $p * \overline{p}$ — but it might also go up.

Figure 2.8.2. HERMAN'S-RING VARIANT

With those definitions, for proof of termination we simply refer to Lem. 2.7.1:

- the state space is finite (so that we need not bound our variant),

- the invariant *Inv* is trivially maintained and

- the variant V decreases strictly with probability at least $p*\overline{p}$, which is nonzero since p itself is neither zero nor one — let the trailing (least-clockwise) processor in the shortest segment decide to *pass* while the leading (most-clockwise) processor decides to *keep*.

Thus Lem. 2.7.1 has given us almost-certain termination: that with probability one eventually only one processor contains tokens $(\overline{G})$, and that it has all N of them (Inv).[34]

---

[34]In Herman's original algorithm each processor has at most one token — and when two tokens collide, both disappear. The argument for termination in a state where exactly one processor has (now only) a single token is as above, but with invariant "the number of tokens is odd, and each process has at most one." (Thus the number of tokens must be odd initially, to establish the invariant.)

See Footnote 67 on p. 74 where we give the exact expected (quadratic) number of steps to termination for the case where there are initially three tokens. Elsewhere, we prove an expected worst-case steps-to-termination result of $O(N^2)$ [MM04a]; together with the quadratic result for three tokens, that gives expected complexity of $\Theta(N^2)$.

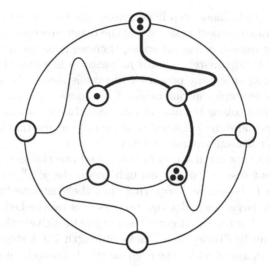

For each node, the probability that it passes all of its tokens to any directly connected neighbour (including itself) is at least some constant $p > 0$.

The variant is the smallest number of edges in any connected sub-graph that contains all tokens. It decreases with probability at least $p^N = p^{N-1} * p$, a lower bound on the probability that all nodes keep their tokens except one, which instead passes all its tokens to its unique neighbour in the graph.[35]

With probability one, all tokens collect in a single processor.

Figure 2.8.3. HERMAN'S-GRAPH VARIANT

2.8.1   Variations on the ring

The variant technique makes it easy to see that the use of a ring above is not essential for correctness: there are many other possibilities. In Fig. 2.8.3, for example, we see that in an arbitrary *undirected* connected graph, the variant is simply the smallest number of edges of any connected sub-graph containing all tokens.[35]

For a *directed* and strongly connected graph,[36] a suitable variant would be the length of a shortest directed path containing all tokens. (Our original ring is a special case of this; in general, however, there could be repeated edges.) Call a *min-straggler* any processor at the trailing end of some directed all-token-covering path of minimum length: with nonzero prob-

---

[35]That sub-graph must be acyclic, since otherwise its size could be decreased, without losing any tokens, by breaking a cycle; if it is acyclic then it must contain a univalent node; any univalent node must contain tokens, else it and its edge could be removed.

[36]That is, there is a directed path from any processor to every other.

ability, on any synchronous step it decreases the variant by passing all its tokens to the processor next along, while the other processors retain theirs.

If we consider *asynchronous* networks, then we must include an assumption of fairness. Unfortunately, *almost fairness* — in which the probability of the scheduler ignoring any processor forever is zero — is not enough in this case.[37] For example, in our original but now asynchronous directed ring, with just two tokens the almost-fair "schedule each processor until it passes its tokens one step clockwise" could avoid termination by circulating the tokens at a constant separation forever.[38]

But there are many variations of fairness, and one can simply choose one — or even invent one — "strong enough to do the job." In this case we propose so-called "k-fairness," requiring that there is some fixed integer k (no matter how large) such that no processor is overlooked more than k times in a row.[39] For termination we then argue intuitively that any execution sequence can be divided into blocks of length $k+1$ steps, and within each block the argument as for the original synchronous case applies: there is a nonzero probability that during the block the size of some separation measure will decrease.

More precisely, let the graph be directed and strongly connected. Say that the *priority* of a processor is the number of steps since it was last scheduled; and define

$$V_0 \quad := \quad \text{the smallest length of any directed path containing all tokens,}$$
$$\text{and} \quad V_1 \quad := \quad \text{the maximum priority among all min-stragglers.}$$

Now V_0 can take only finitely many values; and V_1, bounded below by zero, is bounded above by k due to our k-fairness scheduling policy. Thus construct the overall variant $V := (k+1)V_0 - V_1$,[40] and observe that on each step the processor scheduled is either

- a min-straggler, in which case with nonzero probability it passes its tokens along the min-path, and the expression $(k+1)V_0$ decreases by $k+1$ (and $-V_1$ can increase by no more than k); or the processor is

- not a min-straggler, in which case with nonzero probability it keeps its tokens so that $-V_1$ decreases and V_0 is unchanged.[41]

---

[37] It is however sufficient for the *Dining Philosophers* of Sec. 3.2.

[38] See however Footnote 17 on p. 93 concerning the dependence of the scheduler's almost-fairness on the algorithm it is scheduling.

[39] Unlike most forms of fairness, this k-fairness is a safety property.

[40] Using two or more variants like this, "layered" one within the other, is discussed more generally in Sec. 3.1.4.

[41] Since in this case the processor keeps its tokens, the overall configuration is unchanged and — in particular — whether or not some other processor is a min-straggler is unchanged also. Thus since no min-straggler was scheduled, V_1 must increase (by one).

Thus in an arbitrary finite, directed and strongly connected token-graph, if each processor when scheduled passes all its tokens to some directed neighbour (including possibly itself), with some nonzero probability for each neighbour, then even under demonic asynchronous scheduling the tokens are almost certain to end up all together in one processor — provided that the scheduling is k-fair.[42]

Finally, we mention *probabilistic* fairness, where on each asynchronous step there is a constant nonzero lower-bound probability that any given processor will be scheduled. This gives a very direct and simple argument: the nonzero probability of variant decrease is just the probability that the min-straggler will be selected *times* the probability that it passes its tokens upstream.

Although probabilistic fairness might be expensive to implement, for "real-world" scheduling caused by natural phenomena it might be a reasonable assumption.

2.9 Uncertain termination

In the arguments of Sec. 2.5, the cases where termination was "uncertain," *i.e.* occurred with guaranteed probability strictly less than one, were made by inspection. We are not aware of a general "variant-style" approach in such circumstances, and in many cases informal arguments suffice anyway. For completeness, however, we set out here the details of a straightforward and rigorous inductive argument based on the fact that wp for loops is defined as a least fixed-point.

We know for our usual G, *body* loop that we have

$$wp.loop.1 \quad\equiv\quad (\sqcup k\colon \mathbb{N} \bullet T_k) \quad {}^{43}$$

$$\text{where} \quad T_0 \;:=\; \lceil \overline{G} \rceil \qquad\qquad (2.14)$$
$$T_{k+1} \;:=\; 1 \textbf{ if } \overline{G} \textbf{ else } wp.body.T_k \;,$$

---

[42]In Sec. 3.2.1 we call demonic scheduling "adversarial."

[43]We write $\mathbb{N}$ for the NATURAL NUMBERS, the non-negative integers including zero.

The comprehension $(\sqcup k\colon \mathbb{N} \bullet T_k)$ gives the quantifier $\sqcup$, then the bound variable k and the set $\mathbb{N}$ from which it is drawn, then — following the " $\bullet$ " separator — the expression T_k, evaluated over all values of the bound variable within the given set, to which the quantifier is applied. The parentheses are always used: they give the scope of the bound variable.

a fact derived directly from the expression of least fixed-points as
ℕ-suprema, which in turn follows from the fact that programs denote continuous expectation-transformers.[44] Each T_k term is a lower bound for the
probability of termination in no more than k executions of the loop body;
and, in practice, using the above amounts to a formalisation of the inductive argument that would have been applied by inspection — as we now
illustrate.

2.9.1 Example: an inductive termination argument

We consider the program

$$\textbf{do } n \neq 0 \to \quad n := 0 \ {}_{1/n^2}\oplus\ n := n+1 \quad \textbf{od} ,$$

whose probability of termination lies strictly between zero and one. By
inspection we guess that the loop *fails* to terminate from initial state $n = N$
with probability given by the infinite product

$$(\prod_{n:=N}^{\infty} 1 - 1/n^2) ,$$

and in Case $N = 2$ we find we are looking at

$$3/4 * 8/9 * 15/16 * 24/25 * \cdots , \tag{2.15}$$

whence we notice that the partial products

$$
\begin{array}{cccc}
3/4 & 2/3 & 5/8 & 3/5 \cdots \\
\uparrow & \uparrow & \uparrow & \uparrow \\
\text{term} \quad 2 & 3 & 4 & 5 ,
\end{array}
$$

suggest $1/2 + 1/2n$ for the nth term — a fact easily verified by induction
over n and giving just $1/2$ in the limit as n tends to infinity.

In Case $N = 3$ the limit would be $4/3$ of that (to remove the first
multiplicative term $3/4$ in the product (2.15) above), *i.e.* $2/3$ and which
suggests $1 - 1/N$ in general; again this is easily verified by induction (but
this time on N).

Thus we guess that the probability of termination from the initial state in
which $n = N$ is just $1/N$, and that therefore for any $n > 0$ the probability
p of termination within k steps satisfies the equation

$$1/n \quad = \quad p + \overline{p}/(n+k) ,$$

expressing that termination overall $(1/n)$ is either termination within k
steps (p) or termination at step k or after $(\overline{p}/(n+k))$ — which gives
$k/(n*(n+k-1))$ for p.

---

[44]From (1.19) we have $wp.loop.1 \equiv (\mu Q \bullet wp.body.Q$ **if** G **else** $1)$, whose right-hand
side as a function of Q generates the inductive definition of T_k above when applied
repeatedly starting from zero.

So far, the reasoning is informal (though convincing).

If we need absolute formality, we would use the above to hazard that for $k \geq 1$ we have

$$T_k \;\equiv\; k/(n*(n+k-1)) \text{ if } n \neq 0 \text{ else } 1 \;, \qquad (2.16)$$

which — rigorously — we can prove by induction over k using (2.14). The equality $T_0 \equiv [n=0]$ comes from (2.14) directly; note that for $k \geq 1$ the conditional $n \neq 0$ avoids our having to consider the indeterminate $1/0$ in (2.16).

For $k+1$ we now reason

$$T_{k+1}$$
$$\equiv \quad 1 \text{ if } n = 0 \text{ else } wp.body.T_k \qquad\qquad (2.14)$$

$$\equiv \qquad\qquad\qquad\qquad\qquad\qquad \text{inductive hypothesis (2.16)}$$
$$1 \text{ if } n = 0 \text{ else } wp.body.(k/(n*(n+k-1)) \text{ if } n \neq 0 \text{ else } 1)$$

$$\equiv \qquad\qquad\qquad\qquad\qquad \text{definition } wp.body\,;\, \text{arithmetic}\,[45]$$
$$1$$
$$\text{if } n = 0 \text{ else}$$
$$1 \;_{1/n^2}\!\oplus\; (k/((n+1)*(n+k)) \text{ if } n+1 \neq 0 \text{ else } 1)$$

$$\equiv \quad 1 \text{ if } n = 0 \text{ else } 1 \;_{1/n^2}\!\oplus k/((n+1)*(n+k)) \qquad n \geq 0 \text{ is invariant}$$
$$\equiv \quad 1 \text{ if } n = 0 \text{ else } (k+1)/(n*(n+k)) \;, \qquad \text{expanding } _{1/n^2}\!\oplus$$

as required to prove (2.16) for all k. Thus — from (2.14) — we have

$$(\sqcup k\colon \mathbb{N} \bullet \;\; k/(n*(n+k-1)) \text{ if } n \neq 0 \text{ else } 1) \;,$$

that is

$$1/n \text{ if } n \neq 0 \text{ else } 1$$

as a lower bound for the overall termination probability $wp.loop.1$.

2.10 Proper post-expectations

Thus far we have concentrated mainly on probabilistic correctness, that is the (worst-case) probability that a given program will establish some postcondition. It is clear, though, that our reasoning tools can be applied to what we will call *proper* post-expectations, the more general case where

---

[45]For convenience — and by analogy with programs — we write $exp_1 \;_p\!\oplus exp_2$ for $p * exp_1 + \overline{p} * exp_2$.

they are not of the form $[post]$ — and that in turn allows us to ask more general questions about programs' expected behaviour.

For example, if we were interested in a program's running time, in terms of the number of loop iterations, we could include an extra variable n in the program text and arrange for it to be incremented on each iteration.[46] Its final value then gives the number of iterations required, calculated "in parallel" with whatever calculation the program was designed to perform.[47] To prove an upper-bound N for that complexity, we would write $n \le N$ for the postcondition.

In the probabilistic setting the same technique applies — but we have extra opportunities as well. A post-expectation of $[n \le N]$ delivers as a pre-expectation the probability that no more than N steps are taken — but a post-expectation of just n itself will deliver a lower bound for the *expected* number of steps, which is something we could not do before.[48]

As our first example of proper post-expectations, we return to the martingale program of Sec. 2.5.1.

2.10.1   The martingale revisited

We modify our earlier program (Fig. 2.5.1) so that termination is certain, occurring either when the gambler finally wins a bet or when he runs out of money; the new version is shown in Fig. 2.10.1.

Note informally that on each iteration the gambler's expected win is exactly zero: with probability $1/2$ he gains b; and with probability $1/2$ he loses b. Our demonic/pessimistic wp establishes easily than indeed his expected win is no less than zero:

$$
\begin{array}{lll}
& c & \\
\cdot \equiv & c + 2b \;\; {}_{1/2}\oplus \;\; c & \text{applying } wp.(c, b := c + 2b, 0 \;\; {}_{1/2}\oplus \;\; b := 2b) \\
\cdot \equiv & c + b \;\; {}_{1/2}\oplus \;\; c - b & \text{applying } wp.(c := c - b) \\
\equiv & c \;. &
\end{array}
$$

That is, by showing $c \equiv wp.body.c$ we verify that his expected capital c after a bet is at least its value before the bet.

The "at least" in the previous sentence comes not from an inequality in the calculation — for all the relations were $\equiv$ — but rather from the

---

[46]This is the approach advocated for example by Hehner [Heh89] for treatment of program termination in general.

[47]In fact this is an example of the well-known technique of program SUPERPOSITION, where an extra "monitoring" layer is superposed on the original computation without disturbing it. Superposition is, in turn, an example of *data refinement*, which we discuss in Chap. 4.

[48]One could argue that an upper bound is more likely to be useful, but lower bounds are what wp delivers and so — at least for the moment — we are stuck with that. But at (2.21) further below we show how to adapt our techniques to determine upper and exact bounds as well.

init	$\rightarrow$	$c, b := C, 1;$
loop	$\rightarrow$	**do** $0 < b \leq c \rightarrow$
body	$\rightarrow \quad \big\{$	$\quad c := c - b;$
		$\quad c, b := c + 2b, 0 \;\; {}_{1/2}\oplus\; b := 2b$
		od

The program *prog* is the whole of the above.

The gambler's capital c is initially C, and his intended bet b is initially one.

On each iteration, if his intended bet does not exceed his capital, he is allowed to place it and has $1/2$ chance of winning.

If he wins, he receives twice his bet in return, and sets his intended bet to 0 to indicate he is finished; if he loses, he receives nothing and doubles his intended bet — hoping to win next time.

The gambler stops either because his bet has been set to zero (because he won) or because his intended bet has become more than he can afford.

Figure 2.10.1. THE MARTINGALE, REVISITED

---

inherent pessimism built-in to *wp*, which resolves demonic choice towards lower pre-expectations. In this *deterministic* program, however, there is no demonic choice, and so the "at least" is overly cautious: knowing that we have calculated the exact greatest pre-expectation, and that the program is deterministic, we can take the result as exact. We formalise that as follows.

To verify — with less hand-waving — that his expected winnings are indeed no *more*- (as well as no less) than zero, as a coding trick we take post-expectation $C + 1 - c$, that is how far he has to go towards his goal of an overall increase of one unit. Almost the same arithmetic as above shows

$$C + 1 - c \quad \equiv \quad wp.body.(C + 1 - c) \,, \tag{2.17}$$

which as before establishes a lower bound of zero, this time on the expected change in $C + 1 - c$. But that becomes an *upper* bound of zero on the expected change in c itself, since $C+1$ is constant and c occurs in a negative sense — and so we have that his expected winnings in one iteration are *exactly* zero.

Now there are two issues raised by the technique illustrated at (2.17): the first $(\dagger)^{49}$ is that tricky expectations like the $C + 1 - c$ above must still

---

[49] We employ symbols $\dagger, \ddagger$ for "local" cross-references; the target is indicated with a $\dagger$ in the margin, and occurs within a few paragraphs of the point ($\dagger$) of reference.

be non-negative and bounded above;[50] and the second (‡) is that we don't want to repeat trivial calculations as at (2.17) if we have effectively done them already.

†   For the first issue, in fact we "know" that $C + 1 - c$ is non-negative because of the standard loop-invariants $c + b = C + 1$ (recall p. 45) and $b, c \geq 0$ (easily verified here). This appealing style of "reasoning within a standard invariant" is an extension of the modular reasoning we saw earlier in Sec. 1.7, but now applied to the multiple iterations of a loop rather than to a single program fragment. Both standard predicates are now the single *Inv*, which we call an *auxiliary* invariant. Its use is justified by the following lemma, whose proof is similar to the proof of Lem. 1.7.1:

Lemma 2.10.2 REASONING WITHIN A LOOP INVARIANT   For our usual loop G, *body* let both $[Inv]$ and I be invariants, separately. That is, suppose we have

$$
\begin{aligned}
G \wedge Inv &\quad\Rightarrow\quad wp.body.Inv \\
\text{and} \quad [G \wedge Inv] * I &\quad\Rightarrow\quad wp.body.I . \qquad\quad \text{[51]}
\end{aligned}
$$

Then in fact $[Inv] * I$ is an invariant of the loop as well.   □

The importance of Lem. 2.10.2 is that it justifies our using the two invariants separately: we use invariant $C+1-c$ to investigate the post-expectation of primary interest; and separately we prove invariance of an auxiliary, "well-formedness" condition for it, in this case the predicate

$$
c + b = C + 1 \quad \wedge \quad b, c \geq 0
$$

which guarantees $0 \leq (C + 1 - c) \leq C + 1$. (See p. 70 however for further comments on this calculation.)

‡   For the second issue, we note that *deterministic* programs satisfy the stronger property of *linearity* (as opposed to sublinearity alone): if *prog* is (pre-)deterministic, then we have

$$
\begin{aligned}
&wp.prog.(c_1 * postE_1 + c_2 * postE_2) \\
\equiv\quad &c_1 * wp.prog.postE_1 \;+\; c_2 * wp.prog.postE_2 ,
\end{aligned} \qquad (2.18)
$$

for non-negative reals c_1, c_2 and expectations $postE_1, postE_2$.[52] A sufficient — but not necessary[53]— condition for a program to be deterministic is of course that it contains no occurrence of $\sqcap$.

---

[50]We have not emphasised "bounded above" so far, though it is one of our assumptions about expectations generally. Below we will see that it is sometimes crucial.

[51]Note how we use the auxiliary *Inv* to help prove the invariance of I.

[52]We return to this, strengthening it, at Def. 8.3.2.

[53]Program $x{:=}\,1 \sqcap x{:=}\,1$ is deterministic, for example.

For deterministic *and terminating* programs *prog* we can therefore avoid repeating calculations like (2.17); for if

$$preE \quad \equiv \quad wp.prog.postE \quad ^{54} \tag{2.19}$$

and if — as above — we can appeal to a standard invariant *Inv* assuring $postE \leq H$ for some constant H, then we reason for any state satisfying *Inv* that

$$
\begin{array}{lll}
& wp.prog.(H - postE) & \\
=^{55} & wp.prog.H \;-\; wp.prog.postE & \text{refer (2.18)} \\
= & H * wp.prog.1 \;-\; wp.prog.postE & \text{scaling} \\
= & H \;-\; wp.prog.postE & \textit{prog} \text{ is terminating} \\
\\
= & H \;-\; preE \,. & \text{refer (2.19):} \\
& & \text{note that} \equiv \text{there is necessary} \\
& & \text{to avoid inappropriate} \leq \text{here}
\end{array}
$$

Thus from (2.19) we know directly that the expected final value of *postE* is *at least preE*; and from the subsequent calculation we know that the expected final value of $H - postE$ is at least $H - preE$, *i.e.* that the expected final value of *postE* is *at most preE* (because H is constant). Thus the expected final value of *postE* is exactly *preE*.

To summarise, we can say that (2.19) means that *preE* is the exact expected final value of *postE* whenever *prog* is deterministic and terminating, and *postE* is bounded above (possibly within some auxiliary invariant).

For loops as a whole, unfortunately, we do not usually have an exact pre-expectation like (2.19) to work from — even when the loop is deterministic. That is because the conclusion of our loop rule Lem. 2.4.1 is intrinsically an inequality $preE \Rightarrow \cdots$. We however reach similar conclusions if our invariant is *exact* in this sense, that whenever the guard G holds (and possibly some auxiliary invariant as well) we have

---

[54]Note that the equality $\equiv$ does *not* mean "the expected final value of *postE* after *prog* is exactly *preE*" — that is, after all, precisely what we are trying to establish. Rather it means "the GREATEST LOWER BOUND of the possible expected final values of *postE* is exactly *preE*". The significance of the $\equiv$ is that it is necessary for the conclusions we are about to draw.

$$I \quad = \quad wp.body.I \ . \quad ^{55} \tag{2.20}$$

For an exact and bounded invariant I, provided the loop body is deterministic and terminating, we therefore have that

> in any initial state from which loop termination is guaran-
> teed, an exact invariant I gives the exact expected value of   (2.21)
> any expression $postE$ that agrees with I everywhere on $\overline{G}$. [56]

Thus finally we can return to Fig. 2.10.1. We have shown that c is an exact invariant, and that it is bounded implicitly within the accompanying standard invariant; and it is easy to see that termination of the loop is certain (because variant b is set to zero with probability $1/2$ on each iteration). Thus we conclude that the gambler's expected final capital c is in fact C exactly, which is just what he started with.

What he had hoped for, of course, was that $C + 1 \Rrightarrow wp.prog.c$ — but that is not what he got.

2.11   Bounded *vs.* unbounded expectations

We conclude the chapter by looking more closely at the importance — or not — of expectations' being non-negative and bounded, constraints which we have so far accepted because they assure well-definedness of wp in a straightforward way. At the core of our approach is the operation of *taking the expected value*, for us of a so-called post-expectation over a distribution of final states realised by executing a program. Even if the program reaches a potentially infinite set of final states, that expected value is well defined for all non-negative and bounded post-expectations — which is exactly why we impose those conditions.

---

[55] As noted in Footnote 51 on p. 33, ordinary "=" is appropriate in these two places because the equality is contingent on the preceding text, "whenever Inv holds" or "whenever G holds."

In the latter case, we must avoid the stronger $[G] * I \equiv wp.body.I$ that first comes to mind, since it would not usually be true: equality outside G is irrelevant, because at the beginning of every iteration G holds unconditionally. In fact what we are using is

$$[G] * I \quad \equiv \quad [G] * wp.body.I \ .$$

[56] Saying the loop body must be deterministic is strictly more demanding than saying that the loop as a whole must be deterministic, since demonic choice "along the way" can be cancelled out by the time the loop terminates. Consider

```
do x > 0 →                      =      x:= x min 0 .
    x:= x−1 ⊓ x−2;
    x:= x max 0
od
```

$$
\begin{array}{lll}
init & \rightarrow & b, n := \text{true}, 0; \\
loop & \rightarrow & \textbf{do } b \rightarrow \\
body & \rightarrow \left\{ \begin{array}{l} \\ \end{array} \right. & \quad b := \text{true } {}_{1/2}\oplus \text{false}; \\
& & \quad n := n + 1 \\
& & \textbf{od}
\end{array}
$$

The program *prog* is the whole of the above; its termination with probability one is shown by variant $[b]$, which termination is used in both $(\dagger, \ddagger)$ of the following analyses.[57]

† To establish the final distribution achieved by *prog*, we use post-expectation $[n = N]$; the pre-expectation will give us the probability that it is achieved, for any N.

The probability of establishing $n = N$ is (we guess) $[n < N]/2^{N-n}$ if b is true, and just $[n = N]$ itself if b is false. That suggests the invariant

$$
[n < N]/2^{N-n} \quad \textbf{if } b \textbf{ else} \quad [n = N] \ ,
$$

which is readily verified and gives overall pre-expectation (using *wp.init*) of $[0 < N]/2^N$ **if** true **else** $[0 = N]$, that is just $1/2^N$ provided $N \geq 1$.

‡ For the expected value of n after running *prog* we use instead the post-expectation n itself, and we guess that its expected final value is $n + K$ for some constant K, provided b is true; if b is false, its expected final value is just n. That suggests invariant $(n + K)$ **if** b **else** n, and to determine K we apply *wp.body*:

$$
\begin{array}{lll}
& n + K \textbf{ if } b \textbf{ else } n & \\
\equiv & n + 1 + K \textbf{ if } b \textbf{ else } n + 1 & \text{applying } wp.(n := n + 1) \\
\equiv & n + 1 + K {}_{1/2}\oplus n + 1 & \text{applying } wp.(b := \text{true } {}_{1/2}\oplus \text{false}) \\
\equiv & n + 1 + K/2 \ . &
\end{array}
$$

For exact invariance we must have $(n + K \textbf{ if } b \textbf{ else } n) = n + 1 + K/2$ whenever the guard b holds, for which $K = 1 + K/2$ is sufficient. That gives $K = 2$, and the expected final value of n is thus $0 + 2$ **if** true **else** 0 — that is, the loop runs for an expected two iterations.

Figure 2.11.1. A GEOMETRIC DISTRIBUTION

For mixed-sign or unbounded post-expectations, however, well-definedness is not assured: such cases must be treated individually. Consider the program of Fig. 2.11.1, for example, which has infinitely many potential final states — albeit with sharply diminishing probability as n increases.

---

[57]We could write the program more simply as

$$
n := 1; \ \textbf{do } 1/2 \rightarrow n := n+1 \ \textbf{od} \ ,
$$

but our current invariant/variant techniques do not apply to "probabilistic guards." Section 7.7 shows how to extend them.

The program is well defined, terminating in state $n = N$ with probability $1/2^N$ for $N \geq 1$: it generates an instance of the *geometric distribution* over the final value of n. (See Fig. 7.7.9 however for a more concise way of writing the program.)

Now if we choose the mixed-sign post-expectation $(-2)^n$, its expected value over that distribution would be

$$(\sum_{N:=1}^{\infty} (-2)^N/2^N) \quad = \quad -1 + 1 + -1 + 1 \cdots \tag{2.22}$$

That is, although *prog* itself may be well defined, the greatest pre-expectation $wp.prog.(-2)^n$ is not — and that is a good reason for avoiding mixed signs in general.

Stepping back from the brink, we can consider unbounded but same-sign, say non-negative expectations — and then the indeterminate sum (2.22) cannot occur. In fact, if we adjoin a formal ∞ to the non-negative reals, the operation of taking expected value is well defined even for unbounded expressions.[58] Thus for example in Fig. 2.11.1 we have

$$wp.prog.(2^n) \quad \equiv \quad (\sum_{N:=1}^{\infty} 2^N/2^N) \quad \equiv \quad 1 + 1 + \cdots \quad \equiv \quad \infty .$$

Because of that well-definedness, we can discuss examples such as the martingale (Sec. 2.10.1) with post-expectation c which, though unbounded, still has a well-defined expected value if we allow ∞. As explained in Lem. 2.10.2, we actually use the bounded invariant

$$[c + b = C + 1 \wedge b, c \geq 0] \ast c \tag{2.23}$$

in the loop rule Lem. 2.4.1, reflecting the fact that although c itself is unbounded, in practice it is effectively kept within bounds by other constraints built into the program. Since (2.23) is no more than (the unbounded) c, the pre-expectation calculated for it will also be a pre-expectation for c — which is what we are interested in.

In the case of the geometric distribution program of Fig. 2.11.1, however, we cannot use the same argument for the second analysis (‡) — for in that case n actually does increase potentially without bound, though with exponentially-low probability.

In the next section we look more closely at unbounded invariants and, in particular, whether and when we can use them safely.

---

[58]Unbounded non-*positive* expectations can be similarly treated (using $-\infty$) and we will occasionally make use of them. The advantage of non-negative expectations, however, is that we know more about their algebra — for example that they satisfy sublinearity Def. 1.6.1 and its consequences (Sec. 1.6).

$$
\begin{array}{lll}
init & \rightarrow & n := 1; \\
loop & \rightarrow & \textbf{do } n \neq 0 \rightarrow \\
& & \quad n := n+1 \;_{1/2}\!\oplus\; n-1 \\
& & \textbf{od}
\end{array}
$$

The program *prog* is the whole of the above; it is an example of a *random walk* over the integers. It can be shown (Sections 3.3 and 10.4) that this program terminates with probability one.

Expectation n is trivially an invariant of *loop*; thus we might reason

$$
\begin{array}{lll}
& wp.prog.n \\
\Lleftarrow & wp.prog.([n=0]*n) & \text{monotonicity of } wp.prog \\
\Lleftarrow & wp.init.n & \text{loop rule Lem. 2.4.1 } applied\ inappropriately\ (\dagger) \\
\equiv & 1\,,
\end{array}
$$

concluding that the expected value of n on termination is at least one, obviously incorrect since on termination n is exactly zero.

† The use of the loop rule is inappropriate because the invariant n is not bounded.

Figure 2.11.2. THE SYMMETRIC RANDOM WALK AS A COUNTER-EXAMPLE

2.11.1 Unbounded invariants: a counter-example

An important limitation remains, in spite of our ability to use unbounded post-expectations as at (2.23) above: it is that the loop rule Lem. 2.4.1 is sound only for bounded non-negative invariants, as the counter-example of Fig. 2.11.2 shows.[59]

Unfortunately, for Fig. 2.11.2 there is no standard invariant of the random walk that we can use in the style of Lem. 2.10.2: we say that such invariants are *intrinsically* unbounded.

Yet we do wish to have access to unbounded invariants for performance measures, as in Fig. 2.11.1 where invariant $(n+K)$ **if** b **else** n is intrinsically unbounded as well — as indeed would be the case for any loop containing a "loop counter" variable for determining the expected number of a possibly unbounded number of iterations.[60]

---

[59]In fact it is only Case 2 that requires boundedness explicitly, because Cases 1 and 3 imply it.

[60]Such "unboundedly iterating" loops do not occur in elementary treatments of standard sequential programs, where it is common to observe that from their terminating initial states all programs are "image-finite," or have "only bounded nondeterminism" if everyday programming constructs are used [Dij76, Chap. 9]; semantically, that amounts to saying that such programs are *continuous* (here our Lem. 5.6.6, and corresponding to a similar property for standard programs [Dij76, Property 5 p. 72]). That is the reason, for example, that on p. 21 in Sec. 1.5.2 we insisted that the abbreviation for demonic choice require finiteness of the choice.   . . .

It can be shown that

> a sufficient condition for an unbounded invariant to be used
> soundly in the loop rule (Case 2) is that the greatest ex-
> pected value of $[G] * I$, that is of the invariant evaluated   (2.24)
> only "within the loop" as the body is about to execute, is
> guaranteed to tend to zero as the loop continues to iterate.[61]

For the geometric distribution we observe informally that the probability
of Fig. 2.11.1's exceeding N iterations is no more than $1/2^N$; that in N
iterations the value of n cannot exceed N; and thus that the expected value
of $[G] * I$ in this case is no more than $N/2^N$, which indeed tends to zero as
N increases.

A more interesting example is given in Fig. 2.11.3. Because $a + b + c$ is
invariant in that program, the "original" state space is finite,[62] a subset
of the possible triples $\{a, b, c \mid 0 \le a, b, c \le A + B + C\}$.[63] Only the
"superposed" counter n is unbounded.

It is thus a form of bounded two-dimensional random walk, and its termination is almost certain.[64] Indeed, since $A + B + C$ consecutive losses for
any player is guaranteed to cause termination, the probability of continuing
to play decreases exponentially.[65]

[...][60] A significant and distinguishing feature of probabilistic programs is that unbounded
probabilistic choice is common, and does not lose continuity.

The geometric-distribution program Fig. 2.11.1 is a simple example of that, since it
can set n finally to any positive integer. Attempting however to replace its $_{1/2}\oplus$ with
the closest standard equivalent $\sqcap$, to produce an unboundedly nondeterministic *standard*
program, simply results in **abort** instead.

[61] The "is guaranteed to" turns out to be important: if the loop body is demonic, then
the limit of zero must apply for all its deterministic refinements. (If it's deterministic
itself, then the "for all refinements" is trivial of course.)

A sketch proof of (2.24) is given in Footnote 9 on p. 330.

[62] The same applies to the geometric-distribution program.

[63] We use the same syntax for set comprehensions as we do for comprehensions in
general, except that the brackets $\{\cdots\}$ do "double duty," both giving the scope of the
bound variables and indicating that we are constructing a set. In this case the bound
variables a, b, c vary over their implicit type $\mathbb{N}$ but are further constrained by the predicate $0 \le a, b, c \le A + B + C$. Since no " $\bullet$ " appears, the default expression is used, the
triple (a, b, c) of bound variables as given.

[64] The invariant $a + b + c$ gives it only two degrees of freedom (rather than three): the
walker moves randomly over a triangle, until eventually — with probability one — he
falls off one of the edges.

Interestingly, $(a - b) \bmod 3$ *etc.* are also invariant, which makes it clear that he can
reach only a subset of the triangle.

[65] The probability that a fixed player will not lose every one of a block of $A+B+C$
consecutive rounds is $p := 1 - (2/3)^{A+B+C}$; the probability that the player survives N
such blocks is no more than p^N.

$$a, b, c := A, B, C;$$
$$n := 0;$$
$$\textbf{do } a \neq 0 \wedge b \neq 0 \wedge c \neq 0 \rightarrow$$
$$a, b, c := a{-}1, b{-}1, c{-}1;$$
$$\left|\begin{array}{l} a := a + 3 \quad @ \; 1/3 \\ b := b + 3 \quad @ \; 1/3 \\ c := c + 3 \quad @ \; 1/3 \, ; \end{array}\right.$$
$$n := n{+}1$$
$$\textbf{od}$$

This program is based on the following game (but simplifies its presentation). Three gamblers start with A, B, C coins respectively, and on each round:

- each player commits one coin;
- the coins are flipped repeatedly until
- one player is "the odd man out,"
- and he collects all three coins.

The game is over when one player is broke; variable n counts the number of rounds until that happens.

Figure 2.11.3. THE "THREE-UP" GAME

Remarkably, the expression $n + abc/(a{+}b{+}c - 2)$ is an exact invariant for the loop.[66] We have

$$n + abc/(a{+}b{+}c - 2)$$

$\cdot \equiv \quad n + 1 + abc/(a{+}b{+}c - 2)$ applying $wp.(n := n + 1)$

$\cdot \equiv \quad n + 1 + \dfrac{(a+3)bc + a(b+3)c + ab(c+3)}{3(a+b+c+1)}$ applying $wp.$ "three-way choice"

$\cdot \equiv$ applying $wp.(a, b, c := a{-}1, b{-}1, c{-}1)$

$\quad n + 1 + \dfrac{(a+2)(b-1)(c-1) + (a-1)(b+2)(c-1) + (a-1)(b-1)(c+2)}{3(a+b+c-2)}$

$\equiv \quad n + 1 + \dfrac{abc - (a+b+c) + 2}{a+b+c-2}$ arithmetic

---

[66]This problem went unsolved for twenty-five years, until the correct invariant was discovered in 1966 [Hon03, p103]. It is presented as *a* solution to a recursion, including the comment "Since recursions can have more than one solution... we need to show that our result is the only [one]... by a standard argument in the theory of absorbing chains [*op. cit.*, p. 107]." Feller is cited (*e.g.* [Fel71]) for the more advanced techniques.

Our theory combining invariance and termination has this extra reasoning "built-in," which may account for some of the intricacy of the proof of Lem. 2.4.1. We take advantage of it here by using *exact* invariants (2.21).

$\equiv \qquad\quad n + abc/(a{+}b{+}c - 2)$.   arithmetic

Now because the invariant increases only linearly (remember a, b, c are bounded above, so only n's increase is important in the long run), while the probability of continuing decreases exponentially, we can appeal to (2.24) above to justify our use of the loop rule to conclude that the expected value of n on termination is given by the calculation

$$
\begin{array}{llr}
 & n + abc/(a{+}b{+}c - 2) & \text{invariant} \\
\cdot \equiv & abc/(a{+}b{+}c - 2) & \text{applying } wp.(n{:}= 0) \\
\cdot \equiv & ABC/(A{+}B{+}C - 2) \,, & \text{applying } wp.(a, b, c{:}= A, B, C)
\end{array}
$$

since the negated loop guard (a prerequisite of termination) zeroes the right summand of the invariant, leaving just n for the overall post-expectation. A game starting with *e.g.* 10 coins each would thus take $10^3/28 \simeq 35.7$ rounds on average.[67]

For the *unbounded* random walk, however, we cannot argue as above to show that the expected value of $[n \neq 0] * n$ tends to zero as the number of iterations increases. Indeed, since n is invariant we know that the expected value of just n itself remains exactly its initial value one, no matter how many iterations have occurred; we know also, from simple arithmetic, that

$$ n \quad \equiv \quad [n = 0] * n \ + \ [n \neq 0] * n \ ; $$

and trivially the left-hand summand $[n = 0] * n$ is identically zero. So the expected value of $[G] * I$, in this case $[n \neq 0] * n$, remains at its initial value of one, and does *not* tend to zero as the loop continues to iterate. Figure 2.11.4 illustrates that for the first few steps.

We return to the random walk in Sec. 3.3, where in a more general setting again we see that bounding the invariant is necessary.

2.12   Informal proof of the loop rule

Finally, the above discussion provides an intuitive explanation for the soundness of the most general Case 3a (at (2.12) on p. 54) of our loop rule Lem. 2.4.1 for bounded invariants, an argument which we now summarise for deterministic programs.

---

[67]This result can be adapted to give an *exact* result for the expected time-to-termination of Herman's Ring (Sec. 2.8) in the case where the initial number of tokens is three and their separations are A, B, C places: it is $4ABC/(A{+}B{+}C)$ [MM04a].

Variable n (columns) is the random walker's distance from his goal, back at the origin where $n = 0$; his steps (rows) however can move away from as well as towards his goal.

He starts in Step 0, located at $n = 1$ with probability one.

$n =$	0	1	2	3	4	5	6	7
Step								
0		1						
1	.5		.5					
2	.5	.25		.25				
3	.625		.25		.125			
4	.625	.125		.1875		.0625		
5	.6875		.1563		.125		.0313	
6	.6875	.0781		.1406		.0781		.0156

Probability distribution, over n.

For $n \neq 0$,
the expected value
of n remains one...

$* 1$		$* 3$		$* 5$		$* 7$	
↓		↓		↓		↓	
.0781	+	.4219	+	.3906	+	.1094	$= 1.0$

...even though the total probability
of being in this region where $n \neq 0$
tends to zero as the iterations continue...

...because the probability of termination tends to one.

After 6 steps, the walker has reached his goal $n = 0$ with probability .6875 and, as the number of steps increases, that probability will tend to one — so that the probability he is still in the region $n \neq 0$ tends to zero.

In spite of that, the expected value of n over the region $n \neq 0$ remains one no matter how many steps are taken; above for example we calculate the expected value $.0781 + .4219 + .3906 + .1094 = 1$ at Step 6.

This "paradoxical" fact — that the expected value of an expression can remain nonzero over a region whose distribution tends to zero in the limit — can occur only if the expression is unbounded. That is what accounts for the potential failure of the loop rule for unbounded invariants.

Figure 2.11.4. THE RANDOM WALK, TABULATED

1. The general condition (2.12) relating invariant I and termination probability T is that we should have $\varepsilon * I \Rightarrow T$ for some $\varepsilon > 0$.[68] Assume that holds.

2. Choose arbitrary $\varepsilon_1, \varepsilon_2 > 0$.

3. Since trivially $T \geq \varepsilon_1$ everywhere in the region $G \wedge (T \geq \varepsilon_1)$, by the Zero-One Law the probability of eventual escape from there is one.[69]

4. Thus by iterating sufficiently often we can make the probability that we have not yet terminated, *but* that there remains a probability ε_1 that we eventually will, as small as we like. That is, from (3) we can choose N sufficiently large so that

$$\mathrm{Pr}_n.(G \wedge (T \geq \varepsilon_1)) \leq \varepsilon_2$$

for all $n \geq N$, where by Pr_n we mean the probability determined by the distribution of states after n steps.

5. We now do some arithmetic: after those $n \geq N$ steps we have

$$\mathrm{Exp}_n.([G] * I)$$

$=$ $\qquad\qquad\qquad\qquad\qquad$ Exp_n distributes $+$

$$\mathrm{Exp}_n.([G \wedge (T < \varepsilon_1)] * I) \;+\; \mathrm{Exp}_n.([G \wedge (T \geq \varepsilon_1)] * I)$$

$\leq$ $\qquad\qquad\qquad$ $I \Rightarrow T/\varepsilon$ from (1); $[T < \varepsilon_1]$ multiplier

$$\mathrm{Exp}_n.([G \wedge (T < \varepsilon_1)] * \varepsilon_1/\varepsilon) \;+\; \mathrm{Exp}_n.([G \wedge (T \geq \varepsilon_1)] * I)$$

$\leq$ $\qquad$ $\varepsilon_1/\varepsilon \;+\; \mathrm{Exp}_n.([G \wedge (T \geq \varepsilon_1)] * I)$ $\qquad$ $[G \wedge (T < \varepsilon_1)] \Rightarrow 1$

$\leq$ $\qquad\qquad\qquad\qquad$ $I \Rightarrow T/\varepsilon \Rightarrow 1/\varepsilon$ from (1); Exp_n scales

$$\varepsilon_1/\varepsilon \;+\; \mathrm{Exp}_n.[G \wedge (T \geq \varepsilon_1)] * (1/\varepsilon)$$

$\leq$ $\qquad$ $(\varepsilon_1 + \varepsilon_2)/\varepsilon \,,$ $\qquad$ expected value of characteristic function
$\qquad\qquad\qquad\qquad\qquad\qquad\qquad$ is probability, and (4)

where by Exp_n we mean the function taking the expected value of its argument over the distribution given by Pr_n.

6. Since ε is fixed, and the positive $\varepsilon_1, \varepsilon_2$ are otherwise unconstrained, we can make $(\varepsilon_1 + \varepsilon_2)/\varepsilon$ arbitrarily close to zero: thus from (5) we

---

[68] Remember that this implies boundedness of I, since $T \Rightarrow 1$; and boundedness is of course exactly what we do not have in the random walk tabulated in Fig. 2.11.4, where n can be arbitrarily large.

[69] This is not a circular argument: we are merely presenting the same result in two different ways. One way is to prove Lem. 2.4.1 in our expectation logic (Chap. 7), and to observe that the Zero-One Law is a consequence of it; the other way is to give an argument in the semantics directly, as we do here, appealing to the Zero-One Law from the probability literature [GS92].

have that

$$(\lim_{n \to \infty} \text{Exp}_n.([G] * I)) \quad = \quad 0 \,.$$

7. Because the invariant's expected value does not decrease, after n steps $\text{Exp}_n.I$ is no less than I's initial value.

8. But in the limit that expected value of I must come entirely from the region $\overline{G}$; we reason

	"initial value of I"	
$\leq$	"expected value of I as $n \to \infty$"	from (7)
$=$	$(\lim_{n \to \infty} \text{Exp}_n.I)$	
$=$		distribution of $+$ through lim
	$(\lim_{n \to \infty} \text{Exp}_n.([\overline{G}] * I)) + (\lim_{n \to \infty} \text{Exp}_n.([G] * I))$	
$=$	$(\lim_{n \to \infty} \text{Exp}_n.([\overline{G}] * I)) + 0$	from (6)
$=$	$(\lim_{n \to \infty} \text{Exp}_n.([\overline{G}] * I)) \,.$	

So we have shown that if $\varepsilon * I \Rrightarrow T$ for some $\varepsilon > 0$ then the expected value of $[\overline{G}] * I$ is in the limit no less than I's initial value — which is essentially Case 3a of the loop rule.

Chapter notes

The invariant technique — first proposed by Floyd [Flo67] and incorporated by Hoare into his programming logic [Hoa69] in the early 1970's — remains the principal intellectual tool for developing and reasoning about loops in program code.[70]

Although many proof frameworks for verification of probabilistic *systems* are not based explicitly on Hoare-logic (exceptions include Harel and Feldman [FH84] and den Hartog and de Vink [dHdV02], as mentioned on p. 36), nevertheless many of them do make use of *fixed points*, which is the fundamental mathematical idea underlying the invariant proof method. The basic temporal operators of *pCTL*, for example, have a fixed-point semantics (see Chapters 9 and 11) — similarly

---

[70]The idea is much used elsewhere as well, for example in transition-style reasoning for state-based concurrency. That suggests a line of research into probabilistic concurrent systems based on expectations as here, perhaps in the style of Back and Kurki-Suonio's *action systems* [BKS83], Misra and Chandy's *UNITY* [CM88] (subsequently extended to probability-one reasoning by Rao [Rao94]) and Abrial's *Event-B* [Abr96b].

de Alfaro and Henzinger [dAH00] use explicit fixed-point expressions to express probabilistic "reachability" and "safety" properties, the latter being similar to a *global invariant* of the whole system (in the sense used here).

Similarly, consequences of Zero-One laws turn up elsewhere, especially in model-checking frameworks where the finiteness of the state space makes them particularly appropriate. Early examples include Vardi's work on model checking probability-one temporal properties [Var85] and Hart, Sharir and Pnueli's results on termination [HSP83] — all of whom use models which contain both probability and demonic nondeterminism. Hart *et al.* formulated the probabilistic variant rule Lem. 2.7.1 (and the related Zero-One Law) and showed it to be sound and finitarily complete [*op. cit.*]; our contribution has been to do the same at the level of program logic.

De Alfaro and Henzinger [*op. cit.*] also treat *angelic* interpretations of nondeterminism within a general theory of concurrent games. Consequently their formulations of "reachability" properties need to be more intricate than the similar "eventually" property defined here.

In fact Zero-One laws in logic appear still more generally — Halpern and Kapron [HK94] for example consider all structures over a given state space: given a formula ϕ they ask in what fraction of those structures is ϕ valid. Zero-One laws are relevant here because it turns out that for almost all of them the formula is true, or for almost all of the them the formula is false.

3

Case studies in termination:

Choice coordination,
the dining philosophers,
and the random walk

3.1   Rabin's choice coordination

Rabin's choice-coordination algorithm is an example of the use of probability for *symmetry breaking*: identical processes with identical initial conditions must reach an asymmetric state collectively, all choosing one alternative or all choosing the other. The simplest example is a coin flipped between two people — each has equal right to win if the coin is fair, the initial conditions are thus symmetric; yet, in the end, one person has won and not the other. In this example, however, the situation is made more complex by insisting that the processes be *distributed*: they cannot share a central "coin."

Although Rabin's article [Rab82] explains the algorithm he invented,[1] it does not give a formal proof of its correctness: we do that here, using the techniques of Chap. 2.

Our argument for partial correctness is entirely standard, and so does not illustrate the new probabilistic techniques. (It is somewhat involved, however, and thus is interesting as an exercise and gives a feeling for the intricacy of the algorithm.)

What we do illustrate however is that the new probabilistic- and familiar standard reasoning can work together and — in particular — that we are not burdened with the probabilistic machinery when dealing only with standard properties. We just treat probabilistic choice as demonic choice and proceed in the traditional style, with standard reasoning, since the theory shows that any *wp*-style property proved of the demonic "retraction" is valid for the original probabilistic program as well.[2]

The termination argument however requires the techniques of Chap. 2.

3.1.1   Informal description of Rabin's algorithm

This informal description is based on Rabin's presentation [Rab82].

A group of tourists are to decide between two meeting places: inside a (certain) church, or inside a museum. But they may not communicate as a group; nor is there a central "authority" (*e.g.* a tour guide) which will make the decision for them.

Each tourist carries a notepad on which he will write various numbers; outside each of the two potential meeting places is a noticeboard on which various messages will be written. Initially the number zero appears on all the notepads and on the two noticeboards.

Each tourist decides independently (demonically) which meeting place to visit first, after which he strictly alternates his visits between them. At each place he looks at the noticeboard, and if it displays "here" goes inside. If it does not display "here" it will display a number instead, in which case the tourist compares that number K with the number k on his notepad and takes one of the following three actions:

if $k < K$ — The tourist writes K on his notepad (erasing k), and goes to the other place.

if $k > K$ — The tourist writes "here" on the noticeboard (erasing K), and goes inside.

---

[1] ... and relates it to a similar problem in nature, where mites must decide whether they should all infest the left or all infest the right ear of a moth.

[2] More precisely, replacing probabilistic choice by demonic choice is an anti-refinement.

if $k = K$ — The tourist chooses K', the next even number larger than K, and then flips a coin: if it comes up heads, he increases K' by a further one. He then writes K' on the noticeboard and on his notepad (erasing k and K), and goes to the other place.[3]

We will show that Rabin's algorithm terminates with probability one; and on termination all tourists will be inside, at the same meeting place.

3.1.2   The program

Here we make the description more precise by giving a *pGCL* program for it. Each tourist is represented by an instance of the number on his pad.

- *The program informally*

We call the two places "left" and "right," and refer to Fig. 3.1.1.

Bag *lout* (*rout*) is the bag of numbers held by tourists waiting to look at the left (right) noticeboard; bag *lin* (*rin*) is the bag of numbers held by tourists who have already decided on the left (right) alternative; number L (R) is the number on the left (right) noticeboard.[4]

Initially there are M (N) tourists on the left (right), all holding the number zero; no tourist has yet made a decision. Both noticeboards show zero.

Execution is as follows. If some tourists are still undecided (so that *lout* (*rout*) is not yet empty), select one: the number he holds is l (r). If some tourist has (already) decided on this alternative (so that *lin* (*rin*) is not empty), this tourist does the same; otherwise there are three further possibilities.

- If this tourist's number l (r) is greater than the noticeboard value L (R), then he decides on this alternative (joining *lin* (*rin*)).

- If this tourist's number equals the noticeboard value, he increases the noticeboard value, copies that value and goes to the other alternative (*rout* (*lout*)).

- If this tourist's number is less than the noticeboard value, he copies that value and goes to the other alternative.

- *Notation*

We use the following notation in the program and in the subsequent analysis.

---

[3]For example if K is 8 or 9, first K' becomes 10 and then possibly 11.

[4]BAGS are like sets except that they can have several copies of each element: the bag $\lfloor 1, 1 \rfloor$ contains two copies of 1, and is not the same as $\lfloor 1 \rfloor$ or $\{1\}$.

- $\lfloor \cdots \rfloor$ — Bag (multiset) brackets.

- $\square$ — The empty bag.

- $\lfloor n \rfloor^N$ — A bag containing N copies of value n.

- $b_0 + b_1$ — The bag formed by putting all elements of b_0 and b_1 together into one bag.

- **take** n **from** b — A program command: choose an element demonically from non-empty bag b, assign it to n and remove it from b.

- **add** n **to** b — Add element n to bag b.

- **if** B **then** *prog* **else** $(\cdots)$ **fi** — A "hybrid" conditional: execute *prog* if B holds, otherwise treat $(\cdots)$ as a collection of *GCL*-style guarded alternatives in the normal way [Dij76].

- $\overline{n}$ — The "conjugate" value $n+1$ if n is even and $n-1$ if n is odd.

- $\widetilde{n}$ — The minimum, $n \min \overline{n}$, of n and $\overline{n}$.

- $\#b$ — The number of elements in bag b.

- *Correctness criteria*

We must show that the program is guaranteed with probability one to terminate, and that on termination it establishes

$$\#lin = M+N \ \land \ rin = \square \quad \lor \quad lin = \square \ \land \ \#rin = M+N . \qquad (3.1)$$

That is, on termination the tourists are either all inside on the left or all inside on the right.

3.1.3   Partial correctness

The arguments for partial correctness involve only standard reasoning; but there are several invariants.

- *First invariant*

The first invariant states that tourists are neither created nor destroyed:

$$\#lout + \#lin + \#rout + \#rin \quad = \quad M + N . \qquad (3.2)$$

It holds initially, and is trivially maintained.

$lout, rout := \lfloor 0 \rfloor^M, \lfloor 0 \rfloor^N;$
$lin, rin := \square, \square;$
$L, R := 0, 0;$

do $lout \neq \square \rightarrow$
    take l **from** $lout;$
    if $lin \neq \square$ **then** **add** l **to** lin **else**
        $l > L \quad \rightarrow \quad$ **add** l **to** lin
     $[\!]\quad l = L \quad \rightarrow \quad L := L + 2 \; {}_{\frac{1}{2}}\oplus \; \overline{(L+2)}; \quad$ **add** L **to** $rout$
     $[\!]\quad l < L \quad \rightarrow \quad$ **add** L **to** $rout$
    fi

 $[\!]\quad rout \neq \square \rightarrow$
    take r **from** $rout;$
    if $rin \neq \square$ **then** **add** r **to** rin **else**
        $r > R \quad \rightarrow \quad$ **add** r **to** rin
     $[\!]\quad r = R \quad \rightarrow \quad R := R + 2 \; {}_{\frac{1}{2}}\oplus \; \overline{(R+2)}; \quad$ **add** R **to** $lout$
     $[\!]\quad r < R \quad \rightarrow \quad$ **add** R **to** $lout$
    fi
od

The multiply-guarded loop executes until all (in this case, both) its guards are false; when some are true, the choice between them is demonically nondeterministic [Dij76]. In any case, only one is selected per iteration.

Thus, in this program, the choice of which tourist moves next is demonic among all those still outside, whether on the left or on the right.[5]

Figure 3.1.1. RABIN'S CHOICE-COORDINATION ALGORITHM

- *Second invariant*

The next invariant is

$$lin, lout \leq R \quad \wedge \quad rin, rout \leq L, \tag{3.3}$$

and expresses that a tourist's number never exceeds the number posted at the *other* place.[6] To show invariance we reason as follows:

- It holds initially;

- Since L, R never decrease, it can be falsified only by adding elements to the bags;

- Adding elements to lin, rin cannot falsify it, since those elements come from $lout, rout;$

---

[5]This is "adversarial" scheduling, which we discuss in more detail in the next section, for example at Footnote 14 on p. 89.

[6]By $b \leq K$ we mean that no element in the bag b exceeds the integer K.

- The only commands adding elements to *lout, rout* are

$$\textbf{add } L \textbf{ to } rout \quad \text{and} \quad \textbf{add } R \textbf{ to } lout ,$$

and they maintain it trivially.

- *Third invariant*

Our final invariant for partial correctness is

$$\textsf{max } lin > L \quad \text{if } lin \neq \square \quad \text{and} \quad \textsf{max } rin > R \quad \text{if } rin \neq \square , \qquad (3.4)$$

expressing that if any tourist has gone inside, then at least one of the tourists inside there must have a number exceeding the number posted outside.[7]

By symmetry we need only consider the left (*lin*) case. The invariant holds on initialisation (when $lin = \square$); and inspection of the program shows that it is trivially established when the first value is added to *lin* since the command concerned,

$$l > L \rightarrow \textbf{add } l \textbf{ to } lin ,$$

is executed when $lin = \square$ to establish $lin = \lfloor\!\lfloor l \rfloor\!\rfloor$ for some $l > L$.

Since elements never leave *lin*, it remains non-empty and $\textsf{max } lin$ can only increase; finally L cannot change when *lin* is non-empty.

- *On termination*

On termination we have $lout = rout = \square$, and so with invariant (3.2) we need only

$$lin = \square \quad \lor \quad rin = \square . \qquad (3.5)$$

Assuming for a contradiction that both *lin* and *rin* are non-empty, we then have from invariants (3.3) and (3.4) the inequalities

$$\overbrace{L \geq \underbrace{\textsf{max } rin}_{(3.4)} > R}^{(3.3)} \geq \overbrace{\underbrace{\textsf{max } lin}_{(3.4)} > L}^{(3.3)} ,$$

---

[7]By max b we mean the maximum value held in bag b, undefined if b is empty.[8]

[8]We take the convenient but legitimate view — promoted *e.g.* by Abrial — that all syntactically well-formed expressions denote a value, even those that we might call "undefined." Because the laws for those operators have as antecedents that their operands are within the "usual" ranges, we are prevented from constructing paradoxical (*i.e.* unsound) arguments.

An alternative (and equally legitimate) approach is to use say a three-valued logic, including a special *undefined* value. For us the issue occurs so rarely that the extra trouble is not worth it.

whose conclusion $L > L$ gives us the required impossibility. Thus on termination we have

$$lin = lout = rout = \square \quad \vee \quad rin = lout = rout = \square \;,$$

which together with (3.2) gives us (3.1) as required.

3.1.4 Showing termination: the variant

For termination we need probabilistic arguments, since it is easy to see that no standard variant will do. (Recall that the standard variant technique is incomplete for probabilistic programs.) Suppose for example that the first $M + N$ iterations of the loop take us to the state

$$\begin{aligned} lout, rout \; &= \; \lfloor 4 \rfloor^M, \lfloor 4 \rfloor^N \\ lin, rin \; &= \; \square, \square \\ L, R \; &= \; 4, 4 \;, \end{aligned}$$

differing from the initial state only in the use of 4's rather than 0's. (All coin flips came up heads, and each tourist had exactly two turns.) Since the program contains no absolute comparisons,[9] we are effectively back where we started — and because of that, there can be no standard variant that decreases on every step we take.

So indeed it is not possible to prove termination using a standard invariant whose strict decrease is guaranteed. Instead we appeal to the probabilistic variant rule Lem. 2.7.1.

To find our variant, we note that the algorithm exhibits two kinds of behaviour: the shuttling back-and-forth of the tourists, between the two meeting places (small-scale); and the pattern of the two noticeboard numbers L, R as they increase (large-scale). Our variant therefore will be "layered," one within another: the small-scale *inner* variant will deal with the shuttling, and the large-scale *outer* variant will deal with L and R.[10]

- *Inner variant — tourists' movements*

The aim of the inner variant is to show that the tourists cannot shuttle forever between the sites without eventually changing one of the noticeboards. Intuition suggests that indeed they cannot, since every such movement increases the number on some tourist's notepad and, from invariant (3.3), those numbers are bounded above by $L \max R$.

---

[9]The program checks only whether various numbers are greater than others, not what the numbers actually are.

[10]Layered variants are also called "lexicographic," because the LEXICOGRAPHIC, or dictionary, order of words is a layering of the order a–z between their individual letters. Above we have an order "between two-letter words," with the first letter being the outer invariant and the second being the inner.

The inner variant (increasing) is based on that idea, with some care taken however to make sure that it is bounded above and below by fixed values, independent of L and R.[11] We define V_0 to be

$$
\begin{aligned}
V_0 \quad := \quad & \#\lVert x\colon lout \;+\; rout \mid x \geq L \rVert \\
+ \quad & \#\lVert x\colon lout \;+\; rout \mid x \geq R \rVert \\
+ \quad & 3 * \#(lin \;+\; rin) \ .
\end{aligned}
$$

It is trivially bounded above by $3(M+N)$ and below by zero; and since the outer variant will deal with changes to L and R, in checking the increase of V_0 we can restrict our attention to those parts of the loop body that leave L, R fixed — and we show in that case that the variant must increase on every step:

- If $lin \neq \square$ then an element is removed from $lout$ (V_0 decreases by at most 2) and added to lin (but then V_0 increases by 3); the same reasoning applies when $l > L$.

- If $l = L$ then L will change; so we need not consider that. (It will be dealt with by the outer variant.)

- If $l < L$ then V_0 increases by at least one, since l is replaced by L in $lout+rout$ — and (before) $l \not\geq L$ but (after) $L \geq L$.

The reasoning for $rout$, on the right, is symmetric.

- *Outer variant — changes to L and R*

For the outer variant we need two further invariants; the first is

$$
\widetilde{L} - \widetilde{R} \quad \in \quad \{-2, 0, 2\} \ , \tag{3.6}
$$

stating that the notice-board values can never be "too far apart." It holds initially; and, from invariant (3.3), the command

$$
L := \ L + 2 \ _{\frac{1}{2}} \oplus \overline{(L+2)}
$$

is executed only when $L \leq R$, thus only when $\widetilde{L} \leq \widetilde{R}$, and has the effect

$$
\widetilde{L} := \widetilde{L} + 2 \ .
$$

Thus we can classify L, R into three sets of states:

- $\widetilde{L} = \widetilde{R} - 2 \ \vee \ \widetilde{L} = \widetilde{R} + 2$ — write $L \not\cong R$ for those states.

- $\overline{L} = R$ (equivalently $L = \overline{R}$) — write $L \cong R$.

- $L = R$.

---

[11]The independence from L, R is important, given our variant rule, because L and R can themselves increase without bound.

Then we note that the underlying iteration of the loop induces state transitions as follows. (We write $\langle L = R \rangle$ for the set of states satisfying $L = R$, and so on; demonic choice is indicated by our usual $\sqcap$; the transitions are indicated by $\rightarrow$.)

$$
\begin{array}{lcl}
\langle L \ncong R \rangle & \rightarrow & \langle L \ncong R \rangle \;\sqcap\; \langle L = R \rangle \;{}_{\frac{1}{2}}\!\oplus\; \langle L \cong R \rangle \\
\langle L = R \rangle & \rightarrow & \langle L = R \rangle \;\sqcap\; \langle L \ncong R \rangle \\
\langle L \cong R \rangle & \rightarrow & \langle L \cong R \rangle
\end{array}
$$

However to explain the absence of a transition leaving states $\langle L \cong R \rangle$ we need a second extra invariant: it is

$$\overline{L} \notin rout \;\wedge\; \overline{R} \notin lout . \tag{3.7}$$

It holds initially, and cannot be falsified by the command **add** L **to** $rout$, because $\overline{L} \neq L$. That leaves the command $L := L + 2 \;{}_{\frac{1}{2}}\!\oplus\; \overline{(L+2)}$; but in that case from (3.3) we have

$$rout \;\leq\; L \;<\; L + 2, \overline{(L+2)} \;=\; \overline{(L+2)}, \overline{(L+2)} \,,$$

so that in neither case does the command set L to the conjugate of a value already in $rout$.

Thus with (3.7) we see that execution of the only alternatives that change L, R cannot occur if $L \cong R$, since for example selection of the guard $l = L$ implies $L \in lout$, impossible if $L \cong R$ and $\overline{R} \notin lout$.

For the outer variant we therefore define

$$
V_1 \;:=\; \begin{array}{ll}
2, & \text{if } L = R \\
1, & \text{if } L \ncong R \\
0, & \text{if } L \cong R \,,
\end{array}
\tag{3.8}
$$

and note that whenever L or R changes, the variant V_1 decreases with probability at least $1/2$.

- *The two variants together*

If we put the two variants together lexicographically, forming two layers with the outer variant V_1 being the more significant, then the composite satisfies all the conditions required by the probabilistic variant rule.[12] In particular it has probability at least $1/2$ of strict decrease on *every* iteration of the loop.

Thus the algorithm terminates with probability one — and so from Lem. 2.4.1, Case 1, we have that (3.1) is established almost certainly by Rabin's algorithm of Fig. 3.1.1.

---

[12] Actually the inner variant increases rather than decreases; the two variants put together, "squashing" the layered structure, would be $V_1 * (1 + 3(M + N)) - V_0$.

3.2   The dining philosophers [13]

The *dining philosophers* problem was proposed by Dijkstra [Dij71] as an allegory for the issues of resource allocation among competing concurrent processes. It runs as follows.

A number of philosophers sit at a round table, looking inwards towards a plate of spaghetti. Between each adjacent pair of philosophers is a single fork; but a philosopher needs two forks to eat. When he is hungry, he tries to pick up the forks on either side — but one or both might already be held by his neighbour(s), in which case he either must wait, or try again later.

The principles brought out in Dijkstra's abstraction are these: the processes are concurrent; they are identical; there is no central control; and there is the possibility of "deadlock" and "livelock," as we see below. Figures 3.2.1 and 3.2.2 illustrate the case of five philosophers, and the problem is to "program" them so that they can share the forks without starving each other.

In the most straightforward approach, a hungry philosopher tries to pick up his left fork and, if it is not available, he waits for his neighbour to put it down; once he has the left fork, he waits similarly for his right fork. This can lead to *deadlock* however, where all philosophers end up holding left forks, and each waits forever for his right neighbour to put the other fork down.

To avoid deadlock, the philosophers might be programmed instead to seek their first fork on the left and on the right alternately, without waiting; but in that case they could suffer *livelock*, where they are forever busy but still make no progress towards eating.

Both deadlock and livelock lead eventually to starvation. And in both cases the philosophers could be said to be the victims of an *adversarial scheduler*, our metaphor for anticipating the worst case: we imagine that a demon manipulates the philosophers' rates of progress in order to cause deadlock or livelock if he can — and we try to program the philosophers so that they *avoid* deadlock and livelock, no matter what the demon does.

---

[13]The argument of this section was presented by Morgan to the IFIP WG2.3 at its Napa Valley meeting in January 1997, and was subsequently used by McIver to estimate the expected number of rounds until eating [McI02].

Normally however one assumes the adversary, no matter how bad, is at least "fair" (Sec. 3.2.1) in the sense that he cannot ignore any single philosopher forever.[14]

Between the two approaches above we might expect programs where the philosophers choose probabilistically between their initial forks.[15] Rabin and Lehmann proposed just such a solution [LR94], shown here in Fig. 3.2.3, which guarantees that if some philosopher is hungry, then with probability one eventually some philosopher will eat, provided the scheduler is fair.[16]

Our goal in this section is to verify that claim, using our technique of probabilistic variants: we allow any finite number of philosophers, because the variant technique is complete for finite state spaces (Footnote 28 on p. 55; and Sec. 7.6). Thus we know that if eating occurs eventually with probability one then there must be a variant that shows that to be so.

In fact the variant we present is highly structured, having an outer layer of size $N + 2$, where N is the number of philosophers, with each variant value representing a "phase" in the computation. (See Diagram (3.13) on p. 96 for a look ahead at the phases.) Within each value in the outer layer there is an inner variant that is based on the pattern of state changes in the philosophers' programs during that phase.

---

[14]The correspondence between "adversary" and "demon" is very strong. Demonic choice in our programs represents an adversary controlling the program flow in order to make our postcondition false (standard case) or to make the expected value of our post-expectation as low as possible (probabilistic case).

The effect is described as *defiance probability* by Aspnes and Herlihy [AH90]. In their terms, by

$$p \quad \equiv \quad wp.prog.[post] \qquad\qquad (3.9)$$

we would mean that

in defiance of the adversary who "wants" to avoid *post* on termination, the program *prog* still manages to achieve the postcondition with probability at least p. (3.10)

Writing "$p \equiv$" above, rather than the weaker "$p \Rightarrow$", expresses that in (3.9) we have found the greatest p for which the statement (3.10) is true.

[15]Although there are many non-probabilistic ways to avoid both deadlock and livelock, Rabin and Lehmann showed that all involve either some form of centralised control, built-in differences between the philosophers' programs that break the symmetry of the problem or stronger assumptions than we have made about the "good behaviour" of the scheduler.

[16]Their second solution [*op. cit.*] established the stronger property that if some philosopher is hungry then eventually that philosopher will eat.

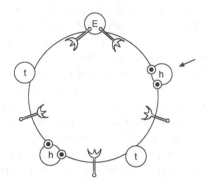

- The philosopher in state E (eating) holds both his forks.

- The philosophers in state h (hungry) are looking at both of their forks, but neither has yet decided which fork to request first.

- The philosophers in state t (thinking) are not interested in the forks at all.

There is no contention here, because no philosopher is waiting for a fork that another philosopher holds. Figure 3.2.2 shows a later stage in which contention has occurred.[a]

---

[a]Footnote 17 on p. 93 below discusses an adversarial scheduling with respect to the arrowed philosopher.

Figure 3.2.1. DIJKSTRA'S DINING PHILOSOPHERS

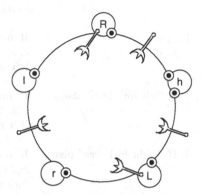

- The philosopher in state R (right-holding) holds his right fork, and is looking at his left fork.

- The philosopher in state L (left-holding) holds his left fork, and is looking at his right fork.

- The philosopher in state l (left-looking) is looking at his left fork, which he does not yet hold. If scheduled he will not be able to pick it up, because it is already held by his left neighbour.

- The philosopher in state r (right-looking) is looking at his right fork, which he does not yet hold. If scheduled he will not be able to pick it up, because it is already held by his right neighbour.

- The philosopher in state h (hungry) is potentially interested in both forks. If scheduled, he will move to state l or r depending on which fork he decides to pick up first; when scheduled again, he will attempt to pick up the fork on that side.

We label fork-holding philosophers with upper-case state-names.

There is contention in this configuration, between philosophers l and R, and between philosophers r and L: in each case the "upper-case" philosopher is holding a fork that the "lower-case" philosopher needs.

Figure 3.2.2. THE DINING PHILOSOPHERS IN CONTENTION

Current state		$\mapsto$	New state	Comment
thinking	t	$\mapsto$	t $\sqcap$ h	If thinking, then either remain thinking or become hungry.
hungry	h	$\mapsto$	l $_{1/2}\oplus$ r	If hungry, then flip a coin and look either left or right for a fork.
left-looking	l	$\mapsto$	L if "left fork free" else l	If (hungry and) looking left, then take left fork if it is free.
right-looking	r	$\mapsto$	R if "right fork free" else r	If (hungry and) looking right, then take right fork if it is free.
Left-holding	L	$\mapsto$	E if "right fork free" else h	If holding left fork, then take right fork if it is free; otherwise drop left fork and begin again.
Right-holding	R	$\mapsto$	E if "left fork free" else h	If holding right fork, then take left fork if it is free; otherwise drop right fork and begin again.
Eating	E	$\mapsto$	E $\sqcap$ t	If eating, then either remain eating or resume thinking.

While thinking (t) a philosopher might become hungry (h). A hungry philosopher flips a coin to decide whether to look left (l) or right (r) for his first fork. If it is held by his neighbour, he waits until it is free and then takes it himself (L or R). He then tries to take his other fork: if it is free, he eats; but if not, he puts down the fork he holds and begins again (h). Once he has eaten enough, he resumes thinking.

Each of the seven actions is atomic.

Deadlock cannot occur, because a philosopher never waits for one fork while holding another. We show that livelock cannot occur either (except with probability zero).

Figure 3.2.3. RABIN AND LEHMANN'S DINING-PHILOSOPHERS ALGORITHM

3.2.1   Fairness, and adversarial scheduling

The definition of *fair scheduler* that we take for the dining philosophers is that with probability one every philosopher is scheduled infinitely often.[17] More precisely we would say that the collective probability of those potential execution paths in which no philosopher eats, and yet every philosopher is infinitely often scheduled, is zero.

But we must take several preliminary steps to adapt this fairness to our probabilistic-variant style of proof.

† The first step is to extend the variant rule itself, Lem. 2.7.1: rather than requiring that the variant decrease on each step with at least some fixed probability $\varepsilon > 0$, we now use a weaker property, that

the strict decrease of the variant occurs *eventually* with at least some fixed probability $\varepsilon > 0$.   (3.11)

Much the same reasoning as before allows us to conclude that with probability one the variant is eventually forced below L, which causes termination. We proceed as follows.

Let the low and high bounds for the variant again be L, H. The crucial step, in the sketched proof of Lem. 2.7.1, is that the probability of eventual termination cannot be less than $\varepsilon^{H-L} > 0$, since that is the probability of the $H - L$ consecutive decrements of the variant. Our revised argument is simply that if the probability of *eventual* decrease is no less than ε, then still the probability of eventual decrease by $H - L$ is no less than $\varepsilon^{H-L} > 0$. The Zero-One Law (Sec. 2.6) then assures us, just as before, that termination in fact occurs with probability one.

---

[17]This is another example of "almost-fair" scheduling. (Recall the remark before Footnote 37 on p. 60.)

Note that like Herescu and Palamidessi [HP01] we therefore allow schedules with phases like "repeatedly schedule a philosopher — for example the one marked ✓ in Fig. 3.2.1 — until he enters state r." Because a philosopher's 50/50 choice between l and r ensures with probability one that the phase will end, the scheduler is sure eventually to get on with other philosophers as fairness requires.

If a schedule seeks to exploit the outcome of the philosophers' choices — as a fully adversarial schedule may — then its fairness might be guaranteed by those very choices.

We return to the issue of "composition of eventualities" in Chap. 10, where Lem. 10.1.2 allows us to treat them rigorously.[18]

‡   The second step is to allow us to reason locally, about various subsets of the philosophers, rather than all of them at once. Say that an individual philosopher's state is *ready* if it must change the very next time that philosopher is scheduled, independently of the states of any other philosophers. That is, States t,E,l,r are not ready: in the first two the philosopher might decide to remain where he is;[19] in the last two he might be blocked by his neighbour. But States h,L,R are ready: in the first, the philosopher must change to l or to r; in the last two, he must change either to E (if he is not blocked) or to h (if he is).

Say that a *group* of philosophers is *ready* if at least one of them must change state when next scheduled, independently of the states of any (other) philosophers not in the group; call that philosopher "ready with respect to the group." For example, although the (singleton) group r is not ready, the group rr is ready: because the right-hand philosopher is in state r (and so is not holding a fork), we know that the left-hand philosopher will move from state r to state R if scheduled. That is, the left-hand r is ready with respect to the group rr, and so the group as a whole is ready.

We now formulate our fairness criterion in terms of readiness.

Lemma 3.2.4 READY PHILOSOPHERS MUST ACT   If every philosopher is scheduled infinitely often with probability one then, for any fixed group of philosophers,

> if the group is ready then eventually one of its members must change state.[20]

Proof   Assume that group G is ready, and let p be a philosopher within it that is ready with respect to G.

---

[18] With the techniques of Chap. 10 we would write $p * [P] \Rightarrow \Diamond [Q]$ for "from any state in P, with probability at least p eventually Q is reached"; and we would write $q * [Q] \Rightarrow \Diamond [R]$ for a similar probabilistic eventuality leading on from Q to R. Lem. B.6.1 (p. 342) will then allow us to reason

$$
\begin{array}{lll}
 & \Diamond R & \\
\equiv & \Diamond\Diamond R & \text{Lem. B.6.1} \\
\Leftarrow & \Diamond(q * [Q]) & \text{second eventuality; } \Diamond \text{ is monotonic} \\
\equiv & q * \Diamond [Q] & \Diamond \text{ is scaling} \\
\Leftarrow & q * p * [P] \ , & \text{first eventuality}
\end{array}
$$

that R is eventually reached from P with probability at least $q * p$.

Above, we applied this $H - L$ times to conclude from $\varepsilon [V < N] \Rightarrow \Diamond [V < N - 1]$ that in fact $\varepsilon^{H-L} [V < H] \Rightarrow \Diamond [V < H - (H - L)] \equiv \Diamond [V < L]$.

[19] We have assumed that philosophers cannot remain in t,E forever; but that does not mean they must move on any individual step.

[20] From now on in this chapter "eventually" will mean "eventually with probability one."

Assume also, for a contradiction, that no philosopher in G ever changes state.

Because every philosopher is scheduled infinitely often, eventually p will be scheduled. At that point p must be ready: it was ready originally (our first assumption) and, because no philosopher in G has changed state (our second assumption, for a contradiction), it is still ready now.

But if it is ready, and is scheduled, then it will change state — which violates our second assumption.   □

Our overall strategy for showing termination is to put the above two steps (†, ‡) together: we identify several successive phases, with the last one being "some philosopher is eating," and we show that eventually each phase must lead to a later one.[21]

We use the second step by formulating within each stage a variant function V and a ready group G; we show for every philosopher in G that if it changes state then V is decreased with probability at least $1/2$. Since fairness (in the form of Lem. 3.2.4) guarantees that eventually some member of G must change, we have that eventually V must decrease with probability $1/2$.

We then use our first step, appealing to the extended variant rule (3.11), to conclude that eventual escape from that stage is certain.

Finally, composition of eventualities tells us that if each stage eventually escapes to a later one,[22] then eventually we must reach the last stage — some philosopher eats.

3.2.2   Proof of termination

We consider the overall system to be an iteration

> "Initially not all philosophers are thinking";
> **do** "No philosopher is eating" →
>
>   "Schedule one of the philosophers fairly"   (3.12)
>
> **od** ,

where "fairly" represents any instance of additional code that ensures every philosopher is infinitely often scheduled, with probability one, thus guaranteeing that the loop body has the property of Lem. 3.2.4.

We note before starting that "not all philosophers are thinking" is an invariant of the loop, since the only way a non-thinking philosopher can resume thinking is via eating, and the loop guard prevents that.

---

[21] This too is composition of eventualities, as explored in Footnote 18 but now in the special case $p = q = 1$: that if P leads to Q with probability one, and similarly Q to R, then in fact P leads to R with probability $1 * 1 = 1$.

[22] ... and there are only finitely many stages.

Given that not all philosophers are thinking, we will see that as a whole the philosophers must be in one of the phases A, $B_{\{L,R\}}$ or $C_{\{1..N\}}$, ordered

$$A \underset{\searrow}{\overset{\nearrow}{}} \begin{array}{c} B_L \\ \\ B_R \end{array} \underset{\nearrow}{\overset{\searrow}{}} C_N \to C_{N-1} \cdots \to C_1 \qquad (3.13)$$

where N is the number of philosophers. In Phase C_1, some philosopher is eating; the other phases are described in detail below.

We now show that each phase eventually reaches a strictly later one, thus guaranteeing a steady progress from left to right in (3.13) towards termination at C_1.

- *Phase A: no forks held*

The only possible individual states in Phase A are t,h,l,r; but because not all states are t (invariant), and no forks are held, the system as a whole is ready; hence eventually some philosopher must change state. The only possibilities are as follows (from Fig. 3.2.3):

$$\begin{array}{c} \text{decreasing} \\ \textit{sum-of-distances} \\ \text{variant} \end{array} \downarrow \left| \begin{array}{ccc} t & \mapsto & h \\ h & \mapsto & l_{1/2} \oplus r \\ l & \mapsto & L \\ r & \mapsto & R \end{array} \right. \begin{array}{l} \ldots \textit{distance 4} \\ \ldots \textit{distance 3} \\ \ldots \textit{distance 2} \\ \ldots \textit{distance 1} \end{array} \qquad (3.14)$$

L, R   We have already left this phase.   ... *distance 0*

Note the effect of "must change state," guaranteed by Lem. 3.2.4, which means that in (3.14) we need not consider "useless" transitions such as the $t \mapsto t, l \mapsto l, r \mapsto r$ which are allowed by the full Fig. 3.2.3.[23]

The variant for this phase is the sum of the philosophers' distances from the end of the list above, where any "have left this phase" states at the bottom are assigned zero: for example, if all five (say) philosophers were hungry, the variant would be $5 * 3 = 15$; if one of them changed state to r, the variant would decrease to 13.

From inspection of the list we see that in fact the variant strictly decreases each time any one of the philosophers changes state — and thus eventually they must leave this phase.[24]

---

[23]A more direct approach, arguing "within any group some philosopher eventually must be scheduled," would not allow us this important simplification. That is why we took such care to formulate Lem. 3.2.4.

[24]Since the state space is finite, we do not need to give explicit bounds for the variant — but in fact they are $L, H := N, 4N$.

- *Phases B_L, B_R: only left, or only right forks held*

Invariant for this phase is that some forks are held (which thus prevents a return to Phase A) — for the only transitions decreasing the number of held forks (refer Fig. 3.2.3) are E $\mapsto$ t and L, R $\mapsto$ h: the former cannot occur because of our guard (no philosopher is eating); and, if the latter occurs, it can only be because some other philosopher is still holding a fork.

Again, the system as a whole is ready.

Assume without loss of generality that only left forks are held (Phase B_L), so that no philosopher is in state R (or E).[25] In this phase we order the possible state changes as follows, again with variant decreasing down the list:

$$
\begin{array}{rcl}
\text{t} & \mapsto & \text{h} \\
\text{l} & \mapsto & \text{L} \\
\text{L} & \mapsto & \text{E} \sqcap \text{h} \\
\text{h} & \mapsto & \text{r} \qquad \text{with probability } 1/2 \\
\text{r} & \mapsto & \text{R}
\end{array}
$$

R   We have left this phase.
E   We will leave the main loop.

Although the transition h $\mapsto$ r is not certain, it does not need to be. Still the variant eventually decreases with nonzero probability $1/2$, and so eventually the philosophers must leave this phase.

- *Phases C_n: a left and a right fork are held, but no less than n apart*

Invariant for these phases is that both a left- and a right fork are held (preventing a return to A or B phases), but possibly by different philosophers. To see the invariance, note without loss of generality that if a philosopher is in state L and its transition L $\mapsto$ h removes an L, then its right neighbour must (still) be in state L — for otherwise the transition would have been from L to E rather than to h.

In these phases we concentrate on a particular group, *a smallest contiguous block of philosophers with L at the clockwise end and R at the anti-clockwise end*, and we let n in the phase name C_n be the length of that group. For example, the bracket in Fig. 3.2.5 shows that the system of Fig. 3.2.2 is in Phase C_3.

Because the group is smallest, it contains no L, R, E in its interior; and so both endpoints are ready with respect to the group — thus the group

---

[25]By *without loss of generality* we mean that the argument would be essentially the same for Phase B_R.

In Fig. 3.2.2 the group of philosophers L-h-R, of length 3 (philosophers), shows
that the system is in Phase C_3, as we illustrate here:

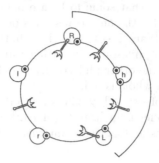

The variant within that phase has value $1 + 4 + 1 = 6$.

Because the group is ready, eventually one of its three philosophers must change
state; if it is either of the endpoints L or R, then that philosopher will eat immediately. If instead it is the centre h, then he will move to either l or r, the variant
decreases to 5 or to 4, and the argument is repeated. If once more the centre
philosopher changes first, then he will move to either L or R and the system enters Phase C_2 — where the above again applies but now for a smaller group, of
size two.

Figure 3.2.5. THE DINING-PHILOSOPHERS' VARIANT FUNCTION

---

itself is ready. Our transition ranking in this phase is thus

$$
\begin{array}{rcl}
\text{t} & \mapsto & \text{h} \\
\text{h} & \mapsto & \text{l}\ _{1/2}\oplus\ \text{r} \\
\text{l} & \mapsto & \text{L} \\
\text{r} & \mapsto & \text{R} \\
\text{endpoint L,R} & \mapsto & \text{E} \quad \text{because the interior contains no L,R,E}
\end{array}
$$

interior L,R   We have moved to Phase $C_{n'}$ for $n' < n$.
       E   We will leave the main loop.

The transitions $l, r \mapsto L, R$ that decrease n do so because they can occur
only on the interior of the group; whichever of L,R appears, the distance n'
from there to the endpoint of opposite sense is strictly less than the size n
of the current group — thus we move to Phase $C_{n'}$, with $n' < n$.

The ordering of the C_n phases is decreasing in n — in fact the last one
C_1 occurs when a single philosopher holds both a left- and a right fork, and
so is in State E.

That concludes the proof of termination of Program (3.12), and hence
that Rabin and Lehmann's algorithm avoids livelock.

3.3   The general random "jump"

For our final case study we consider a random-walk process acting over the integers $\mathbb{Z}$, *i.e.* both positive and negative, and which can take possibly large steps in either direction. We saw a special case of this earlier, the symmetric random walk $n := n+1 \,_{1/2}\oplus n-1$ in Fig. 2.11.2 which was restricted to just the natural numbers $\mathbb{N}$ and took steps of size one.

We now show that the more general "jumping" process is recurrent in the sense that from any position on the number line it eventually reaches or passes any other position, provided the assumptions below are met.

3.3.1   Assumptions and claim

We consider a program fragment *jump* over integer variable n, which variable can take both positive and negative values, and we assume the following:

1. The jumper *jump* is guaranteed with some nonzero probability to change position, *i.e.* to move strictly up or down: it cannot remain forever where it is. Thus for all N we have

$$[n = N] \quad \Rightarrow \quad \lceil wp.jump.[n \neq N] \rceil . \quad {}^{26} \tag{3.15}$$

   This condition rules out the trivial jumper $n := n$, for example.

2. The length of its jump is uniformly bounded: there is a K such that for every N we have

$$[n = N] \quad \Rightarrow \quad wp.jump.[N - K \leq n \leq N + K] . \quad {}^{27}$$

---

[26]We write $\lceil \cdot \rceil$ for the CEILING function such that $\lceil x \rceil$ is the least integer no less than the real x.

On the left we have $[n = N]$, nonzero only when n is N so that the implication is trivially true otherwise; that in effect makes N a "logical constant" that captures n's initial value. And when the left-hand side is one, the $\lceil \cdot \rceil$ on the right simply requires that the enclosed expression is not zero. Thus in general we have

$$[pred] \quad \Rightarrow \quad \lceil exp \rceil$$

as a convenient idiom for "*exp* is nonzero on *pred*."

[27]Note that this implies termination of *jump* as well, *i.e.* that it does not abort, since it is guaranteed to establish $N-K \leq n \leq N+K$.

3. The *expected jump* is everywhere non-negative, so that

$$ n \;\Rrightarrow\; wp.jump.n \, . \qquad [28,29] $$

In other words, the *expected* final position is no less than (is "to the right of") the initial position.

Our treatment is more general than usual (*e.g.* [GS92, Sec. 5.10]): we do not make the normal assumption that the jumps are *independent and identically distributed*, represented by so-called *iid* random variables; that is, we do not assume that the distribution of distances jumped is the same everywhere and at every time. Instead we are considering a "non-homogeneous" walker [Sti94, p. 326] that can even demonically use different distributions at the same n on different visits.

We claim that under the above assumptions the jumper eventually moves up beyond any position of our choice, no matter how distant.

3.3.2   Proof of the claim

Let our program *prog* be as in Fig. 3.3.1. The process begins at $n = N$, and every execution of the loop body *jump* causes a non-negative expected movement of n; we show under the stated conditions that, for all constants H, eventually we have $H \leq n < H + K$. [30]

---

[28] Although we seem to be flirting with negative expectations (recall the warnings of Sec. 2.11), we see at Footnote 31 below that in our detailed reasoning we can shift the calculations up into positive values, if we want to.

[29] In Sec. 3.3.3 we consider the complementary case of the "non-positive" jump; either way, the point is that the expected movement must be everywhere one or the other, and not a mixture.

[30] Assumption 2 above is necessary, as we can now see by considering the instance

$$
\begin{array}{lll}
init & \rightarrow & n := -2; \\
loop & \rightarrow & \textbf{do } n < 0 \rightarrow \\
& & \qquad n := n^2 \;\;_{1/n^2}\!\oplus\; n := n - 1 \quad \leftarrow \quad \text{Violates Assumption 2} \\
& & \textbf{od}
\end{array}
$$

of Fig. 3.3.1. In discussing its termination, we can assume the guard $n < 0$ (and so avoid in particular any concerns about the "undefined" probability $1/0^2$).

Its body satisfies Assumptions 1 and 3 when the guard $n < 0$ holds, but the potentially unbounded up-jump $n := n^2$ fails Assumption 2 even then. A glance back at Sec. 2.9.1 shows that the loop has probability $1/2$ from its starting position $n = -2$ of continuing to decrement n forever, of never terminating and so never establishing $n \geq 0$.

$$
\begin{array}{lll}
init & \rightarrow & n := N; \\
loop & \rightarrow & \textbf{do } n < H \rightarrow \\
& & \quad jump \qquad \leftarrow \text{ Satisfies conditions of Sec. 3.3.1} \\
& & \textbf{od}
\end{array}
$$

The program *prog* is the whole of the above. It is an example of a *random walk* over the integers, with loop body *jump* increasing or decreasing n according to a distribution with (we stipulate) non-negative expected change in n.

We show that if $N < H$ then the program terminates with probability one in a state satisfying $H \le n < H + K$, where K is the largest possible jump.

Figure 3.3.1. THE GENERAL RANDOM WALK PROGRAM

We do this by showing termination of certain related programs $prog_L$, indexed by an arbitrary lower bound L below H; they are

$$
\begin{array}{lll}
init & \rightarrow & n := N; \\
loop_L & \rightarrow & \textbf{do } L \le n < H \rightarrow \\
& & \quad jump \\
& & \textbf{od },
\end{array}
$$

where $L \le N < H$.

Showing termination of $loop_L$ amounts to proving n's "escape" from the interval $[L, H)$.[31] In fact it is easy to show that escape is certain, because this modified program is a *bounded* random walk; but — more importantly — we show that there is always an L_* low enough (depending on N, and "far away enough" from it) that the escape of $loop_{L_*}$ from $[L_*, H)$ occurs *above H* — rather than below L_* — with probability at least $1/2$.

Since there is such an L_* for each N, we will conclude that the original *loop* also has the property that escape above H occurs with probability at least $1/2$ — and that the $1/2$ applies no matter which N we choose. The Zero-One Law then tells us that in fact escape towards H occurs with probability one.

†   Here are the details. We see that $loop_L$ differs from *loop* only in that its stronger guard possibly forces early termination in a state *not* satisfying $H \le n < H + K$, and so we will not have "cheated," *i.e.* increased our probability of achieving that postcondition, if we use $prog_L$ instead of *prog*.

---

[31] Similarly, showing termination of the original *prog* is just proving n's escape from $(-\infty, H)$. Note however that for these subsidiary programs our reasoning is confined to the interval $[L - K, H + K]$ and so can be shifted by $K - L$ up into positive values if we are worried about the use of negative expectations. (Recall Footnote 28 on p. 100 above.)

Written in terms of wp, the above informal observation becomes this inequality between the two programs' behaviours:

$$wp.prog_L.[H \leq n < H + K] \quad \Rightarrow \quad wp.prog.[H \leq n < H + K] \; . \quad ^{32}$$

Keeping that in mind for later, we begin by identifying these two invariants of $loop_L$:

- Since the guard is $L \leq n < H$ we have the standard invariant

$$Inv \quad := \quad L - K \leq n < H + K$$

  from Assumption 2 above.

- The (non-negative) expectation

$$I \quad := \quad (n \ominus L)/(H + K - L) \tag{3.16}$$

  is invariant as well, which we see as follows:

$$
\begin{array}{ll}
& wp.jump.I \\
\equiv & wp.jump.((n \ominus L)/(H + K - L)) \\
\Leftarrow & \qquad\qquad\qquad\qquad\qquad\qquad\qquad \text{sublinearity Def. 1.6.1} \\
& (wp.jump.n)/(H + K - L) \quad \ominus \quad L/(H + K - L) \\
\\
\Leftarrow & \qquad\qquad\qquad\qquad\qquad\qquad\qquad \text{Assumption 3 above} \\
& n/(H + K - L) \quad \ominus \quad L/(H + K - L) \\
\\
\equiv & (n \ominus L)/(H + K - L) \\
\equiv & I \; .
\end{array}
$$

Although this expectation is potentially unbounded, we appeal to the standard invariant Inv (recalling Lem. 2.10.2) to see that it is bounded over the region we are dealing with.

---

[32]To show this rigorously, we would rely on the least-fixed-point definition of the pre-expectation of a loop (recall (2.3) and Footnote 3 on p. 38): pre-expectations $preE$ and $preE_L$ are the least fixed-points of functions F and F_L defined

$$
\begin{array}{lll}
& F.preE & := \quad [H \leq n < H + K] \text{ if } n \geq H \text{ else } wp.jump.preE \\
\text{and} & F_L.preE_L & := \quad [H \leq n < H + K] \text{ if } n < L \vee n \geq H \text{ else } wp.jump.preE_L
\end{array}
$$

respectively. To show $preE_L \Rightarrow preE$ it is sufficient to check that $F_L.preE \Rightarrow preE$, easily verified from the definitions immediately above, and then to appeal to the general least-fixed-point property that $f.x \leq x$ implies $\mu.f \leq x$, where μ takes the least fixed-point of its argument function.

For termination of $loop_L$ we take (increasing) variant n, because the loop guard bounds it by L, H, and the probability of its increase is everywhere positive within that finite interval.[33] Thus termination is certain.

And on termination, the loop guard is false — we have $\overline{L \leq n < H}$ — so that either

- $n < L$, in which case I in (3.16) is zero, or

- $H \leq n < H + K$ (with the upper bound coming from the standard invariant Inv), which implies that I is still no more than one.

Thus, either way, the negated guard and invariant(s)

$$\left[\overline{L \leq n < H} \right] \; * \; [Inv] \; * \; I$$

together imply ($\Rightarrow$) the post-expectation $[H \leq n < H + K]$, as required.

Now from our loop rule Lem. 2.4.1 Case (2) we have that a sufficient pre-expectation for $prog_L$ to establish $H \leq n < H + K$ is therefore given by the calculation

---

[33]Informally, we argue that for every n in $[L, H)$ at least some of the potential movement guaranteed by Assumption 1 must be strictly positive — else Assumption 3 would fail for that n.

We can also give a rigorous proof of this, always worth doing but not necessarily worth *reading* the first time through.[34,35]

[34]We give two independent arguments for this, as examples. The first is more intuitive, but the second is more direct.[35] In both cases we argue in a state where n is N.

For the first argument our strategy is to reason over deterministic programs, which have more intuitive behaviour than demonic programs, and where (as a result) we can appeal to linearity.

Suppose for a contradiction that $wp.jump.[n > N] = 0$ (when $n = N$ initially). By Fact B.3.5 there is a deterministic refinement det of $jump$ with $wp.det.[n > N] = 0$ also, and we concentrate on that. In the current state $n = N$ we have

$$wp.det.n$$

$=$ $\quad wp.det.(n * [n < N] + n * [n = N] + n * [n > N])$ $\qquad$ arithmetic

$=$ $\qquad\qquad\qquad\qquad\qquad\qquad\qquad\qquad\qquad\qquad\qquad$ det is linear, (2.18)
$\quad wp.det.(n * [n < N]) + wp.det.(n * [n = N]) + wp.det.(n * [n > N])$

$=$ $\quad wp.det.(n * [n < N]) + wp.det.(n * [n = N])$ $\qquad$ assumption $wp.det.[n > N] = 0$

$\leq$ $\quad wp.det.((N{-}1) * [n < N]) + wp.det.(N * [n = N])$ $\qquad$ monotonicity

$=$ $\quad (N{-}1) * wp.det.[n < N] + N * wp.det.[n = N]$ $\qquad$ scaling

$\leq$ $\quad (N{-}1) * wp.det.[n \neq N] + N * wp.det.[n = N]$ $\qquad$ monotonicity

$<$ $\qquad N .$ $\qquad\qquad$ $wp.det.[n \neq N] > 0$
$\qquad\qquad\qquad\qquad$ and $\quad wp.det.[n \neq N] + wp.det.[n = N] \leq 1$

Thus $wp.det.n < N$ in this state, and since $jump \sqsubseteq det$ we have $wp.jump.n < N$ also, violating Assumption 3.

Therefore $wp.jump.[n > N]$ cannot be zero when $n = N$.

‡

$$wp.prog_L.[H \le n < H + K]$$
$\Leftarrow$ $wp.init.preE_L$
$\equiv$ $wp.init.([Inv] * I)$ above
$\equiv$ $([L - K \le n < H + K] * (n \ominus L)/(H + K - L)) \langle n \mapsto N \rangle$
$\equiv$ $[L - K \le N < H + K] \; * \; (N \ominus L)/(H + K - L) \,,$

and so by choosing a "low enough" L, specifically $L_* := 2N - H - K$ to
make $(N \ominus L_*)/(H + K - L_*) = 1/2$, we can continue

$$wp.prog_{L_*}.[H \le n < H + K]$$ repeating (‡) above, for $L := L_*$
$\Leftarrow$ $\cdots$ as before
$\equiv$ $[L_* - K \le N < H + K]\,/2$ choice of L_*
$\Leftarrow$ $[N < H \wedge K \ge 0]\,/2\,.$ again

But now we can remove our dependence on L_* by remembering that $prog_L$'s
probability of termination in $H \le n < N + K$ is if anything worse than
$prog$'s. That is, we have

$$[N < H \wedge K \ge 0]\,/2$$
$\Rightarrow$ $wp.prog_{L_*}.[H \le n < H + K]$ immediately above
$\Rightarrow$ $wp.prog.[H \le n < H + K]\,,$ recall (†) further above

and the outer inequality gives us that the constant lower bound of $1/2$ is
sufficient for any H, any $K \ge 0$ and $N < H$ (the N being hidden in the
initialisation of $prog$). Most importantly, the $1/2$ is independent of L.

Thus we may now finish the argument, over N, H, K alone, by appealing
to the Zero-One Law (Sec. 2.6): if for all n initially N below H we have
that the probability of escape to $H \le n < H + K$ is at least $1/2$, then in
fact that escape probability must be one — which concludes the proof. We
have shown that

$$[n < H] \quad \Rightarrow \quad wp.prog.[H \le n < N + K]$$

under the assumptions of Sec. 3.3.1.

---

[35] For the alternative, second argument we use sublinearity directly to construct our
$[n > N]$ from "the pieces we know about," i.e. from n and from $[n \ne N]$; we continue to
argue at initial state $n = N$. We have

$$wp.jump.((K+1) * [N < n])$$

$\ge$ Assumption 2 and Lem. 1.7.1 with its Footnote 52; let interior operators associate left
     $wp.jump.(n \ominus (N-1) + [n \ne N] \ominus 1)$

$\ge$ $wp.jump.n \ominus (N-1) + wp.jump.[n \ne N] \ominus 1$ sublinearity twice
$\ge$ $n \ominus (N-1) + \varepsilon \ominus 1$ Assumptions 1,3; some $\varepsilon > 0$
$\ge$ ε operators associating left; in current state $n = N$
$>$ $0\,,$ assumption about ε

whence by scaling we have $0 < wp.jump.[N < n]$ also.

3.3.3 *Variations and consequences*

In the previous section we showed that non-negative expected movement of n guarantees that eventually $H \leq n < H+K$ with probability one, for all H; thus by symmetry a similar process but with non-*positive* expected movement will achieve the same $H \leq n < H+K$ with probability one, again for all H.[36] And if the expected movement is exactly zero (as it is in the simple ± 1-style symmetric random walk), then that interval $[H, H+K)$ is visited infinitely often, for any H.

If K is one, then the interval collapses to just the singleton $[H, H+1)$, *i.e.* just $\{H\}$ on its own: therefore we know in that case that any H we choose is visited infinitely often, even under the very liberal Assumptions 1 and 2 above.

Chapter notes

Termination with probability one is fundamental to probabilistic correctness and, not surprisingly, there are many techniques for its verification. From a model-checking perspective, Vardi [Var85] and Courcoubetis and Yannakakis [CY95] have studied the complexity of determining whether a temporal logic formula is satisfied with probability one, termination being a special case. There is also an extensive collection of concrete case studies using Kwiatkowska's *PRISM* model checker [PRI].

Similarly, fairness often plays an important role in termination arguments and there are specialised and sophisticated methods to deal with such situations [Fra86]. Pnueli and Zuck [PZ93, AZP03] for example have studied termination in the context of several kinds of fairness. Closer to us is J.R. Rao [Rao94], who uses a generalised *wp* for reasoning about probabilistic programs within a *UNITY* setting [CM88] which also assumes fair executions between "competing" threads of control.

As we have just seen, probabilistic variants can often "cut through" seeming complexities to produce startlingly succinct yet rigorous proofs of termination; indeed the point of the chapter was to show that.[37]

Duflot *et al.* [DFP01] also argue that some of the original termination arguments for probabilistic self-stabilisation were unnecessarily complicated. They

---

[36]Use $H' := H + K - 1$ in the non-positive argument.

[37]That is not to say of course that there are not convincing informal proofs as well: see for example Motwani and Raghavan's "stepwise" treatment [MR95] of Choice Coordination, our Sec. 3.1.

too have improved earlier proofs, in their case by applying more sophisticated results from Markov-process theory to the special properties of the problem domain. However in most cases, and certainly in all finite cases, our simple variant technique here will suffice in principle.

Finally, there are many variations of the dining philosophers (as an example of resource-access control) — for example, again Duflot *et al.* [DFP02] remove the fairness condition altogether, and Herescu and Palamidessi [HP01] consider the "generalised" dining philosophers problem applied to network topologies other than rings.

4

Probabilistic data refinement: [1]
The steam boiler

4.1 Introduction: refinement of datatypes

One *datatype* is said to be refined by another if the second can replace the first in any program without detection.[2] Techniques for proving such refinements are well established,[3] and involve setting up a correspondence between the (abstract) states in the specification datatype and the (concrete) states in the implementation datatype. The standard techniques however are inadequate when the datatypes (and the programs that use them) contain probabilistic choice — for that we need a correspondence between *probability distributions* over abstract states and *probability distributions* over concrete states, rather than between individual states.

---

[1] An earlier version of this chapter [MMT98] was co-authored with Elena Troubitsyna.

[2] That makes the definition of refinement sensitive, of course, to what we are able to detect. Here we limit ourselves to functional properties only (not including running time, for example).

[3] They have been comprehensively surveyed by de Roever and Engelhardt [dRE98].

In this chapter we show how the methods we have seen so far can be applied to probabilistic data refinements. Probability is of particular relevance when a quantified analysis of safety is required — thus our example is the "steam boiler," which is widely known as a test case for formal system specification [Abr96a].

In Sec. 4.2 we set out the standard definitions for data refinement in the probabilistic context. Section 4.3 contains the first technical contribution of this chapter — a completely worked example of a probabilistic datatype, consisting of a single operation which chooses probabilistically either to skip or to abort. It is of the kind lying strictly outside the domain of standard programming, and its behavioural simplicity belies the subtleties of the calculations required to prove data refinement. That then sets the scene for Sec. 4.4 in which we look at the steam boiler proper and, though much simplified, it does demonstrate how probabilistic datatypes are relevant in a realistic setting.

In general, datatype refinements can be viewed in two ways. One way is to accept the probabilistic behaviour of the components and then to determine by calculation what is the best specification that can be achieved of a system built from them; we call this *supplier-oriented*. The other way is to deduce from a specification the minimum acceptable behaviour of the components from which it is to be built: this is *customer-oriented*. In the steam-boiler study we take the supplier's point of view.

4.2   Data refinement and simulations

Data refinement is a generalised form of program refinement — an "abstract datatype" is replaced by a more "concrete datatype" in a program while preserving its algorithmic structure, the difference being that the two datatypes operate over different (local) state spaces and thus cannot be related directly using ordinary program refinement. In this context we refer to a *datatype* (in the style of Hoare, He and Sanders [HHS87]) as a triple (*init, OP, fin*) where *init* and *fin* are programs (respectively the initialisation and the finalisation) and *OP* is a set of programs indexed by operation name — the operations of the datatype.

Typically *init* initialises the local variables, possibly but not necessarily referring to global variables in doing so; operations in *OP* may refer to and change both local and global variables; and *fin* simply "discards" the final values of the local variables, projecting the state (local and global) back onto its purely global component.

In the activity of data refinement the programs *init, fin* and those in *OP* are replaced by corresponding concrete programs, perhaps changing the local state space in the process. The effect of the replacement on a program using the datatype is *proper* program refinement (*i.e.* not *data* refinement)

between that using the abstract type and that using the concrete, provided
the local state cannot be observed directly — an assumption that is vital
for the data refinement technique.

Since we are considering datatypes whose operations may have probabilistic choices, we must apply the above ideas to our current context.
We start by defining data refinement by appeal to program refinement, as
above.

Definition 4.2.1 DATA REFINEMENT AS SUB-COMMUTATION
We say $(init, OP, fin)$ is *data refined* by $(init', OP', fin')$ iff for all program
contexts $\mathcal{C}$ over the operations OP of the datatype we have

$$init;\ \mathcal{C}(OP);\ fin \quad \sqsubseteq \quad init';\ \mathcal{C}(OP');\ fin'\ .\quad ^4$$

□

Note that the effect of prepending and appending respectively the initialisation and the finalisation make a composite that refers only to global
variables — the local variables are hidden in the sequential compositions
of Def. 4.2.1. Hence the refinement relation shown is ordinary program
refinement between programs over the globally observed state space.

The fragment $\mathcal{C}$ is formed from the programming language constructs
(and takes arguments from the appropriate datatypes) — naturally, we use
$pGCL$ here.

Rather than using the cumbersome Def. 4.2.1 directly — for which we
would need to check refinement *for all* fragments $\mathcal{C}$ that could ever be written in $pGCL$[5] — to verify data refinements we use instead the traditional
technique of simulations [HHS87]. For simplicity we assume that fin is indeed just a projection back onto the global state, and we omit it from here
on.

---

[4]By the notation $\mathcal{C}(OP)$ we mean an "implicit substitution"

$$\mathcal{C}\ \langle X_0, X_1 \cdots \mapsto op_0, op_1 \cdots \rangle$$

within program fragment $\mathcal{C}$ of actual operation texts op_i for place-holder variables (operation names) X_i whose names we suppress. This gives us the convenient and uncluttered
notation $\mathcal{C}(OP')$ for the same fragment except that primed operations appear instead
of the unprimed ones.

[5]In fact we do just that for the very simple example tabulated in Fig. 4.3.4 below.

Definition 4.2.2 SIMULATION FOR DATA REFINEMENT
We say that datatype $(init', OP')$ *simulates* datatype $(init, OP)$ if there is a terminating program *rep* satisfying

$$init; rep \quad \sqsubseteq \quad init'$$
$$op; rep \quad \sqsubseteq \quad rep; op' \ ,$$

where the second refinement holds for all corresponding pairs op, op' in OP, OP'.[6]   □

It is well known [GM91] that a simulation of datatypes implies data refinement Def. 4.2.1; the same holds in our probabilistic model, provided *rep* is suitably defined.[7]

We now use the simulation technique in a worked example.

4.3   Probabilistic datatypes: a worked example

The example we consider in this section is set out in Fig. 4.3.1, with Fig. 4.3.2 showing a special case. The abstract — or specification — datatype has a single operation which skips with probability p and aborts with probability $\bar{p}$. The concrete (implementation) datatype has a local variable i; the operation op'_N first checks whether i is at least N, aborting if it is and skipping otherwise, which behaviour is achieved by the assertion command $\{i < N\}$.[8] After that it either increments i (probability p'_N) or sets it to 0 (probability $\overline{p'_N}$). Thus in the concrete datatype abortion is possible only after the left branch has been selected N times in succession.

The qualitative external behaviour of the two datatypes is identical — each provides a single operation which acts either like **skip** or like **abort**. Our task however is *quantitative* — to determine the relation between p and p'_N which ensures that op'_N is no more likely to abort than op is.

The object of this section is to work carefully through the calculation of probabilities required to establish a simulation between the datatypes and, since our aim is merely to illustrate the technique, we treat the simplest

---

[6]The usual third requirement is *fin* $\sqsubseteq$ *rep; fin'*; but if *rep* terminates and *fin, fin'* are just projections, it is satisfied trivially.

[7]For *rep* to be a simulation that implies data refinement between datatypes it must satisfy two conditions. The first is continuity (as a function over the program domain); and the second is that it must not modify the global state space. The proof then of the soundness of the simulation technique follows exactly that explained by Gardiner [GM91] for standard programs, save for distribution of *rep* through external demonic nondeterminism. However distribution through even that operator can be established provided the locality of variables is modelled explicitly [Gro].

[8]ASSERTION commands $\{pred\}$ [Mor94b, BvW98] act as **skip** if *pred* is true and as **abort** otherwise. Thus the above is **skip if** $i < N$ **else abort**.

$$
\begin{array}{ll}
init & \textbf{skip} \\
op & \textbf{skip}\ {}_p\!\oplus \textbf{abort}
\end{array}
$$

$$
\begin{array}{ll}
init' & i\!:=0 \\
op'_N & \{i < N\};\ (i\!:=i+1\ {}_{p'_N}\!\oplus i\!:=0)
\end{array}
$$

Figure 4.3.1. A FAULTY N-SKIPPER

$$
\begin{array}{ll}
init & \textbf{skip} \\
op & \textbf{skip}\ {}_p\!\oplus \textbf{abort}
\end{array}
$$

$$
\begin{array}{ll}
init' & i\!:=0 \\
op' & \{i < 2\};\ (i\!:=i+1\ {}_{p'}\!\oplus i\!:=0)
\end{array}
$$

In this $N = 2$ instantiation of Fig. 4.3.1, for the concrete probability we write just p' instead of p'_2.

Figure 4.3.2. A FAULTY 2-SKIPPER

---

interesting case — when $N = 2$ — as is set out in Fig. 4.3.2. We return to the general case at the end of the section.

For the simple 2-skipper, according to Def. 4.2.2, we need a program *rep* satisfying the refinements

$$
\begin{aligned}
\textbf{skip};\ rep &\ \sqsubseteq\ i\!:=0 \\
op;\ rep &\ \sqsubseteq\ rep;\ op'
\end{aligned}
\tag{4.1}
$$

The program *rep* can be thought of as calculating concrete states for the corresponding abstract states.[9]

We note first that a successful execution of *op* must correspond to a successful execution of *op'*, which means i initially takes the values 0 or 1 — the initial assertion $\{i < 2\}$ in *op'* ensures that — and thus 0, 1 or 2 finally in some distribution. Bearing in mind that *rep* "converts" abstract states into concrete states, we might imagine that it sets i to some distribution of those outcomes. Thus we suppose that *rep* has the form

$$
\left|
\begin{array}{ll}
i\!:=0 & @\ q \\
i\!:=1 & @\ r \\
i\!:=2 & @\ s\ ,
\end{array}
\right.
$$

for some one-summing q, r, s to be determined. (They must sum to one because *rep* is terminating.)

---

[9]Of course in this case we have simplified things so much that there is only one (anonymous) abstract state — or two, if you count the "$\perp$" we will introduce in Chap. 5 (p. 130) to model aborting behaviour.

Now we substitute into the simulation equations (4.1) above to get

$$
\begin{array}{l}
\mathbf{skip}\,_p\oplus\mathbf{abort};\\
\left|\begin{array}{ll}
i\colon=0 & @\ q\\
i\colon=1 & @\ r\\
i\colon=2 & @\ s
\end{array}\right.
\end{array}
\quad\sqsubseteq\quad
\left|\begin{array}{ll}
i\colon=0 & @\ q\\
i\colon=1 & @\ r\\
i\colon=2 & @\ s;\\
\{i<2\};\ i\colon=i+1\ _{p'}\oplus\ i\colon=0
\end{array}\right.
\tag{4.2}
$$

Then we simplify by computing the various sequential compositions along the probabilistic branches, multiplying probabilities as we go (and appealing to Law 21 from the list of algebraic laws given in Sec. B.1); on the left the result is

$$
\left|\begin{array}{lll}
\mathbf{skip}; & i\colon=0 & @\ pq\\
\mathbf{abort}; & i\colon=0 & @\ \overline{p}q\\
\mathbf{skip}; & i\colon=1 & @\ pr\\
\mathbf{abort}; & i\colon=1 & @\ \overline{p}r\\
\mathbf{skip}; & i\colon=2 & @\ ps\\
\mathbf{abort}; & i\colon=2 & @\ \overline{p}s\ .
\end{array}\right.
$$

Simplifying the program fragments and, as a result, collapsing the three aborting cases gives

$$
\left|\begin{array}{ll}
i\colon=0 & @\ pq\\
i\colon=1 & @\ pr\\
i\colon=2 & @\ ps\ ,
\end{array}\right.
$$

where (recall p. 20) when the probabilities sum to less than one the deficit represents aborting behaviour. Similar calculations on the right show us that (4.2) is equivalent to

$$
\left|\begin{array}{ll}
i\colon=0 & @\ pq\\
i\colon=1 & @\ pr\\
i\colon=2 & @\ ps
\end{array}\right.
\quad\sqsubseteq\quad
\left|\begin{array}{ll}
i\colon=0 & @\ \overline{p'}(q+r)\\
i\colon=1 & @\ p'q\\
i\colon=2 & @\ p'r
\end{array}\right.
$$

from which we must extract a relation between p and q,r,s,p'.

The refinement relation $\sqsubseteq$ applied between probabilistic programs requires that any outcome guaranteed on the left with a certain probability is guaranteed on the right with a probability at least as great.[10] In this case we have that when i is assigned a particular value on the left-hand side the associated probability must be at least as great as for the same assignment on the right-hand side. So matching the cases $i=0,1,2$ separately we end up with the inequations

---

[10]In fact it requires more than this in general — recall our earlier discussion of compositionality (p. 15) — but the programs here are pre-deterministic, and so checking at points is sufficient. (See Sec. A.1 for an example of when it is not.)

$$pq \leq \overline{p'}(q+r) \qquad (4.3)$$
$$pr \leq p'q \qquad (4.4)$$
$$ps \leq p'r , \qquad (4.5)$$

and these are the constraints we seek.

As mentioned in the introduction, there are now two approaches we could take. In the "customer-oriented" approach the specification is fixed — the customer demands a certain reliability — and the supplier has to find components that are themselves reliable enough to meet it; in our current example we would fix p and calculate how high p' would then have to be.

In the "supplier-oriented" approach we imagine an implementor working with "off the shelf" components and calculating the impact from the customer's point of view. That is the view we take here: thus we fix p' and find the largest p for which the above inequations can be satisfied by some one-summing q, r, s.

We note first that if q, r, s solve (4.3–4.5), for fixed p, p', then we can multiply all three by any positive constant and they will still be a solution. In particular, if $q+r+s > 0$ then we can scale them up or down so that they sum to one — and so the constraint $q+r+s = 1$ can effectively be ignored.

Without the one-summing constraint, we next notice that s appears only in (4.5), and so we can set it to zero; then (4.3) can be rewritten

$$(p - \overline{p'})\, q \quad \leq \quad \overline{p'}\, r$$

so that together with (4.4) we see that there are nonzero solutions for q, r exactly when $p\,(p - \overline{p'}) \leq p'\,\overline{p'}$. For fixed p', *i.e.* customer-oriented analysis, the largest p satisfying that inequality is given by

$$p := \frac{\overline{p'} + \sqrt{1 + 2p' - 3p'^2}}{2} , \qquad (4.6)$$

which gives us $q = \overline{p'}/p$, $r = p'q/p$ and $s = 0$. (Note that with (4.6) we have $q + r + s = 1$ for these values.)

Now we return to the initialisation: from (4.1) above, we must have **skip**; $rep \sqsubseteq i := 0$ — which clearly is not true for the rep we have chosen. To deal with that we use the standard technique of composing data refinements [Mor88a], *i.e.* doing a second data refinement after the first: since reducing the value of i will make abortion less likely in the concrete datatype, and is what we must do to get from the **skip**; rep we have to the $i := 0$ we want, we propose a second representation function $i: \leq i$ — that is, a demonic assignment that cannot increase the value of i.[11]

From (4.1) our "intermediate" $init'$ must be rep itself; and so in the second data refinement, for the initialisation we must show

$$rep; i: \leq i \quad \sqsubseteq \quad i:= 0 \ ,$$

which is trivial. But we must also show that our second data refinement allows us to keep the op' we have chosen already — i.e. that it is data-refined by $i: \leq i$ to itself. Again using program algebra, we reason as follows:

$$
\begin{array}{ll}
& op'; \ i: \leq i \\
= & \{i < 2\}; \ (i:= i+1 \,_{p'}\oplus i:= 0); \ i: \leq i \\
= & \hspace{5cm} \text{Law 13 on p. 323} \\
& \{i < 2\}; \ ((i:= i+1; \ i: \leq i) \,_{p'}\oplus (i:= 0; \ i: \leq i)) \\
= & \{i < 2\}; \ (i: \leq i+1 \,_{p'}\oplus i:= 0) \\
= & \{i < 2\}; \ (i: \in \{0..i+1\} \,_{p'}\oplus i:= 0) \\
\sqsubseteq & \{i < 2\}; \ (i: \in \{1..i+1\} \,_{p'}\oplus i:= 0) \quad \text{refine away the zero case} \\
= & \{i < 2\}; \ ((i: \leq i; \ i:= i+1) \,_{p'}\oplus i:= 0) \\
= & \{i < 2\}; \ ((i: \leq i; \ i:= i+1) \,_{p'}\oplus (i: \leq i; \ i:= 0)) \\
\sqsubseteq & \{i < 2\}; \ i: \leq i; \ (i:= i+1 \,_{p'}\oplus i:= 0) \quad \text{Law 16 on p. 324} \\
= & i: \leq i; \ \{i < 2\}; \ (i:= i+1 \,_{p'}\oplus i:= 0) \\
= & i: \leq i; \ op' \ .
\end{array}
$$

Thus the final form for our representation program is given by the sequential composition

$$(i:= 0 \,_{q}\oplus i:= 1); \ i: \leq i \ ,$$

which is just $i:= 0 \,_{\geq q}\oplus 1$. Although using it for rep directly would have allowed us to show in one step that both refinements in (4.1) held, in fact the calculations are often easier if done in stages.

When p' is $1/2$, the corresponding abstract p given by (4.6) is $(\sqrt{5}+1)/4$, obtained by setting q to the value $(\sqrt{5}-1)/2$; thus we are using a rep defined as

$$i:= 0 \quad _{\geq(\sqrt{5}-1)/2}\oplus \quad i:= 1 \qquad (4.7)$$

to show the refinement of Fig. 4.3.3.

Informal reasoning about data refinement can be treacherous. A very quick analysis might read "the concrete datatype aborts only if it increments i twice in a row, with probability $(0.5)^2 = 0.25$, but the abstract datatype can fail every time, with probability $\overline{p}$ — thus p should be 0.75."

---

[11]By this we mean $i: \in \{0..i\}$. In general, we write a generalised assignment $x: \odot \ expr$ for "choose fresh x' demonically such that $x' \odot \ expr$ and then execute $x:= \ x'$", a notation borrowed from Abrial.

init   **skip**
op   **skip** $_{(\sqrt{5}+1)/4}\oplus$ **abort** $\leftarrow$ $(\sqrt{5}+1)/4 \simeq 0.81$

init' $i := 0$
op' $\{i < 2\}$; $i := i + 1$ $_{0.5}\oplus i := 0$

Figure 4.3.3. A FAULTY SKIPPER FOR $p := 1/2$

This reasoning is incorrect, however, because it is comparing one invocation of *op* with two invocations of *op'*. (The two datatypes must be compared with respect to the *same* context.) Moreover, the failure of *op'* occurs only on the *third* invocation, after i has been set to 2 by the second.

Thus chastened, we might say "what we do know is that three consecutive invocations of *op'* will fail with probability at least 0.25, and the corresponding probability for *op* is $1 - p^3$ — thus p cannot be more than $\sqrt[3]{0.75}$." That value is shown at the end of Row 3 in Fig. 4.3.4, and it is correct that we must have p less than 0.9086.

In fact it must be *much* less than that, since assuming that i is 0 is too optimistic. Indeed, when $i = 1$ failure will occur with probability 0.5 after only two more invocations of *op'*'s which, by similar reasoning, gives the too-pessimistic constraint that $p \leq 0.7071$.

The problem is that we have no way of knowing what the value of i is, since it is hidden inside the datatype and has been subject to probabilistic assignments that we cannot observe directly.[12] But for such a simple program, we can fall back on the definition Def. 4.2.1 of data refinement and tabulate all possible contexts. (Recall Footnote 5 above.) The resulting values in the final column of Fig. 4.3.4 tend in the limit to our exact value $(\sqrt{5}+1)/4$ calculated earlier.

---

[12]Standard techniques for Markov processes can show that in the long run the conditional probability that i will be 0 rather than 1 (*i.e.*, given it is not 2 and that abortion has not already occurred) is the $(\sqrt{5}-1)/2$ given by our *rep* in (4.7). This reinforces the intuitive view that *rep* estimates the distribution assumed by the hidden state.

In fact, simple arithmetic based on the recurrence

$$p'_0 := (p_0 + p_1)/2$$
$$p'_1 := p_0/2$$

calculated in the first two columns of Fig. 4.3.4 shows that $p_0, p_1 = (F_n, F_{n-1})/2^n$, where F_n is the nth *Fibonacci number*, and so the quotient p_0/p_1 should approach the *Golden Ratio* $(1 + \sqrt{5})/2 \simeq 1.618$ in the limit.

From there we can argue that the long-term probability that *op'* will abort on the following invocation — given that it has not aborted already — is $1/2$ *times* the (conditional) probability that i is 1 now. That calculation will lead to the correct value for the abstract p.

n	Probable effect of $init'$; $(op')^n$				Corresponding abstract probability p
	$i := 0$	$i := 1$	$i := 2$	**abort**	
0	1.0	0.0	0.0	0.0	
1	0.5	0.5	0.0	0.0	1.0000
2	0.5	0.25	0.25	0.0	1.0000
3	0.375	0.25	0.125	0.25	0.9086
4	0.3125	0.1875	0.125	0.375	0.8891
5	0.25	0.1563	0.0938	0.5	0.8706
6	0.2031	0.125	0.07813	0.5938	0.8606
7	0.1641	0.1016	0.0625	0.6719	0.8528
8	0.1328	0.0820	0.0508	0.7344	0.8473
9	0.1074	0.0664	0.0410	0.7852	0.8429
10	0.0869	0.0537	0.0332	0.8262	0.8395

$$\vdots$$

n	$i := 0$	$i := 1$	$i := 2$	**abort**	p
29	0.0015	0.0010	0.0006	0.9969	0.8193
30	0.0012	0.0008	0.0005	0.9975	0.8190

$$\vdots$$

The table shows that the concrete program $init'$; $(op')^6$, for example, will abort with just over 59% probability; but from Fig. 4.3.2 we see that the corresponding abstract program — the specification — allows $init$; op^6 to abort with probability no more than $1 - p^6$. Thus for the implementation to abort *no more often* than the specification, we must have

$$0.5938 \quad \le \quad 1 - p^6 ,$$

that is $p \le \sqrt[6]{0.4062} \simeq 0.8606$ as given in the rightmost column.

Similar reasoning applies for all other n, as well, and the estimate for p grows steadily worse the more times op is supposed to have been invoked. It approaches just under 0.81 asymptotically, however — and that greatest lower bound is the value predicted by our data refinement.

Figure 4.3.4. TABULATION OF 2-SKIPPER PROGRAM

Finally we generalise the above ideas to treat the N-skipper of Fig. 4.3.1. Following the same reasoning as above we use

$$rep \quad := \quad (\llbracket\, n \mid 0 \le n \le N \cdot i := n \,@\, q_n); \; i :\le i , \qquad (4.8)$$

where $q_0, q_1, \cdots, q_N$ are probabilities summing to 1. As before, when the refinement equations are simplified we recover a set of constraints

$$\begin{aligned} p\, q_0 \quad &\le \quad \overline{p'_N}\, \overline{q_N} \\ p\, q_{n+1} \quad &\le \quad p'_N\, q_n \quad \text{for } 0 \le n < N . \end{aligned} \qquad (4.9)$$

Notice that this time the assignment to i is chosen from $0, \cdots, N$ since if op'_N does not abort, i must have one of these values finally. Once more we maximise for p, the result of which is summarised in the following lemma.

Lemma 4.3.5 The program *rep* defined in (4.8) is a simulation between the datatypes in Fig. 4.3.1 provided $p \le \widehat{p}$, where $\widehat{p}$ is the largest solution in the interval $[0, 1]$ of the equation

$$\widehat{p}^N \quad = \quad \overline{p'_N} \left(\widehat{p}^{N-1} + \widehat{p}^{N-2} p'_N + \cdots + (p'_N)^{N-1} \right) . \qquad (4.10)$$

Proof   As with (4.3–4.5) earlier we rewrite $\overline{q_N}$ in the first inequation as $q_0 + \cdots + q_{N-1}$ and then use a scaling argument to see that we can ignore one-summing. Thus we take $q_N := 0$ and substitute the inequalities for $q_1 \cdots q_{N-1}$ into the inequality for q_0; after multiplying both sides by p^{N-1}/q_0, we have

$$p^N \quad \le \quad \overline{p'_N} \left(p^{N-1} + p^{N-2} p'_N + \cdots + (p'_N)^{N-1} \right) , \qquad (4.11)$$

for which p is zero gives a trivial solution.

But when p is one, the right-hand side of (4.11) is no more than one; thus its maximum solution $\widehat{p}$ in $[0, 1]$ must occur at an equality.

Finally, re-inspection of the original equations (4.9) shows that if $\widehat{p}$ satisfies them for some one-summing $q_0, \ldots, q_N$, then so does any $p \le \widehat{p}$.
□

In the general case there is a unique solution of the equation (4.10) lying in the interval $[0, 1]$. (We are interested in that interval since we look for probabilities.) A careful analysis shows that it always allows us to prove a data refinement with p somewhat larger than $1 - (p'_N)^N$, the probability that i is incremented N times in a row.

4.4   A safety-critical application: the probabilistic steam boiler

In this section we consider the problem of calculating the reliability of a regulator for a steam boiler. An accurate model would be much more complicated than the one presented here — but we concentrate on only a small number of details, for our primary aim is to illustrate how the data-refinement techniques developed in the previous sections might apply to a realistic analysis of a safety-critical system.

4.4.1   System safety

The steam boiler consists of a reservoir of water supplied by pumps. During operation the water level continually fluctuates, in reaction to steam loss and to varying pump- and water pressure. If the water level becomes either

steam discharge	pump	simplified effect δ
disabled	off	0
disabled	on	+1
enabled	on	0
enabled	off	−1

The above shows the possible change in the steam-boiler water level during a single unit of time.

Figure 4.4.1. CHANGE IN STEAM-BOILER WATER LEVEL

---

too low or too high, the boiler could be seriously damaged; thus we describe our system reliability in terms of the level's remaining between a low (L) and a high (H) boundary. We use the standard safety-engineering definition of *unreliability* [Sto96] — it is the proportion of observed failures in a given interval of time (*reliability* is the complement). Thus in this application our unreliability measures appear as a probability of failure per unit time; and each unit is a single iteration of one of the programs given below.

The steam boiler has a *regulator*, which we write as a datatype specifying the overall tolerance on system reliability,[13] and which otherwise has very simple behaviour — it abstracts from the sensors and the fault tolerance mechanism which would appear in the more complicated implementation. By proving refinement between specification and implementation we aim to establish a link (as in standard safety analysis) between high-level "system reliability" and low-level component failure rate. We concentrate on the two system components that together contribute to unreliability by independently affecting the water level: they are the *environment*, and the regulator.

4.4.2 The environment

Our intended environment is a very general one, for it can influence the water level in either direction. Disturbances in the ambient temperature, for example, impact on the rate of steam production and can cancel the effect of pump action. In Fig. 4.4.1 we illustrate the possible changes in water level during a single unit of time.

In a more elaborate system model we might abstract specific environment reactions by a demonic choice over a wide range of pump- and steam pressures. Here we simplify matters by assuming that the rate of water movement δ is either constant or zero; normalisation then makes it −1, 0

---

[13]In fact we present only the operations of the datatypes — it is a routine exercise to include the initialisation.

$Regulator(w, \delta)$:=

$$
\begin{array}{lll}
\textbf{if} & w = H & \rightarrow & w := w - 1 \;_{\geq e}\oplus\; w := w + \delta \\
\square & L < w < H & \rightarrow & w := w + \delta \;_{\geq f}\oplus\; \textbf{fail} \\
\square & w = L & \rightarrow & w := w + 1 \;_{\geq e}\oplus\; w := w + \delta \\
\textbf{fi}; \\
\{L \leq w \leq H\}
\end{array}
$$

Figure 4.4.2. REGULATOR SPECIFICATION

---

or $+1$. With that abstraction the environment's behaviour is very simple: it is a demonic choice over the possibilities of water-level movement δ in one time unit: we write it

$$\delta :\in \{-1, 0, 1\} \; .$$

4.4.3 Safety specification for the regulator

The behaviour of the steam-boiler system is to alternate between the environment action and the regulator reaction. A single step of such a cycle is therefore given by

$$
\begin{array}{l}
\delta :\in \{-1, 0, 1\}; \\
Regulator(w, \delta); \\
\{L \leq w \leq H\} \; .
\end{array}
$$

The task of the procedure $Regulator(w, \delta)$ is to keep the water level w between the boundaries H and L, possibly counteracting the environment to do so — and if it fails, the assertion command $\{L \leq w \leq H\}$ will cause the whole system to abort.

The specification of the regulator set out in Fig. 4.4.2 describes both correct and faulty behaviour, defining the overall tolerance placed on the regulator's (and hence the system's) reliability by its users. The specification says that when the water level is dangerously high ($w = H$) the regulator will lower it (executing $w := w - 1$) with probability at least e. When the water level lies strictly between the boundaries, in the "safe" region, the regulator does not need to intervene: but with probability at least f the water level is updated by the environment alone; failure here is modelled by the statement **fail**, which we explain below.

The $w = L$ case is similar to the $w = H$.

$$RegRef(w, \delta) \ := \ \begin{array}{llll} \textbf{if} & w \geq H & \rightarrow & w := w - 1 \ {}_e\oplus \ \textbf{abort} \\ \square & L < w < H & \rightarrow & w := w + \delta \ {}_f\oplus \ \textbf{abort} \\ \square & w \leq L & \rightarrow & w := w + 1 \ {}_e\oplus \ \textbf{abort} \\ \textbf{fi} \end{array}$$

We replace $\geq_e\oplus$ *etc.* by the simpler ${}_e\oplus$ without loss of generality, because the right-hand alternatives are all **abort**.

Figure 4.4.3. REGULATOR REFINEMENT

To begin our analysis, we use program algebra to move the boundary check $\{L \leq w \leq H\}$ into the branches of the alternation. In the worst case for $w = H$ (assuming pessimistically that δ could be 1) the result is the statement

$$w := w - 1 \ {}_{\geq e}\oplus \ \textbf{abort} .$$

(Again the $w = L$ case is similar.)

Here we are reserving **abort** to model *catastrophic failure*: for us, that is when w escapes the boundaries. The statement **fail** on the other hand is deemed a lesser failure — immediate catastrophe is not its consequence.[14] However an occurrence of **fail** would still require invocation of some back-up procedure while necessary repairs were carried out; we call that *maintenance failure*. The different levels of severity mean different reliabilities: maintenance failure would be tolerated at a much higher rate than catastrophic failure would be and thus, in general, the catastrophic reliability e would be expected to be much higher than maintenance reliability f. In any case $\overline{e}$ should be considerably less than the failure rate of the components (see below) — that, as we will see, can be achieved by the design of the implementation.

Having distinguished between failure types, for simplicity we drop that distinction — modelling both with **abort** in our formal treatment. We do, however, preserve the brief association with the notion of "maintenance failure" in that we allow f to be much less than e: we can tolerate maintenance failures at a higher rate. With that and a final simplification the regulator specification becomes the program fragment in Fig. 4.4.3.

4.4.4 The regulator implementation

In reality a regulator would not have direct access to the water level, but rather would rely on the sensors to relay that information — it is the sensors that cause the overall failure, and our task is to minimise that risk. Thus

---

[14]The **fail** occurs for example when the regulator ignores δ because an internal sensor has failed for some time — its estimate of the water level is so inaccurate that it acts as if the boundaries might be violated even when they cannot be.

$$
\begin{array}{llll}
RegImp(w,\delta) \quad := & \textbf{if} & g+i \geq H & \rightarrow \quad w := w - 1 \\
& \Box & L+i < g < H-i & \rightarrow \quad w := w + \delta \\
& \Box & g-i \leq L & \rightarrow \quad w := w + 1 \\
& \textbf{fi}; \\
& i := i+1 \ _s\oplus \ i, g := 0, w
\end{array}
$$

Figure 4.4.4. REGULATOR IMPLEMENTATION

the implementation set out in Fig. 4.4.4 must not refer directly to w in its guards, but rather uses variables g (for "guess," the last successful sensor reading of the water level) and i (the length of time that the sensor has continuously failed).

In each time unit, we assume the sensor fails with probability s. If $i = 0$ the sensor is working and the water-level reading g coincides with the actual water level w. If on the other hand $i > 0$, then g only holds the last accurate reading — for g is *not* updated along with an increase in i — so the regulator must "guess," calculating bounds on the actual water level using the (invariant) relation

$$
g - i \leq w \leq g + i \, .
$$

The probabilistic assignment to i and g models a simple time-independent failure/repair rate of s and $\bar{s}$ respectively. And now the extent to which the regulator can avert catastrophe by operating under the invariant above is to be measured by the relationship between e and s (which we determine below).

Before we discover formally what that relationship is, by arguing informally we can see that, when $g + i \geq H$ and $g - i \leq L$ do not hold simultaneously, the regulator can safely push the water level away from one boundary without running the risk of crossing the other. After some time however (and in the worst case when $i > (H-L)/2$) that situation is no longer tenable, and at that point the regulator can only assume that the water level is both too low and too high, an impossibility that can be resolved (in the specification at least) only by abortion. Thus we would expect that e in Fig. 4.4.2 is roughly at least $1-s^{(H-L)/2}$, the probability that the sensor fails $(H-L)/2$ times in a row. As discussed above we now discover the precise quantitative relation between s, f and e by determining the conditions under which $RegRef(w,\delta)$ is data-refined by $RegImp(w,\delta)$ (Figs. 4.4.3 and 4.4.4).

4.4.5 Proof of data refinement

As in Sec. 4.3, we take a supplier-oriented view of the problem — fixing the reliability $\bar{s}$ of the sensor in the implementation, we wish to find the greatest values for e and f which will allow a data refinement. Recall that

$$
\begin{array}{llll}
\textbf{if} & w \geq H & \rightarrow & \{i < (H-L)/2\};\ w:=w-1 \\
\square & L < w < H & \rightarrow & \{i = 0\};\ w:=w+\delta \\
\square & w \leq L & \rightarrow & \{i < (H-L)/2\};\ w:=w+1 \\
\textbf{fi}; &&& \\
\end{array}
$$
$$i:=i+1 \ _s\!\oplus\ i:=0$$

Figure 4.4.5. INTERMEDIATE DATA-REFINEMENT STEP

---

overall reliability of the regulator depends on the reliability of the sensors, and establishing a data refinement is contingent on that dependency.

We prove the refinement in two stages: starting from the specification in Fig. 4.4.3 we first add the variable i, to reach the intermediate program set out in Fig. 4.4.5; only then do we introduce the variable g. We sketch the details of the formal proofs here, as well as appealing freely to standard data-refinement results — for example, as before, that the relation of data refinement is transitive, and that ordinary program refinement is simply a special case of it [Mor88a].

Lemma 4.4.6 $RegRef(w, \delta)$ is data-refined by the datatype set out in Fig. 4.4.5 provided that f is no greater than $\bar{s}$ and that e is no greater than the solution $\widehat{e}$ of

$$\widehat{e}^N = \bar{s}(\widehat{e}^{N-1} + s\widehat{e}^{N-2} + \cdots + s^{N-2}\widehat{e} + s^{N-1}), \qquad (4.12)$$

where $N := (H - L)/2$. This is essentially (4.10) for s rather than p'_N.

Proof   The lemma follows by noting first that each guarded statement, together with the assignment to i that follows, is essentially an instance of the faulty skipper in Fig. 4.3.1. To prove a data refinement, we use a representation program

$$rep := (\|\, n \mid 0 \leq n \leq (H-L)/2 \cdot i := n @ q_n);\ i:\leq i\,, \qquad (4.13)$$

where the q_n are defined analogously to (4.9) of Sec. 4.3. The required data refinement can now be proved via (4.13): we have that rep distributes through the guards (they do not involve i and rep does not mention w); moreover despite the q_n being solutions to the faulty N-skipper when N is $(H-L)/2$ (thus applicable to the boundaries) we note that the demonic decrease of i in (4.13) gives us the refinements (4.1) for the intermediate $L < w < H$ case as well. $\square$

Lemma 4.4.7 The "intermediate" datatype in Fig. 4.4.5 is data refined by the $RegImp(w, \delta)$ datatype of Fig. 4.4.4.

Proof   We add the variable g to the intermediate datatype in Fig. 4.4.5 using the standard technique of coupling invariants [Mor88a]. We observe first that the real water level w is always maintained between the estimated boundaries $g - i$ and $g + i$, hence we define the coupling invariant

$$g - i \leq w \leq g + i\,.$$

Using this, we can (routinely) replace the guards and augment the assignment. □

Composing the data refinements of Lem. 4.4.6 and Lem. 4.4.7, we can finally conclude that using a sensor of reliability $\bar{s}$ guarantees overall system reliability of at least (and in fact strictly greater than) $1-s^{(H-L)/2}$ since, as remarked earlier (p. 117), the solution of (4.12) exceeds $1 - s^N$.

For example, if the waterflow/sampling rate were such that $H - L$ was 10, and the probability of sensor failure was 1% per sample interval, our specification would be met with values

$$e \ := \ 1 - 10^{-10}$$
$$f \ := \ 0.99 \ ,$$

indicating extremely low chance of catastrophic failure, but rather high chance of maintenance failure.

4.5 Summary

In this chapter we have illustrated how data refinements within our probabilistic domain may be proved using the technique of simulations; and we have argued that establishing probabilistic data refinements provides a quantitative link between system- and component-level reliabilities.

It is well known that representation programs (our *rep*) are not necessarily unique, and one might conjecture that a non-probabilistic representation might be found that would satisfy the refinements in (4.1) and thus avoid some of the calculations above. That the probabilistic nature of our example needs the extended techniques can be seen by considering the 2-skipper — for we may assume a finite state space and thus only finitely many possibilities (nine in all[15]) for a non-probabilistic *rep*, and it can be checked that none of them satisfies the refinements in (4.1).

More generally we can explain the choice of *rep* by considering how accurately an observer can guess the value of the local variable at "run time" of a program using the 2-skipper. We assume that he knows the implementation's code, but not the *actual* run-time value of i (since he cannot observe i directly). After several calls to *op'* at best he can only infer a probability distribution for i; from this he can calculate the chance that the 2-skipper will abort next time *op'* is called — it is the probability that the actual value of i is 2. Hence *rep* encodes the observer's most accurate knowledge of i's value — if *rep* was non-probabilistic that would correspond either to perfect knowledge (for a deterministic *rep*) or to complete lack of knowledge (for a demonic *rep*); in neither case is this appropriate for the 2-skipper as the observer's knowledge lies somewhere in between and can

---

[15]The possible initial values for i are 0 and 1; the possible final values are 0, 1 and 2.

only be described by a probability distribution. More details concerning the general treatment of probabilistic assignments to local variables can be found elsewhere [Gro].

Naturally one would hope to extend the techniques of this chapter to the application of quantified analyses of safety-critical systems in general. For example our modelling all types of system failures as **abort** is certainly unacceptable for a quantified assessment of risk, for which we would need to continue to distinguish between the possible identified hazards (for risk measures the cost modified by likelihood of all the hazards [Sto96]).

Chapter notes

The steam-boiler specification problem has received a great deal of attention, including an entire workshop devoted to case studies based on it [ABL96]. It typifies those situations where the environment endows a probabilistic context — in this case the known failure rates of the hardware components. The question of *quantifying* the reliability (studied in this chapter) has also been considered by Andriessens and Lindner [AL96] whose solution appeared at the workshop.

The approach we have taken is based on the formalisation of data refinement as described by Gardiner and Morgan [MG90]. But the *representation functions* critical for proofs of data refinement are similar to the *abstraction relations* which underlie abstract interpretation, and which are crucial for extending the applicability of model checking; Monniaux [Mon01] has explored this topic in the context of probability. Moreover Huth [Hut03] considers the wider issues of probabilistic abstraction and has reviewed many methods in an extensive and informative survey.

More generally, data refinement is just one way to formulate the problem of how to compare programs' behaviours when they operate over different state spaces — and in fact such a comparison only makes sense when it is restricted to prescribed "observable" states or events. There are, of course, many other formulations of this problem, in particular concurrency theory's *bisimulation*, of which there are a number of treatments for probabilistic systems. One of the earliest formalisations is by Larsen and Skou [LS91] who characterise *bisimilar processes* in terms of testing, taking the view that user *observations* are the crucial ingredient for judging comparisons. Other work along these lines includes Stoelinga and Vaandrager [SV03] — they suggest that tests based on random variables may be able to distinguish processes.

One technical difference between bisimulation and data refinement is that the former makes *exact comparisons* of events, whereas the latter, which is formulated in terms of *program refinement*, does not. Some treatments of probabilistic bisimulation also avoid making exact comparisons by formalising the "distance" between process behaviour with the introduction of appropriate (pseudo-)metrics, an approach taken by Ying and Wirsing [YW00, Yin02], by Desharnais *et al.* [DGJP02] and by van Breugel *et al.* [vBMOW03] for example. Ying [Yin03] has also suggested a quantitatively approximate notion of (ordinary) refinement: a possible line of research would be to extend this to data refinement.

Part II

Semantic structures

128

5

Theory for the demonic model

In Part I we saw how probabilistic sequential programs are interpreted as expectation transformers satisfying certain "healthiness" conditions (Sec. 1.6), the principal nontrivial one being sublinearity. At the same time, we introduced a simple probabilistic programming language *pGCL* and gave it an informal operational interpretation as a gambling game (Sec. 1.3).

The two views — expectation transformers *vs.* gambling games — are strongly related, and indeed that relationship is part of what gives the entire approach its legitimacy. In this chapter we fill in the details, defining the operational model more mathematically and stating precisely how it is related to the transformer approach.

This chapter is based on an earlier work *Probabilistic Predicate Transformers* [MMS96] © ACM, 1996.

In so doing we see how the sublinearity condition is generated, and why it is important.

Except for Chap. 7, in this part we use a more mathematical notation than in Part I, working "in the semantics" rather than "in the logic" — thus (recall p. 25) we treat expectations *etc.* as functions rather than as expressions.[1] We use lower-case Greek letters α, β for random variables, and upper-case Greek letters Δ, Δ' for distributions.

5.1   Deterministic probabilistic programs: the operational view

We now formalise our operational model for deterministic probabilistic behaviour, the "gambling game" of Sec. 1.3.

A *standard* sequential program begins execution in an *initial* state and terminates (if it does) in a *final* state. For standard programs, *deterministic* means that for every initial state from which termination is guaranteed there is exactly one corresponding final state; from an initial state that can lead to nontermination, however, any final state is possible as well.[2]

We formalise that view as follows: given a set S of states, a standard deterministic program denotes a total function of type $S \to S_\perp$, where we write $S_\perp$ for the set $S \cup \{\perp\}$ so that nontermination can be indicated explicitly by a special (improper) final state $\perp$.[3]

In this chapter we require the *state space* S to be finite.[4]

A *probabilistic* deterministic program takes an initial state to a fixed final probability "sub-distribution" over S, rather than to some single element of $S_\perp$:

Definition 5.1.1 SUB-DISTRIBUTION   For state space S, the set of *sub-distributions* over S is

$$\overline{S} \quad := \quad \{\Delta \colon S \to [0,1] \mid \textstyle\sum \Delta \leq 1\} \,,$$

---

[1] Chap. 7 for consistency continues the expression-style notation used in Chap. 2, of which it is a continuation.

[2] Some writers call that PRE-DETERMINISTIC, others LIBERALLY DETERMINISTIC [Hes92]: it means "deterministic if terminating."

[3] We write $A \to B$ for the set of total functions between sets A and B.

[4] Infinite state spaces interfere not so much with definitions as with proofs: Lem. 5.7.4 to come would then be argued in an infinite dimensional Euclidean space, where Farkas' Lemma (B.5.2) does not apply directly. In Chap. 8 we return to the infinite case, where the results are shown to apply provided continuity is imposed on the expectation transformers.

the set of functions from S into the closed interval of reals $[0,1]$ that sum to no more than one. (We write $\sum \Delta$ to abbreviate $\sum_{s:S} \Delta.s$.)   □

Thus sub-distributions over S assign a probability to every element individually of S, and those probabilities must sum to no more than one. Usually, for simplicity we will drop the "sub-." [5]

There is a complete partial order over $\overline{S}$, inherited pointwise from $[0,1]$:

Definition 5.1.2 ORDERING OVER DISTRIBUTIONS   For $\Delta, \Delta' \in \overline{S}$ we define

$$\Delta \sqsubseteq \Delta' \quad := \quad (\forall s\colon S \cdot \Delta.s \le \Delta'.s)\,.$$

□

The least element of $\overline{S}$ is $(\lambda s\colon S \cdot 0)$, a constant function which we write $\underline{0}$; and Δ in $\overline{S}$ is maximal when $\sum \Delta = 1$, in which case Δ indicates certain termination.[6]

Thus we define a space of probabilistic deterministic programs together with a refinement order:

Definition 5.1.3 DETERMINISTIC PROBABILISTIC PROGRAMS   For state space S the space of deterministic probabilistic programs over S is defined

$$\mathbb{D}S \quad := \quad (S \to \overline{S}, \sqsubseteq)\,,$$

where for programs f, f' in $S \to \overline{S}$ we define $\sqsubseteq$ pointwise (again): [7]

$$f \sqsubseteq f' \quad := \quad (\forall s\colon S \cdot f.s \sqsubseteq f'.s)\,.$$

The order $\sqsubseteq$ of $\mathbb{D}S$ is called the *refinement* order.   □

It is easy to see that $\mathbb{D}S$ is a complete partial order; and its least element is the nowhere-terminating program given by

$$\textbf{abort}.s := \underline{0} \quad \text{or equivalently} \quad \textbf{abort}.s.s' := 0\,, \tag{5.1}$$

for any initial state s and final state s'. (Recall that **abort** *is* deterministic in our sense; more precisely, it is pre-deterministic.) Because it is least, we have for every program f in $\mathbb{D}S$ that $\textbf{abort} \sqsubseteq f$.

In general, if $\sum f.s = 1$ then $f.s.s'$ is the exact probability that f takes s to s'; but if $\sum f.s < 1$ then $f.s.s'$ is only a lower bound for that probability,

---

[5]In Kozen's terminology [Koz81, p.331], these would be called DISCRETE SUB-PROBABILITY MEASURES.

[6]We use λ-notation for explicit functions, this one taking argument s and returning the constant value zero; as usual we include parentheses to indicate the scope of the bound variable s.

[7]We write "f" for such programs to remind us that they are functions, *i.e.* single-valued, even though they are in the "relational" semantics in the sense of Sec. 5.4 below.

since the aborting component of weight $1 - \sum f.s$ is unpredictable. Thus for **abort** itself, where that lower bound is zero, we are told nothing about the probability of reaching any particular final s'.

For state s in S we write $\overline{s}$ for the "point" distribution "certainly s":

Definition 5.1.4 POINT DISTRIBUTION   For state s the *point distribution* at s is defined

$$\overline{s}.s' \quad := \quad 1 \text{ if } (s = s') \text{ else } 0 \;.$$

□

Point distributions allow us to embed standard deterministic programs in the probabilistic space as follows:

Definition 5.1.5 EMBEDDED DETERMINISTIC PROGRAM   For every standard deterministic program f in $S \to S_\bot$ there is a corresponding deterministic probabilistic program $\overline{f}$ in $\mathbb{D}S$, defined

$$\overline{f}.s \quad := \quad \begin{array}{ll} \overline{f.s} & \text{if } f.s \neq \bot \\ \underline{0} & \text{otherwise,} \end{array}$$

where s is the initial state. Equivalently we could define for arbitrary final state s' that

$$\overline{f}.s.s' \quad := \quad \begin{array}{ll} 1 & \text{if } f.s = s' \\ 0 & \text{otherwise,} \end{array}$$

noting in both cases that s, s' are restricted to proper elements (not $\bot$).

□

As an example, consider the properly probabilistic program over state space $\{1, 2, 3\}$ written in $pGCL$ as

$$s := 1 \; {}_{\frac{1}{3}}\oplus \; s := 2 \;. \tag{5.2}$$

We can give its operational interpretation in the terms above, so that it will be in the space $\mathbb{D}\{1, 2, 3\}$, and is f_1 where

$$f_1.s.s' \quad := \quad \begin{array}{ll} 1/3 & \text{if } s' = 1 \\ 2/3 & \text{if } s' = 2 \\ 0 & \text{otherwise.} \end{array}$$

(The absence of s on the right reflects only that (5.2) is insensitive to its initial state.)

We do not give here the definitions of program operators for $\mathbb{D}S$, since we treat them later in a more general setting (Definitions 5.4.5–5.4.7). But note for example that since

$$s := 1 \ {}_{\frac{1}{3}}\oplus \ \textbf{abort} \tag{5.3}$$

denotes f_2 where

$$
f_2.s.s' \quad := \quad \begin{array}{ll} 1/3, & \text{if } s' = 1 \\ 0, & \text{otherwise,} \end{array}
$$

we can see that $f_2.s$ delivers a *sub*-distribution: that is, $f_2.s$ does not sum to one. Thus we now have $\textbf{abort} \sqsubset f_2 \sqsubset f_1$, illustrating both nonstrictness of ${}_{1/3}\oplus$ and proper (nontrivial) refinement.[8],[9] Note also that terminating programs in $\mathbb{D}S$ are maximal in the refinement order, which by analogy with standard programs justifies their being called deterministic.[10]

5.2 The sample space, random variables and expectations

Since we are concerned with distributions over a state space S, that S will be our *sample space* in the sense of Sec. 1.4. And, as we saw there, *random variables* are functions from that space into the non-negative reals $\mathbb{R}_{\geq}$; as we discussed in Sec. 2.11, and for reasons we will revisit here, we usually require the random variables to be bounded.[11]

---

[8]By $\sqsubset$ we mean "$\sqsubseteq$ but not $=$".

[9]A function f is *strict* just when $f.\bot = \bot$; for functions of several arguments the terminology is analogous.

Here, we are therefore using $\bot$ in its more general sense of "bottom element" in a partial order, that is the one which is below ($\sqsubseteq$) all others. Because **abort** is refined by all other programs, it is the "bottom program," and strictness of a binary operator $\odot$ on programs would therefore mean that

$$prog \odot \textbf{abort} \quad = \quad \textbf{abort} \odot prog \quad = \quad \textbf{abort} \ .$$

Most program operators have that property: conditional (in its program arguments) and probabilistic choice are the main exceptions.

[10]Singleton result sets, delivered by terminating standard deterministic programs, are $\supseteq$-maximal among the non-empty subsets of S.

[11]Over a finite state space that happens automatically of course.

Definition 5.2.1 RANDOM VARIABLE   A *random variable* is a non-negative real-valued function on the sample space which, for us, is the state space over which our programs operate.[12]   □

Given a probability distribution Δ (in $\overline{S}$) we define the expected value of random-variable α as follows [GW86]:

Definition 5.2.2 EXPECTED VALUE   For bounded random variable α in $S \to \mathbb{R}_\geq$ and distribution $\Delta \in \overline{S}$, the *expected value of α over Δ* is defined

$$\int_\Delta \alpha \ := \ \sum_{s:S} (\alpha.s * \Delta.s) .$$

□

We use the integral notation — rather than our earlier Exp (p. 5) — for compatibility with expectations in general.[13] Note that as a special case we have $\int_{\overline{s}} \alpha = \alpha.s$.

With Def. 5.2.2 we can write the expectation of the square of the final value delivered by a program f in $\mathbb{D}S$, if executed in initial state s, as

$$\int_{f.s} (s^2 \, \mathrm{d}s) , \tag{5.4}$$

where we read $(s^2 \, \mathrm{d}s)$ as the function $(\lambda s \cdot s^2)$ over (that finite subset of) the natural numbers forming the state space. Note that s in the function application $f.s$ is free (it is the initial state), and that $f.s$ is a distribution; but s in the body, the random variable $(s^2 \, \mathrm{d}s)$, is bound. Taking Program (5.2)

---

[12]The usual constraint on random variables is only that they be "measurable" over the probability distributions with respect to which their expected values will be calculated [GS92, p. 19]. Boundedness assures that; but if we relax that constraint then we rely instead on their being non-negative (and if necessary adjoin an infinity element).

However non-negativity is important to us for another reason as well, *viz.* that with mixed-sign expectations it might not be clear what quantity our demonic choice was to minimise.

[13]For example, the expectation (or average) of the function $f \in \mathbb{R} \to \mathbb{R}$ over the interval $[a, b]$ is given in analysis by

$$\frac{1}{b-a} * \int_a^b f(x) \, \mathrm{d}x ,$$

which in the above style one could write $\int_{\langle a, b \rangle} f$, inventing the notation $\langle a, b \rangle$ for this example only to denote a uniform distribution on $[a, b]$.

We are in fact dealing with integration over measures. But generally for measure μ we prefer to write $\int f$ rather than the $\int f \mathrm{d}\mu$ of normal mathematical practice, because the latter greatly $^\mu$ confuses the role of bound and free variables: we are avoiding the notational inconsistency where in $\int_a^b f(x) \, \mathrm{d}x$ the variable x is bound, but in $\int f \mathrm{d}\mu$ the measure μ is free.

as f in (5.4) gives us $(1/3)1^2 + (2/3)2^2$, or an expected value of 3 for the square of the final value of s.

More significantly, Def. 5.2.2 can be used to apply a probabilistic program to a *distribution* of initial states (rather than only to a "definite" single state) — and importantly — to the (intermediate) distribution yielded by the program immediately before it in a sequential composition. Following Jones [Jon90, Sec. 4.2] we have

Definition 5.2.3 KLEISLI COMPOSITION OF PROGRAMS   For $f \in \mathbb{D}S$, (initial) distribution $\Delta \in \overline{S}$ and (final) state $s' \in S$ we define f^*, an element of $\overline{S} \to \overline{S}$, as follows:

$$f^*.\Delta.s' \ := \ \int_\Delta \left(f.s.s' \, \mathrm{d}s \right) .$$

$\square$

Thus we may, if we wish, regard probabilistic deterministic programs as homogeneous functions (*i.e.* from $\overline{S}$ to itself), taking initial distributions to final distributions — the distributions then represent "probabilistic states," as for example explained by Kozen [Koz81], and in execution the deterministic program simply moves from one probabilistic state to the next.[14] The program is deterministic because for a given initial distribution Δ there is only a single possible final distribution $f^*.\Delta$.

The specialization to our earlier view ($S \to \overline{S}$) is obtained for initial state s by taking the point distribution $\overline{s}$ for Δ initially; we then observe that from Def. 5.2.3 (with change of bound variable from s to s'') and the remark following Def. 5.2.2 we have

$$f^*.\overline{s}.s' \ = \ \int_{\overline{s}} f.s''.s' \, \mathrm{d}s'' \ = \ f.s.s' ,$$

so that $f.s$ is recovered as $f^*.\overline{s}$.

5.3   Probabilistic deterministic transformers

The logic of Kozen [Koz85] and later Jones [Jon90, Chap. 7] treats deterministic probabilistic programs at the level of Hoare triples [Hoa69] and Dijkstra-style weakest preconditions [Dij76]. We summarise it below.

For a probabilistic and deterministic program f, and random variables α, β, we write

$$\alpha \ \Rightarrow \ wp.f.\beta \tag{5.5}$$

---

[14]But our programs f^* satisfy Kozen's property [Koz81, Thm. 6.1] by construction: they are completely determined by their behaviour on the point distributions $\overline{s}$ of $\overline{S}$.

to mean that for every initial state s we have

$$\alpha.s \leq \int_{f.s} \beta . \qquad (5.6)$$

We are saying that α gives, in any initial state s, a lower bound for the expected value of β in the final distribution $f.s$ reached via execution of f from s.

We have said this before (1.15)... but now in (5.6) we can write it mathematically. And it is consistent with our earlier usage: whereas formerly $\Rrightarrow$ acted between real-valued *expressions* over the program variables, now it acts as pointwise $\leq$ between real-valued *functions* of the state space defining those variables' values. The random variables α, β we write now are the functions of S denoted by the expressions $preE, postE$ we used before, over the variables S defines.

Similarly, function wp formerly acted on program texts $prog$; but in $wp.f$ above it is acting over the function f in $\mathbb{D}S$ that $prog$ denotes.

The overloading we are taking advantage of here is intended to reduce the syntactic clutter that would occur were we to be more precise by using different names for the "syntactic" and the "semantic" wp: things will be arranged so that we get consistent answers either way. (See Fig. 5.9.1 in Sec. 5.9 for a diagram of how it all fits together.)

A more symmetric form of (5.6) comes from taking the homogeneous approach of the last section: it is equivalent to saying that for any initial distribution Δ we have

$$\int_{\Delta} \alpha \;\Rrightarrow\; \int_{f^*.\Delta} \beta , \qquad (5.7)$$

and we will return to this view in Sec. 5.5.

It is as usual instructive to lift the standard approach, where we use predicates, into the probabilistic structure we are dealing with here. Since predicates over the program variables denote subsets of the state space, we represent them as characteristic functions of those subsets, *i.e.* as random variables over S taking values zero or one only. For subsets S_a and S_b of S, we write $[S_a]$ and $[S_b]$ for their characteristic functions, and consider a standard deterministic program $f \in S \to S_\perp$.

Now we show that the standard pre/post specification $S_a \subseteq wp.f.S_b$ of f — in which $\subseteq$ between sets represents implication between predicates — says about f just what the probabilistic

$$[S_a] \;\Rrightarrow\; wp.\overline{f}.[S_b] \qquad (5.8)$$

says about its probabilistic embedding $\overline{f}$. Take any initial state s; then (5.5) becomes for every state $s \in S$ we have

$$[S_a].s \;\leq\; \int_{f.s} [S_b] \;,$$

when written in the equivalent form given by (5.6).

That simplifies first to $[S_a].s \leq [S_b].(f.s)$, and then we have

$$s \in S_a \quad \text{implies} \quad f.s \in S_b \;,$$

for all s, which is just what we expect from $S_a \subseteq wp.f.S_b$.

We call random variables over the state *expectations*, when used like α, β above for reasoning about programs, because the random variable α in (5.5) is in fact giving (a lower bound for) the expected value of β over the distribution delivered by the program — these expectations are our "probabilistic preconditions and postconditions."

If an expectation takes values in $\{0, 1\}$ only (equivalently, if it is $[S']$ for some subset S' of S), then we call it *standard* — and those correspond to (Boolean) predicates.

Jones [Jon90, Sec. 7.8] relates the operational view of Sec. 5.1 to the axiomatic view (this section), showing soundness and completeness. For *deterministic* probabilistic program f and initial state s, the connection she gives is

$$wp.f.\beta.s = \int_{f.s} \beta \quad \text{and} \quad f.s.s' = wp.f.[\{s'\}].s \;. \quad [15]$$

Our "demonic-enabled" wp of Sec. 5.5 below will generalise that.

5.4 Relational demonic semantics

He, McIver and Seidel [HSM97] gave a model for sequential probability that extends the one in Sec. 5.1 above: a program executed in an initial state produces not a (single) final distribution but rather a *set* of them, and the plurality of the set represents demonic choice. We now show that their construction, with a slight modification, provides a "relational" model for our program logic and indeed motivates its extension to include demonic choice as well. In this section we explain He's model, and in the section following we give the connection between it and the probabilistic program logic.

---

[15]Although $[\{s'\}]$ can be written more simply $\overline{s'}$, we reserve the latter notation for distributions only, since we do not wish to confuse them with expectations even when as functions over S they are identical.

Informally (and made precise in Def. 5.4.4 below), our programs — which we now call r to emphasise their properly relational nature — will take an initial state s to a *subset* of the distributions $\overline{S}$; roughly speaking, the operational model becomes $S \to \mathbb{P}\overline{S}$, of which the earlier type is a special case.[16] We call the semantics "relational" because it is isomorphic to the use of relations between S and $\overline{S}$; but the set-valued-function form is more convenient.

Not every set of distributions is appropriate, however, as the result of a probabilistic demonic program; we consider only non-empty, up-closed, convex, and Cauchy-closed sets, constraints which we now discuss.

Non-emptiness is imposed to avoid "infeasible" transformers representing "miracles" [Mor88b, Mor87, Nel89]. Although they simplify the space structurally (see Footnote 37 following Def. 5.7.1 below) the presentation here would be complicated by the adjoined ∞ values (caused by empty infima) in the arithmetic over $\mathbb{R}$.

For *up closure*, we have the following definition.

Definition 5.4.1 UP CLOSURE   A subset $\mathcal{D}$ of $\overline{S}$, a set of distributions, is *up-closed* if it is closed under refinement of its elements — if for all $\Delta, \Delta' \in \overline{S}$ we have

$$\Delta \in \mathcal{D} \text{ and } \Delta \sqsubseteq \Delta' \quad \text{implies} \quad \Delta' \in \mathcal{D} \ .$$

□

As we do elsewhere [MMSS96, Sec. 13.3], we insist on up closure so that refinement (of demonic programs) is expressed by reverse subset inclusion of their result sets: program r is refined by r' just when for every initial state s we have $r'.s \subseteq r.s$.[17]

Recall for example the program **abort**. Whereas the standard **abort** takes every s to $\perp$, in the deterministic probabilistic model of Sec. 5.1 it takes every s to the constant distribution $\underline{0}$; and so in our current model it takes every s to the *up closure* of $\{\underline{0}\}$. But because $\underline{0}$ is the least element of $\overline{S}$, the up closure is $\overline{S}$ itself — thus allowing every possible behaviour just as one expects from **abort**.

Convexity of a set of distributions is defined as follows:

Definition 5.4.2 CONVEXITY   A set $\mathcal{D}$ of distributions is *convex* if for every $\Delta, \Delta' \in \mathcal{D}$ and probability $p \in [0,1]$ we have $\Delta \,_p\!\oplus \Delta' \in \mathcal{D}$ also.[18]

□

---

[16] We write $\mathbb{P}X$ for the powerset of set X.

[17] This is a standard technique when using the Smyth order [Smy78, Smy89], which we effectively are doing here.

[18] We define probabilistic choice between distributions as $(\Delta \,_p\!\oplus \Delta').s := \Delta.s \,_p\!\oplus \Delta'.s$.

We insist on convexity so that demonic choice is refined by any probabilistic choice, a fact which we used long ago in the program algebra at the end of Sec. 1.2. Consider for example the program

$$s := 1 \sqcap s := 2 . \tag{5.9}$$

The result set of distributions will contain the two point-distributions $\overline{1}, \overline{2}$; but by convexity it will contain also the distributions $\overline{1}_p{\oplus}\overline{2}$ for any $p \in [0, 1]$. Thus our notion of refinement allows Program (5.9) to be refined to

$$s := 1 \ _p{\oplus} \ s := 2 ,$$

for any probability p.

For *Cauchy closure*, we note that our distributions Δ in $\overline{S}$ are points in finite-dimensional Euclidean space, where each axis corresponds to an element s of S and the s-coordinate of distribution Δ is $\Delta.s$. (See Chap. 6 still to come for pictures of this "Euclidean view.") We have then

Definition 5.4.3 CAUCHY CLOSURE   A set of distributions over S is *Cauchy closed* if as a subset of N-dimensional Euclidean space $\mathbb{R}^N$ it is closed in the usual sense,[19] where N is the (finite) cardinality of S.   □

We will see below, in Chap. 8, that Cauchy closure is related to the property of continuity (p. 220).

We can now give the model for demonic probabilistic programs.

Definition 5.4.4 DEMONIC PROBABILISTIC MODEL   Given a state space S, the set of non-empty, up-closed, convex and Cauchy-closed subsets of $\overline{S}$ is written $\mathbb{C}S$, and such subsets are said to be *probabilistically closed*. It can be shown that $\mathbb{C}S$ is a *cpo* under $\supseteq$.[20]

The *cpo* of demonic probabilistic programs over S is then defined

$$\mathbb{H}S \quad := \quad (S \to \mathbb{C}S, \sqsubseteq) , \qquad [21]$$

where for $r, r' \in \mathbb{H}S$ we have

$$r \sqsubseteq r' \quad := \quad (\forall s : S \cdot r.s \supseteq r'.s) .$$

□

A full account of the structure of $\mathbb{H}S$ and the definition of the program operators is given by He [HSM97]; here we discuss only probabilistic choice $_p{\oplus}$, demonic choice $\sqcap$, and sequential composition. For well-definedness

---

[19]That is, in the sense that it contains its boundary.
[20]By CPO we mean complete partial order. The infinite intersection of a $\supseteq$-directed collection of non-empty closed subsets of $[0, 1]^N$ cannot be empty; the other conditions on $\mathbb{C}S$ are preserved trivially by arbitrary intersection.
[21]It is $\mathbb{H}$ for (Jifeng) He.

(preservation of the closure conditions) we again refer to He [*op. cit.*] for the first two; we have added Cauchy closure ourselves, and it is discussed following the definitions.

Probabilistic choice between programs is formed by taking the appropriate combinations of the elements of the result sets.

Definition 5.4.5 RELATIONAL PROBABILISTIC CHOICE   For two programs $r, r' \in \mathbb{H}S$, their p-probabilistic combination is defined as [22]

$$(r \, {}_p\!\oplus r').s \quad := \quad \{\Delta : r.s; \; \Delta' : r'.s \cdot \Delta \, {}_p\!\oplus \Delta'\} \; .$$

□

Demonic choice is formed by taking all possible probabilistic combinations. Thus

Definition 5.4.6 RELATIONAL DEMONIC CHOICE   For two programs $r, r' \in \mathbb{H}S$, their demonic combination is defined as

$$(r \sqcap r').s \quad := \quad (\cup p : [0, 1] \cdot (r \, {}_p\!\oplus r').s) \; .$$

□

Note that the extreme values 1,0 for p ensure that $r \sqcap r' \sqsubseteq r$ and $r \sqcap r' \sqsubseteq r'$, as we expect from demonic choice: it is refined by either operand.

Finally, for sequential composition we have

Definition 5.4.7 RELATIONAL SEQUENTIAL COMPOSITION   For two programs $r, r' \in \mathbb{H}S$, their sequential composition is defined as

$$(r; r').s \quad := \quad \{\Delta : r.s; \; f : \mathbb{D}S \mid r' \sqsubseteq f \cdot f^*.\Delta\} \; , \qquad [23]$$

where by $r' \sqsubseteq f$ we mean $(\forall s : S \cdot f.s \in r'.s)$, agreeing with the notion of refinement that would result were $\mathbb{D}S$ to be embedded in $\mathbb{H}S$ in the obvious way.   □

Thus the sequential composition is formed by taking all possible "intermediate" distributions Δ that r can deliver from s and, for each, continuing with all possible deterministic refinements f of r'.

For well-definedness we are obliged to show that for Cauchy-closed arguments each of the above three operators produces a Cauchy-closed result.

---

[22] If p is a function of the state rather than a constant, on the right-hand side use $p.s$ in the definition instead of p.

[23] Here the set comprehension binds the variables Δ and f, then constrains them to their "types" $r.s$ and $\mathbb{D}S$ respectively and imposes the additional requirement $r' \sqsubseteq f$; the elements of the constructed set are then the terms $f^*.\Delta$ that result.

The reason is the same in each case: a continuous function between Euclidean spaces is applied to a closed and bounded — hence compact[24] — set, and so its image is closed. We consider Def. 5.4.6 as an example:

Lemma 5.4.8 If subsets $r.s, r'.s$ of $\overline{S}$ are Cauchy closed then so is

$$(\cup p\colon [0,1] \cdot (r_{\,p}\oplus r').s) \ .$$

Proof   By Def. 5.4.5 the set concerned may be written

$$\{\Delta\colon r.s; \ \Delta'\colon r'.s; \ p\colon [0,1] \cdot \Delta_{\,p}\oplus \Delta'\} \ ,$$

which is the image in $\mathbb{R}^N$, through a continuous function, of the closed and bounded subset formed as the Cartesian product

$$r.s \ \times \ r'.s \ \times \ [0,1]$$

within $\mathbb{R}^{2N+1}$, where N is the cardinality of S.   □

5.5   Regular transformers

Equipped with the probabilistic/demonic relational model of the previous section, we now look for a corresponding expectation-transformer semantics. The key is provided by (5.7), where initial distribution Δ for deterministic program f led uniquely to final distribution $f^*.\Delta$. In the presence of demonic choice, instead an initial distribution can lead potentially to any number of final distributions Δ, and the analogue of (5.7) must hold for them all. That is,

> If a program r satisfies $\alpha \Rrightarrow wp.r.\beta$ then for every pair of distributions Δ_0, Δ in $\overline{S}$ such that (initial) Δ_0 can be taken to (final) Δ by r, we should have

$$\int_{\Delta_0} \alpha \quad \Rrightarrow \quad \int_\Delta \beta \ . \tag{5.10}$$

Thus whereas in the standard case "the truth value of the assertions must not decrease as execution proceeds," in the probabilistic case it is the *expectations* that must not decrease.

In this section we construct a definition of $wp.r$ which satisfies our requirement (5.10).

---

[24]COMPACTNESS is a notion defined for sets in topological (*e.g.* metric) spaces: here we are using that closed and bounded sets are compact, that continuous functions take compact sets to compact sets, and that compact sets are closed in a Hausdorff space [Sut75].

5.5.1   The two models correspond

We recall that a *relational* model for *standard* demonic programs over S is

$$S \leftrightarrow S_\perp \,,$$

where a standard program r relates s to s' just when executing r from initial state s can yield final state s'. A conventional constraint on elements r of $S \leftrightarrow S_\perp$ is that whenever s is taken by r to $\perp$ then it is taken to all of $S_\perp$ as well (thus allowing simple $\supseteq$ to be the refinement order); this is the standard version of up closure.[25]

If an element of $S \leftrightarrow S_\perp$ relates every initial state to at least one final state, we say that it is *total*; and if it relates every initial state to at most finitely many final states, we say that it is *image-finite*.[26] We write $\mathbb{F}^+X$ for the non-empty finite subsets of a set X, so that $S \to \mathbb{F}^+S_\perp$ is the subset comprising the total and image-finite elements from $S \leftrightarrow S_\perp$.[27]

It turns out that the probabilistic analogue for $\mathbb{F}^+S_\perp$ is exactly $\mathbb{C}S$ — for totality corresponds to the non-emptiness imposed by $\mathbb{C}$, and image-finiteness corresponds to Cauchy closure. We now investigate why that is so.

Recall that the *predicate-transformer* model for standard nondeterministic programs is

$$\mathbb{P}S \leftarrow \mathbb{P}S \,,$$

where we write the functional arrow backward just to emphasise that such programs map sets of final states (postconditions) to sets of initial states (weakest preconditions). We relate the four models as shown in Fig. 5.5.1; thus $\mathbb{T}S$ for example is the expectation-transformer model. (It was given earlier at Def. 1.5.2.)

Now the standard relations can be embedded in the standard predicate transformers $\mathbb{P}S \leftarrow \mathbb{P}S$ in such a way that the image of $S \to \mathbb{F}^+S_\perp$ is characterised by *positive conjunctivity* [Hes92, Chap. 6]:

a predicate transformer $t \in \mathbb{P}S \leftarrow \mathbb{P}S$ is the embedding of some relation in $S \to \mathbb{F}^+S_\perp$ iff for all standard postconditions S_a, S_b

---

[25]This up closure is again associated with the Smyth order [*op. cit.*], but over the so-called "FLAT *cpo*" $S_\perp$ made from the structure-free S by adjoining $\perp$ (as we have seen) and by stipulating additionally that $\perp \sqsubseteq s$ for all s. At Footnote 17 we were dealing with the richer set $\overline{S}$ of distributions, but applying the same principle.

[26]That latter happens automatically under our present assumption that S itself is finite.

[27]In so doing, we note the equivalent formulation $S \to \mathbb{P}S_\perp$ of the latter, that is "total functions from S to subsets of $S_\perp$".

Our four principal models can be arranged along two axes, as in this table:

	standard	probabilistic
relational	$S \to \mathbb{F}^+ S_\perp$	$\mathbb{H}S := \quad S \to \mathbb{C}S$
transformer	$\mathbb{P}S \leftarrow \mathbb{P}S$	$\mathbb{T}S := \quad \mathbb{E}S \leftarrow \mathbb{E}S$

The symbols and constructors are summarised as follows:

S — The state space.

$S_\perp$ — The state space with an extra element $\perp$ adjoined.

$\mathbb{F}^+$ — Finite non-empty subsets of...

$\mathbb{C}$ — Up-, convex- and Cauchy-closed sets of discrete distributions over...

$\mathbb{P}$ — Subsets of...

$\mathbb{E}$ — Expectations over...

Figure 5.5.1. FOUR SEMANTIC MODELS

---

within $\mathbb{P}S$ we have

$$t.(S_a \cap S_b) \quad = \quad t.S_a \cap t.S_b \ .$$

Note that it is the finiteness of S that allows us to consider only finite conjunctions.

Thus the standard transformer model is richer than the relational precisely because not all of its programs are positively conjunctive.[28]

In the remainder of this section we construct an analogous embedding for the probabilistic models.

5.5.2 Embedding the relational model in the transformer model

Recall from (5.10) that for program $r \in \mathbb{H}S$, initial s and every possible Δ that r may produce from s, we must have

$$wp.r.\beta.s \quad \leq \quad \int_\Delta \beta \ . \tag{5.11}$$

---

[28]The extra constraints on $S \to \mathbb{F}^+ S_\perp$ induce strictness and continuity on the corresponding predicate transformers.

(Specialise (5.10) to $\Delta_0 := \bar{s}$ and take $\alpha := wp.r.\beta$.) Noting that $wp.r.\beta.s$ should be as large as possible while satisfying (5.11) for all Δ in $r.s$, we minimise over Δ and propose

Definition 5.5.2 RELATIONAL-TO-TRANSFORMER EMBEDDING   The injection $wp \in \mathbb{H}S \to \mathbb{T}S$ is defined

$$wp.r.\beta.s \quad := \quad (\sqcap \Delta : r.s \cdot \int_{\Delta} \beta) \; ,$$

for program $r \in \mathbb{H}S$, expectation $\beta \in \mathbb{E}S$, and state $s \in S$.   □

It is a consequence of Lem. 5.7.2 below that wp is indeed an injection, as an embedding should be.

That we take the minimum $\sqcap$ over result distributions Δ is related to our demonic view of nondeterminism: the demon resolves the choices in a way that makes the greatest pre-expectation as small (as low) as possible. Note that well-definedness of the minimum is guaranteed by non-emptiness of $r.s$, which is why we imposed it; a more general treatment is possible however if ∞ is allowed as a possible value for $wp.r.\beta.s$.

It is again instructive to specialise Def. 5.5.2 to the standard case. Let β there be $[S_b]$ for arbitrary standard postcondition S_b and suppose r is standard, so that $r.s$ is the closure of some set of point distributions; then

$$
\begin{aligned}
&wp.r.[S_b].s \\
={}& (\sqcap \Delta : r.s \cdot \int_{\Delta}[S_b]) \\
={}& \qquad\qquad\qquad\qquad \textit{r is a standard program: see (†) below} \\
&(\sqcap s' : \text{"standard results of } r.s\text{"} \cdot \int_{\underline{s'}}[S_b]) \\[1ex]
={}& (\sqcap s' : \text{"standard results of } r.s\text{"} \cdot [S_b].s') \\
={}& (\sqcap s' : \text{"standard results of } r.s\text{"} \cdot 1 \text{ if } (s' \in S_b) \textbf{ else } 0) \\
={}& [\text{"standard results of } r.s\text{"} \subseteq S_b] \\
={}& [wp.r.S_b.s] \; ,
\end{aligned}
$$

where in the last line we are using the standard set-to-set version of $wp.r$.

† In considering above only the standard results of $r.s$ (*i.e.* ignoring the closures), we note that the minimum over Δ of $\int_{\Delta} \beta$, for any β, is unaffected by leaving out the points generated by closure — up closure, for example, only increases that value, and convex closure generates only values between the extremes.

Thus $wp.r.[S_b]$ is the characteristic function of those initial states all of whose r-produced final states lie in S_b; and that is of course the standard embedding of $S \leftrightarrow S_\perp$ into $\mathbb{P}S \leftarrow \mathbb{P}S$.

Thus wp as used here is an embedding function of type $\mathbb{H}S \to \mathbb{T}S$, converting probabilistic relational- to probabilistic transformer semantics.

With it we define a subset $\mathbb{T}_r S$ of $\mathbb{T}S$: that image through wp of $\mathbb{H}S$ will comprise the transformers which will be of most interest to us initially — since they correspond to operational behaviours in $\mathbb{H}S$.

Definition 5.5.3 REGULAR TRANSFORMER SPACE   The set of *regular expectation transformers* $\mathbb{T}_r S$ over S is the wp-image of $\mathbb{H}S$ in $\mathbb{T}S$, thus defined as

$$\mathbb{T}_r S \quad := \quad \{r\!:\!\mathbb{H}S \bullet wp.r\} \;.$$

□

We say "regular" because $\mathbb{T}_r S$ is supposed to contain just the transformers for the ordinary, "everyday" probabilistic programs found in practice.

In the next section we establish properties of $\mathbb{T}_r S$, and in the section following that we show that they in fact *characterise* it — so that we will have identified a probabilistic analogue of the conjunctivity that characterises relational programs as transformers in the standard case. Dijkstra originally called such properties "healthiness conditions."

5.6 Healthiness conditions for probabilistic programs

The "healthiness conditions" of Dijkstra [Dij76], for the standard predicate transformers $\mathbb{P}S \leftarrow \mathbb{P}S$, are strictness,[29] monotonicity, positive conjunctivity and continuity. Although Dijkstra proved them by structural induction over program texts, in fact they can also be seen as characterising the images of the total and image-finite relations $S \rightarrow \mathbb{F}^+ S_\perp$ under the standard wp-embedding into $\mathbb{P}S \leftarrow \mathbb{P}S$.

In this section we consider *probabilistic* healthiness, and we see how the probabilistic analogues of strictness, monotonicity, positive conjunctivity and continuity give us a characterisation of the regular transformers $\mathbb{T}_r S$.

As we saw in Def. 1.6.1, the analogue of standard positive conjunctivity is probabilistic sublinearity. We restate that property here, in terms of transformers; it is

---

[29] As a pun Dijkstra called this the LAW OF THE EXCLUDED MIRACLE in its standard form $wp.prog.\mathsf{false} \equiv \mathsf{false}$ [Dij76, Property 1 p. 18].

Definition 5.6.1 SUBLINEARITY OF TRANSFORMERS   An expectation transformer $t \in \mathbb{T}S$ is said to be *sublinear* iff for all $\beta_1, \beta_2 \in \mathbb{E}S$ and $c, c_1, c_2 \in \mathbb{R}_\geq$ we have

$$c_1(t.\beta_1) + c_2(t.\beta_2) \ominus \underline{c} \quad \Rightarrow \quad t.(c_1\beta_1 + c_2\beta_2 \ominus \underline{c}) . \qquad {}^{30}$$

$\square$

And we begin by giving the proof that all regular expectation transformers are sublinear.[31]

Lemma 5.6.2 REGULAR TRANSFORMERS ARE SUBLINEAR   All members t of $\mathbb{T}_r S$ satisfy the condition Def. 5.6.1 of sublinearity.

Proof   For any t in $\mathbb{T}_r S$ we have by definition that $t = wp.r$ for some $r \in \mathbb{H}S$; we then proceed using the notation of Def. 5.6.1. For arbitrary $s \in S$, we calculate

$$
\begin{aligned}
& wp.r.(c_1\beta_1 + c_2\beta_2 \ominus \underline{c}).s && \text{call this } (\ddagger) \\
= {} & (\sqcap\Delta\colon r.s \bullet \textstyle\int_\Delta (c_1\beta_1 + c_2\beta_2 \ominus \underline{c})) && \text{Def. 5.5.2} \\
\geq {} & (\sqcap\Delta\colon r.s \bullet \textstyle\int_\Delta (c_1\beta_1 + c_2\beta_2 - \underline{c})) && \text{monotonicity of } \textstyle\int_\Delta \text{ and } \sqcap \\
= {} & (\sqcap\Delta\colon r.s \bullet c_1 \textstyle\int_\Delta \beta_1 + c_2 \textstyle\int_\Delta \beta_2 - c \textstyle\int_\Delta \underline{1}) && \text{distributivity of } \textstyle\int_\Delta \\
\geq {} & (\sqcap\Delta\colon r.s \bullet c_1 \textstyle\int_\Delta \beta_1 + c_2 \textstyle\int_\Delta \beta_2 - c) && \Delta \in r.s \subseteq \overline{S} \text{ implies } \textstyle\int_\Delta \underline{1} \leq 1 \\
= {} & (\sqcap\Delta\colon r.s \bullet c_1 \textstyle\int_\Delta \beta_1 + c_2 \textstyle\int_\Delta \beta_2) - c \\
\geq {} & && + \text{ and } (c*) \text{ monotonic for } c \in \mathbb{R}_\geq \\
& c_1(\sqcap\Delta\colon r.s \bullet \textstyle\int_\Delta \beta_1) + c_2(\sqcap\Delta\colon r.s \bullet \textstyle\int_\Delta \beta_2) - c \\[4pt]
= {} & c_1(wp.r.\beta_1.s) + c_2(wp.r.\beta_2.s) - c , && \text{Def. 5.5.2}
\end{aligned}
$$

and since the first line ($\ddagger$) above is non-negative we may replace the final $(-c)$ by $(\ominus c)$. $\square$

That sublinearity is indeed the probabilistic analogue of conjunctivity may be seen by considering a special case: suppose sublinear $t \in \mathbb{T}S$ takes standard postconditions to standard preconditions. We then have for $S_a, S_b \subseteq S$ that

$$
\begin{aligned}
& t.[S_a \cap S_b] \\
\equiv {} & t.([S_a] \sqcap [S_b]) \\
\equiv {} & t.([S_a] + [S_b] \ominus \underline{1}) \\
\Leftarrow {} & t.[S_a] + t.[S_b] \ominus \underline{1} && \text{sublinearity} \\
\equiv {} & t.[S_a] \sqcap t.[S_b] , && t.[S_a], t.[S_b] \text{ standard}
\end{aligned}
$$

and equality follows from monotonicity of t (below).

---

[30]Recall from p. 28 that $x \ominus y$ is $(x - y) \mathbf{max}\ 0$, and that $\ominus$ has lower syntactic precedence than $+$.

[31]Sec. 6.10 suggests a geometric interpretation of this fact.

The other probabilistic healthiness conditions are all consequences of sublinearity.[32] We restate them here; but for proofs of monotonicity, feasibility and scaling, we refer to their first appearance, in Sec. 1.6.

Lemma 5.6.3 MONOTONICITY OF TRANSFORMERS
If $t \in \mathbb{T}S$ is sublinear then it is monotonic.   □

Lemma 5.6.4 FEASIBILITY OF TRANSFORMERS
If $t \in \mathbb{T}S$ is sublinear, then for all $\beta \in \mathbb{E}S$ we have

$$t.\beta \quad \Rrightarrow \quad \underline{\sqcup\beta} \,,$$

where $\sqcup\beta$ abbreviates $(\sqcup s: S \cdot \beta.s)$.   □

Lemma 5.6.5 SCALING OF TRANSFORMERS
If $t \in \mathbb{T}S$ is sublinear, then for all $\beta \in \mathbb{E}S$ and $c \in \mathbb{R}_{\geq}$ we have

$$t.(c\beta) \quad \equiv \quad c(t.\beta) \,.$$

□

Finally, for continuity we appeal to the finiteness of S.[33]

Lemma 5.6.6 BOUNDED CONTINUITY OF TRANSFORMERS
If $t \in \mathbb{T}S$ is sublinear and $\mathcal{B}$ is a $\Leftarrow$-directed and bounded subset of $\mathbb{E}S$, so that $\sqcup\mathcal{B}$ exists, then

$$t.(\sqcup\mathcal{B}) \quad \equiv \quad (\sqcup\beta: \mathcal{B} \cdot t.\beta) \,.$$

Proof   By monotonicity we need only show $t.(\sqcup\mathcal{B}) \Rrightarrow (\sqcup\beta: \mathcal{B} \cdot t.\beta)$.

Take any c with $c > 1$. For every (final) state s there is a β_s in $\mathcal{B}$ such that $\sqcup\mathcal{B}.s \leq c\beta_s.s$; since $\mathcal{B}$ is directed and S is finite we can thus find a single β_c with $\beta_s \Rrightarrow \beta_c$ for all s, so that $\sqcup\mathcal{B} \Rrightarrow c\beta_c$. For any such c we have

$$
\begin{array}{lll}
& t.(\sqcup\mathcal{B}) & \\
\Rrightarrow & t.(c\beta_c) & \text{monotonicity} \\
\equiv & c(t.\beta_c) & \text{scaling} \\
\Rrightarrow & c(\sqcup\beta: \mathcal{B} \cdot t.\beta) \,, & \beta_c \in \mathcal{B}
\end{array}
$$

which suffices since c can be arbitrarily close to one.   □

---

[32]They are also easily proved from Def. 5.5.2 directly, a useful exercise.

[33]In general, saying that a function is "continuous" amounts to asserting that it distributes limits, although there is some variation in whether that fact is a definition or a theorem. (In analysis, for example, it is conventional to define continuity in the "ε-δ" style, and then to prove that it distributes limits of suitable sequences of numbers.)
For us, a function f is $\sqcup$-CONTINUOUS iff we have

$$f.(\sqcup X) \quad = \quad (\sqcup x: X \cdot f.x)$$

for all $\sqsubseteq$-directed sets X (for which, recall Footnote 38 on p. 25).
In Lem. 5.6.6 immediately below, we refer to "bounded" continuity because of the technical restriction we must make to ensure in our context that the limits exist.

feasibility	$t.\beta \Rrightarrow \sqcup\beta$	for $\beta \colon \mathbb{E}S$
monotonicity	$\beta_1 \Lleftarrow \beta_2$ implies $t.\beta_1 \Lleftarrow t.\beta_2$	for $\beta_1, \beta_2 \colon \mathbb{E}S$
scaling	$t.(c\beta) \equiv c(t.\beta)$	for $\beta \colon \mathbb{E}S$ and $c \colon \mathbb{R}_{\geq}$

$$
\begin{aligned}
&\textit{sublinearity} && c_1(t.\beta_1) + c_2(t.\beta_2) \ominus \underline{c} && \text{for } \beta_1, \beta_2 \in \mathbb{E}S \\
&&& \Rrightarrow \; t.(c_1\beta_1 + c_2\beta_2 \ominus \underline{c}) && \text{and } c, c_1, c_2 \in \mathbb{R}_{\geq}
\end{aligned}
$$

bounded continuity	$t.(\sqcup\mathcal{B}) \equiv (\sqcup\beta \colon \mathcal{B} \bullet t.\beta)$	for bounded $\Lleftarrow$-directed subset $\mathcal{B}$ of $\mathbb{E}S$

Note that in the presence of scaling we can decompose sublinearity into the two properties

$$
\textit{sub-additivity} \qquad t.\beta_1 + t.\beta_2 \Rrightarrow t.(\beta_1 + \beta_2) \qquad \text{for } \beta_1, \beta_2 \in \mathbb{E}S
$$

$$
\ominus\textit{-subdistributivity} \qquad t.\beta \ominus \underline{1} \Rrightarrow t.(\beta \ominus \underline{1}) \qquad \text{for } \beta \in \mathbb{E}S \; ,
$$

and thus we conclude that, over a finite state space, *sublinearity* is equivalent to the three separate (and more intuitive) properties *scaling*, *sub-additivity* and $\ominus$-*subdistributivity*.[34]

Figure 5.6.7. HEALTHINESS CONDITIONS FOR REGULAR TRANSFORMERS

The importance of continuity is usually that it simplifies the treatment of fixed points in a *cpo*.[35] But we recall (p. 25) that neither $\mathbb{E}S$ nor $\mathbb{T}S$ is complete, and so for them a special argument is sometimes needed; the following is typical.

Lemma 5.6.8 EXISTENCE OF FIXED POINT

If continuous $\mathbf{t}$ in $\mathbb{T}S \to \mathbb{T}S$ is such that for any feasible $t \in \mathbb{T}S$ we have that $\mathbf{t}.t$ is feasible also — which property we call *feasibility preserving* — then the least fixed-point $\mu.\mathbf{t}$ exists, and is given by $(\sqcup n \colon \mathbb{N} \bullet \mathbf{t}^n.\mathbf{abort})$.

---

[34]The following counter-example shows that $\ominus$-subdistributivity cannot be eliminated: let S be the two-element state space $\{a, b\}$, and define $t \in \mathbb{T}S$ so that

$$
t.\beta.s \; := \; \beta.a + \beta.b - \sqrt{\frac{(\beta.a)^2 + (\beta.b)^2}{2}} \; .
$$

Then t is scaling and subadditive. But it does not satisfy $\ominus$-subdistributivity (and so is not sublinear): take $\beta.a, \beta.b := 2, 1$ and consider $t.(\beta \ominus \underline{1})$.

[35]More generally, continuity is important because it allows us to infer properties of "infinite" objects from knowledge of their finite approximates. (Referring to Footnote 33 above, we recall that in analysis the value of continuous f applied to the "infinite" limit $(\lim_{n \to \infty} x_n)$ of a sequence, that is $f.(\lim_{n \to \infty} x_n)$, can be approximated by examining its values $f.x_n$ at finite distances along the sequence and taking the limit $(\lim_{n \to \infty} f.x_n)$ of those.)

In program semantics — our context — the continuity of transformers *etc.* allows us to infer properties of iterations, even those allowing infinite loops from some initial states, from knowledge of their behaviour up to some finite (but larger and larger) number of steps.

Proof   Since **abort** is feasible and t is feasibility-preserving, a simple induction over n shows that $t^n.$**abort** is feasible for all n — and so for any (bounded) α in $\mathbb{E}S$ we have $t^n.$**abort**$.\alpha \Rrightarrow \sqcup \alpha$. That is, we have

$$t^n.\textbf{abort} \quad \sqsubseteq \quad (\lambda \alpha \colon \mathbb{E}S \cdot \sqcup \alpha) \ ,$$

for all n, whose constant right-hand side provides a bound beneath which the stated limit must exist (since $\mathbb{R}_\geq$ has the property that any non-empty subset has a least upper bound).

The argument that the limit gives the least fixed-point is then the usual one.   □

We summarise the properties of $\mathbb{T}_r S$ in Fig. 5.6.7 opposite. It remains to show that those properties characterise $\mathbb{T}_r S$ exactly, and for that we will define an inverse to wp.

5.7   Characterising regular programs

We now show that sublinearity characterises the regular expectation transformers $\mathbb{T}_r S$: that is, they are *exactly* the sublinear members of $\mathbb{T}S$.

We begin by defining a map rp from $\mathbb{T}S$ back to $\mathbb{H}S$, to be an inverse to wp on $\mathbb{T}_r S$.[36] Thus for $t \in \mathbb{T}_r S$ in particular the corresponding relation $rp.t \in \mathbb{H}S$ should produce from initial s exactly those final distributions Δ that satisfy (5.11) for all possible post-expectations β:

Definition 5.7.1   TRANSFORMER-TO-RELATIONAL RETRACTION
The function $rp \in \mathbb{T}S \to \mathbb{H}S$ is defined

$$rp.t.s \quad := \quad \{\Delta \colon \overline{S} \mid (\forall \beta \colon \mathbb{E}S \cdot t.\beta.s \leq \int_\Delta \beta)\} \ ,$$

for transformer $t \in \mathbb{T}S$ and state $s \in S$.   □

Concerning definedness of rp, we note that $rp.t.s$ satisfies the closure conditions for $\mathbb{C}S$: in fact for fixed t, β, s the constraint

$$t.\beta.s \leq \int_\Delta \beta$$

on Δ represents a closed upper half-space in the Euclidean $\mathbb{R}^N$ that corresponds to the distributions over our finite state space S, and that half-space satisfies the three closure conditions trivially; furthermore, those conditions are preserved by intersection. (It is an upper half-space because β takes only non-negative values.)

---

[36]The "rp" is a mnemonic for *relational program*, as we view our $\mathbb{H}S$ effectively to be a relational probabilistic model.

If $rp.t.s$ is empty for some s, we consider rp to be undefined at that t; but Lem. 5.7.5 below shows $rp.t$ to be defined whenever t is sublinear.[37]

Our first use of rp is to show that wp is indeed an injection:

Lemma 5.7.2 For any $r \in \mathbb{H}S$ we have

$$rp.(wp.r) \quad = \quad r \ .$$

Proof   Direct from the definitions we have (taking the contrapositive), for arbitrary $\Delta \in \overline{S}$, that

$$\Delta \notin rp.(wp.r).s$$

iff	$\Delta \notin \{\Delta : \overline{S} \mid (\forall \beta : \mathbb{E}S \bullet wp.r.\beta.s \leq \int_{\Delta} \beta)\}$	Def. 5.7.1
iff	$(\exists \beta : \mathbb{E}S \bullet wp.r.\beta.s > \int_{\Delta} \beta)$	
iff	$(\exists \beta : \mathbb{E}S \bullet (\sqcap \Delta' : r.s \bullet \int_{\Delta'} \beta) > \int_{\Delta} \beta)$	Def. 5.5.2
iff	$(\exists \beta : \mathbb{E}S; c : \mathbb{R} \bullet (\forall \Delta' : r.s \bullet \int_{\Delta'} \beta > c) \wedge \int_{\Delta} \beta < c)$	
iff	$\Delta \notin r.s$.	see (†) below for "if"

The last line's "only if" is trivial. For "if," however, we need the *Separating-Hyperplane Lemma* B.5.1, which states that any point not in a closed convex subset of $\mathbb{R}^N$ can be separated from it by a hyperplane.

In our use of the lemma, the point concerned is Δ; and $r.s$ is the closed- and convex subset of $\mathbb{R}^N$ that Δ is not in, as stated on the last line of the proof above. The plane separating Δ from $r.s$ is given to us by Lem. B.5.1, and can be described by its coefficients β and a constant term c — that is, a point Δ' lies on that plane iff

$$c \quad = \quad \int_{\Delta'} \beta \ .$$

We choose the signs of β, c so that "$r.s$ on one side" is expressed

$$(\forall \Delta' : r.s \bullet \int_{\Delta'} \beta > c) \ , \tag{5.12}$$

and "Δ on the other side" is expressed

$$\int_{\Delta} \beta \quad < \quad c \ ,$$

† which two conditions together with the existence of β, c give us our deferred "if" above. Note that the coefficients β are non-negative since, by up closure of $r.s$, Condition (5.12) could not be true otherwise.   □

---

[37]If $\mathbb{H}S$ did not impose non-emptiness (Def. 5.4.4), allowing miracles, then rp would be everywhere defined; in that case however for definedness of wp we would require

$$\mathbb{E}S \quad := \quad S \to (\mathbb{R}_{\geq} \cup \{\infty\}) \ ,$$

and we would have to take some care with the resulting arithmetic.

The Cauchy-closure condition on $r.s$ is essential for Lem. 5.7.2: if $r.s$ were for example

$$\{p\colon (0,1) \bullet \overline{0} \,_p\!\oplus\, \overline{1}\}\;,$$

thus all probabilistic combinations of the point distributions $\overline{0}$ and $\overline{1}$ but *excluding* the endpoints, then $rp.(wp.r).s$ would be

$$\{p\colon [0,1] \bullet \overline{0} \,_p\!\oplus\, \overline{1}\}\;,$$

in which closure has occurred, making it differ from the original $r.s$.

Our next step is to prove an analogous result for the opposite direction; but since rp is not defined for all $\mathbb{T}S$ (for $t \notin \mathbb{T}_r S$ it may be that $rp.t.s$ is empty for some s), we state the lemmas conditionally.

Our first lemma concerns members of $\mathbb{T}S$ generally:[38]

Lemma 5.7.3 For any $t \in \mathbb{T}S$, if $rp.t$ is defined then

$$wp.(rp.t) \quad \sqsupseteq \quad t\;.$$

Proof   Take any $\beta \in \mathbb{E}S$ and $s \in S$; directly from the definitions we have:

$$wp.(rp.t).\beta.s \geq t.\beta.s$$

iff $\quad (\sqcap\Delta\colon rp.t.s \bullet \int_\Delta \beta) \geq t.\beta.s \qquad$ Def. 5.5.2; $rp.t$ defined by assumption

iff $\quad (\forall\Delta\colon rp.t.s \bullet \int_\Delta \beta \geq t.\beta.s)$

iff $\quad (\forall\Delta\colon \overline{S} \bullet (\forall\beta'\colon \mathbb{E}S \bullet t.\beta'.s \leq \int_\Delta \beta') \Rightarrow \int_\Delta \beta \geq t.\beta.s)$

iff $\quad$ true .

$\square$

Our second lemma is restricted to sublinear elements of $\mathbb{T}S$:

Lemma 5.7.4 If $t \in \mathbb{T}S$ is sublinear and $rp.t$ is defined, then

$$wp.(rp.t) \quad \sqsubseteq \quad t\;.$$

Proof   We proceed via the contrapositive: suppose $rp.t$ is defined but $wp.(rp.t) \not\sqsubseteq t$. Then for some $s \in S$ and $\beta \in \mathbb{E}S$, we have

$$wp.(rp.t).\beta.s \quad > \quad t.\beta.s\;,$$

whence $\quad (\sqcap\Delta\colon rp.t.s \bullet \int_\Delta \beta) > t.\beta.s \qquad\qquad$ Def. 5.5.2

implies $\quad (\forall\Delta\colon rp.t.s \bullet \int_\Delta \beta > t.\beta.s) \qquad\qquad$ call this (5.13)

iff $\quad (\forall\Delta\colon \overline{S} \bullet (\forall\beta'\colon \mathbb{E}S \bullet t.\beta'.s \leq \int_\Delta \beta') \Rightarrow \int_\Delta \beta > t.\beta.s) \qquad$ Def. 5.7.1

iff $\quad (\cap\beta'\colon \mathbb{E}S \bullet \{\Delta\colon \overline{S} \mid t.\beta'.s \leq \int_\Delta \beta'\}) \subseteq \{\Delta\colon \overline{S} \mid \int_\Delta \beta > t.\beta.s\}$

iff. . .

---

[38]Sec. 6.10 suggests a geometric interpretation of this lemma.

... iff   for arbitrary sets A, B write $A \subseteq B$ as $A \cap B^C = \emptyset$

$$(\cap \beta' : \mathbb{E}S \cdot \{\Delta : \mathbb{R}^N \mid t.\beta'.s \leq \int_\Delta \beta'\}) \qquad \text{call this (5.14)}$$
$$\cap \quad \{\Delta : \mathbb{R}^N \mid -t.\beta.s \leq \int_\Delta (-\beta)\}$$
$$\cap \quad \{\Delta : \mathbb{R}^N \mid -1 \leq \int_\Delta (\underline{-1})\}$$
$$= \quad \emptyset ,$$

where we have negated the second term to make the inequalities uniform. The third term, also negated, is added because of the retyping of the bound variables Δ: in (5.14) they are taken from all of $\mathbb{R}^N$, rather than the more restrictive $\overline{S}$, and so we must compensate by requiring explicitly

$$\sum \Delta \quad = \quad \int_\Delta \underline{1} \quad \leq \quad 1 .$$

(The remaining constraints $\Delta.s \geq 0$ applying to members Δ of $\overline{S}$, for every $s \in S$, are included already in the first term above: take $\beta' := [\{s\}]$ and recall that $0 \leq t.\beta'.s$.)

Now we argue that some finite sub-collection of the sets (5.14) has empty intersection also. Take the sets determined by $\beta' := [\{s\}]$ for each $s \in S$ (finitely many), and the final set $\{\Delta : \mathbb{R}^N \mid -1 \leq \int_\Delta \underline{-1}\}$: they determine a closed hyper-pyramid in the positive hyper-octant of $\mathbb{R}^N$ (whose apex points downward toward the origin). Since the pyramid is closed and bounded, it is compact — and so the finite-intersection lemma applies within it.[39] Thus we can restrict our attention to a collection of just M, say, of the sets (5.14).

We will shortly appeal to another lemma in the style of Linear Programming (Sec. B.5). The finite M-collection from (5.14) with empty intersection may be regarded as a system of M equations

$$A \cdot x \quad \geq \quad r$$

that has no solution in x, where A is an $M * N$ matrix (of coefficients, representing the β''s, $-\beta$, and $\underline{-1}$), x is an $N * 1$ column vector (representing points Δ in $\mathbb{R}^N$), and r is an $M * 1$ column vector (of constant terms, representing the $t.\beta'.s$'s, $-t.\beta.s$, and -1). The expression $A \cdot x$ denotes matrix multiplication, and the inequality $\geq$ is taken row-wise.

Lem. B.5.2 then gives us a $1 * M$ row-vector C of non-negative reals such that

$$C \cdot A = 0 \quad \text{but} \quad C \cdot r > 0 .$$

---

[39]The FINITE-INTERSECTION LEMMA states that a collection of closed subsets of a compact set has empty intersection only if some finite sub-collection of it does; it is a standard result from analysis (*e.g.* [Roy68]).

Returning to (5.14), we thus have a finite number of non-negative coefficients $c, c_0 \cdots c_{M-2}$ such that

$$c\beta \quad \equiv \quad c_1\beta_1' + \cdots + c_{M-2}\beta_{M-2}' \ominus c_0$$

but

$$c(t.\beta.s) \quad < \quad c_1(t.\beta_1'.s) + \cdots + c_{M-2}(t.\beta_{M-2}'.s) \ominus c_0 \ ,$$

where in both cases we can use $\ominus$ on the right because the left-hand side is non-negative.

We then finish with

$$
\begin{aligned}
& c(t.\beta.s) \\
< \ & c_1(t.\beta_1'.s) + \cdots + c_{M-2}(t.\beta_{M-2}'.s) \ominus c_0 && \text{above} \\
\leq \ & t.(c_1\beta_1' + \cdots + c_{M-2}\beta_{M-2}' \ominus c_0).s \ , && \text{sublinearity of } t \\
= \ & t.(c\beta).s \ , && \text{above}
\end{aligned}
$$

which contradicts t's scaling. Thus t cannot be sublinear. $\square$

Finally we deal with definedness of rp:

Lemma 5.7.5 If $t \in \mathbb{T}S$ is sublinear, then $rp.t$ is defined.
   Proof   The three closure conditions have already been dealt with. For non-emptiness, again we proceed via the contrapositive: suppose for some $s \in S$ that $rp.t.s$ is empty. Then we have immediately

$$(\forall \Delta \colon rp.t.s \bullet \int_\Delta \beta > t.\beta.s) \ ,$$

since we are quantifying universally over the empty set (the body of the quantification is irrelevant). But that is identical to (5.13) in the proof of Lem. 5.7.4, and rp is not mentioned beyond that point. Thus we conclude as before that t cannot be sublinear. $\square$

We have thus established the following:

Lemma 5.7.6 SUBLINEAR TRANSFORMERS ARE REGULAR   If $t \in \mathbb{T}S$ is sublinear, then $rp.t$ is defined and

$$wp.(rp.t) \quad = \quad t \ .$$

   Proof   The result is immediate from Lemmas 5.7.3–5.7.5. $\square$

With Lemmas 5.6.2 and 5.7.6 we have finally our characterization of $\mathbb{T}_r S$:

Theorem 5.7.7 SUBLINEARITY CHARACTERISES
                REGULAR TRANSFORMERS

Transformer t in $\mathbb{T}S$ is sublinear iff it is in $\mathbb{T}_r S$.
   Proof   Lemmas 5.6.2 and 5.7.6. $\square$

Note that Thm. 5.7.7 shows also that the conditions of Fig. 5.6.7 characterise $\mathbb{T}_r S$ collectively, since by Lemmas 5.6.2–5.6.6 they are all implied by sublinearity.

5.8   Complementary and consistent semantics

We have now seen that $\mathbb{H}S$ and $\mathbb{T}_r S$ are placed in 1–1 correspondence by the mutual inverses $wp \in \mathbb{H}S \to \mathbb{T}_r S$ and $rp \in \mathbb{T}_r S \to \mathbb{H}S$. But for the two spaces to give equivalent semantics for regular programs requires further that the program operators preserve the correspondence — that for example Definitions 5.4.5–5.4.7, for relations, correspond with the definitions of Fig. 1.5.3 (p. 26) for expectation transformers. We want the relational- and transformer semantics to be *complementary and consistent* — complementary in that they are two different ways of looking at programs, each with its own advantages; and consistent in the sense that the results agree when translated into the same context.

For sequential composition, for example, we have the following lemma.

Lemma 5.8.1 COMPLEMENTARY AND CONSISTENT SEMANTICS
OF SEQUENTIAL COMPOSITION

Definition 5.4.7 and Fig. 1.5.3 for sequential composition are consistent with respect to the embedding wp; that is, for $r_1, r_2 \in \mathbb{H}S$ we have

$$wp.(r_1; r_2) \quad = \quad wp.r_1 \circ wp.r_2 . \quad ^{40}$$

Proof   The proof is a routine if lengthy application of definitions to both sides, working in towards the middle, with a Skolemised swap of quantifiers to bring things together in the end.[41]

For arbitrary expectation β and state s, and starting from the left-hand side, we have

$$wp.(r_1; r_2).\beta.s$$

---

[40]We write $f \circ g$ for FUNCTIONAL COMPOSITION, so that $(f \circ g).x = f.(g.x)$.

[41]We are using SKOLEMISATION in the sense of "swapping quantifiers by introducing a (Skolem) function." In logic this is usually a transformation of the form

$$(\forall x \bullet (\exists y \bullet \phi)) \qquad \text{becomes} \qquad (\exists f \bullet (\forall x \bullet \phi \langle y \mapsto f.x \rangle)) ,$$

where ϕ is a formula over x and y and (in our notation) every y in ϕ on the left is replaced by $f.x$ on the right: the introduced function f is a *Skolem function* reflecting the ϕ-satisfying y's dependence on x. In effect, the left-hand assertion that "for every x there is some y so that ϕ holds" is replaced by the right-hand assertion that "there is some function f that for each x will give that y."

Thus the order of quantification is swapped — but note that on the right we are quantifying over functions, so that a strictly formal treatment would require higher-order logic.

In our case — and less formally — we will see that it is the quantifiers $\sum$ and $\sqcap$ that are swapped, with the introduced function f giving the value at which the minimum is attained, separately for each term in the summation. (The fact that it *is* attained, rather than merely approached, is assured by Cauchy closure.)

$\equiv \quad (\sqcap \Delta: (r_1; r_2).s \cdot \int_\Delta \beta)$ Def. 5.5.2

$\equiv \quad (\sqcap \Delta: \{\Delta': r_1.s; \; f: \mathbb{D}S \mid r_2 \sqsubseteq f \cdot f^*.\Delta'\} \cdot \int_\Delta \beta)$ Def. 5.4.7

$\equiv \quad (\sqcap \Delta': r_1.s; \; f: \mathbb{D}S \mid r_2 \sqsubseteq f \cdot \int_{f^*.\Delta'} \beta \;)$ set comprehension [42]

$\equiv \quad (\sqcap \Delta: r_1.s; \; f: \mathbb{D}S \mid r_2 \sqsubseteq f \cdot \int_{f^*.\Delta} \beta \;)$ rename bound variable

$\equiv \quad (\sqcap \Delta: r_1.s; \; f: \mathbb{D}S \mid r_2 \sqsubseteq f \cdot \int_{(\lambda s': S \;\cdot\; \int_\Delta (f.s.s' \, ds))} \beta \;)$ Def. 5.2.3

$\equiv$ Def. 5.2.2
$$(\sqcap \Delta: r_1.s; \; f: \mathbb{D}S \mid r_2 \sqsubseteq f \cdot (\textstyle\sum_{s':S} \beta.s' * \int_\Delta (f.s.s' \, ds)))$$

$\equiv$ Def. 5.2.2 again
$$(\sqcap \Delta: r_1.s; \; f: \mathbb{D}S \mid r_2 \sqsubseteq f \cdot (\textstyle\sum_{s,s':S} \beta.s' * f.s.s' * \Delta.s)) \; .$$

Starting from the right-hand side, we have

$$(wp.r_1 \circ wp.r_2).\beta.s$$

$\equiv \quad wp.r_1.(wp.r_2.\beta).s$ functional composition

$\equiv \quad (\sqcap \Delta: r_1.s \cdot \int_\Delta wp.r_2.\beta)$ Def. 5.5.2

$\equiv \quad (\sqcap \Delta: r_1.s \cdot \int_\Delta (\lambda s' \cdot wp.r_2.\beta.s'))$ insert s' argument explicitly

$\equiv \quad (\sqcap \Delta: r_1.s \cdot \int_\Delta (\lambda s' \cdot (\sqcap \Delta': r_2.s' \cdot \int_{\Delta'} \beta)))$ Def. 5.5.2 again

$\equiv \quad (\sqcap \Delta: r_1.s \cdot \int_\Delta ((\sqcap \Delta': r_2.s' \cdot \int_{\Delta'} \beta) \, ds'))$ convert $\lambda s'$ to ds' under $\int$

$\equiv$ Def. 5.2.2
$$(\sqcap \Delta: r_1.s \cdot \int_\Delta ((\sqcap \Delta': r_2.s' \cdot (\textstyle\sum_{s'':S} \beta.s'' * \Delta'.s'')) \, ds'))$$

$\equiv$ Def. 5.2.2 again
$$(\sqcap \Delta: r_1.s \cdot (\textstyle\sum_{s':S} (\sqcap \Delta': r_2.s' \cdot (\textstyle\sum_{s'':S} \beta.s'' * \Delta'.s'')) * \Delta.s'))$$

---

[42]In this comprehension the quantifier is $\sqcap$, the bound variables are Δ' drawn from set $r_1.s$ and f drawn from $\mathbb{D}S$, and there is a further constraint "$r_2 \sqsubseteq f$" on the bound variables (but in fact not affecting Δ'). The expression over which the quantifier operates is $\int_{f^*.\Delta'} \beta$.

As usual the parentheses give the scope.

$\equiv$ 　　　　　　　　　　　　　　distribute $(*\Delta.s')$ inwards through $\sqcap, \sum$
$$(\sqcap\Delta: r_1.s \cdot (\textstyle\sum_{s':S} (\sqcap\Delta': r_2.s' \cdot (\textstyle\sum_{s'':S} \beta.s'' * \Delta'.s'' * \Delta.s'))))$$

$\equiv$ 　　　　　　　　Skolemise $\Delta': r_2.s'$ within $\sum_{s':S}$ as f; see ($\dagger$) below
$$(\sqcap\Delta: r_1.s \cdot$$
$$(\sqcap f: \mathbb{D}S \mid r_2 \sqsubseteq f \cdot$$
$$(\textstyle\sum_{s':S} (\textstyle\sum_{s'':S} \beta.s'' * f.s'.s'' * \Delta.s'))))$$

$\equiv$ 　　　　　　　　　combine doubled $\sqcap$ and $\sum$ quantifiers
$$(\sqcap\Delta: r_1.s; \; f: \mathbb{D}S \mid r_2 \sqsubseteq f \cdot (\textstyle\sum_{s',s'':S} \beta.s'' * f.s'.s'' * \Delta.s')) \; ,$$

which is only a change of bound variable away from where we left the left-hand side.

$\dagger$　　The Skolemisation converts a $\sqcap$-choice of a distribution $\Delta' \in \overline{S}$ from $r_2.s'$, acting within the scope of $\sum_{s':S}$, to an $\sqcap$-choice of a function f of type $S \to \overline{S}$, acting outside of that scope. The interior constraint $\Delta' \in r_2.s'$ on Δ' then becomes the exterior constraint $(\forall s': S \cdot f.s' \in r_2.s')$ on the function f which chooses Δ' given s' — and that is just $r_2 \sqsubseteq f$.　　　　$\square$

For demonic choice the proof is shorter.

Lemma 5.8.2 COMPLEMENTARY AND CONSISTENT SEMANTICS
　　　　　　OF DEMONIC CHOICE

For $r_1, r_2 \in \mathbb{H}S$ we have

$$wp.(r_1 \sqcap r_2) \quad = \quad wp.r_1 \sqcap wp.r_2 \; .$$

Proof　　Again the steps are routine. For arbitrary β, s we calculate

$$wp.(r_1 \sqcap r_2).\beta.s$$
$= \quad (\sqcap\Delta: (r_1 \sqcap r_2).s \cdot \int_\Delta \beta)$ 　　　　　　　　　Def. 5.5.2
$= \quad (\sqcap\Delta_1: r_1.s; \Delta_2: r_2.s; p: [0,1] \cdot \int_{(\Delta_{1p}\oplus\Delta_2)} \beta)$ 　　　Def. 5.4.6
$= \quad (\sqcap\Delta_1: r_1.s; \Delta_2: r_2.s; p: [0,1] \cdot \int_{\Delta_1} \beta \; {}_p\oplus \; \int_{\Delta_2} \beta)$ 　　Def. 5.2.2

$=$ 　　　　　　　minimum occurs at extremes, thus at $p=0$ or $p=1$
$$(\sqcap\Delta_1: r_1.s; \Delta_2: r_2.s \cdot \int_{\Delta_1} \beta \sqcap \int_{\Delta_2} \beta)$$

$= \quad (\sqcap\Delta_1: r_1.s \cdot \int_{\Delta_1} \beta) \quad \sqcap \quad (\sqcap\Delta_2: r_2.s \cdot \int_{\Delta_2} \beta)$
$= \quad wp.r_1.\beta.s \sqcap wp.r_2.\beta.s \; .$ 　　　　　　　　　　　Def. 5.5.2

　　　　　　　　　　　　　　　　　　　　　　　　　$\square$

Except for recursion, it is straightforward with similar calculations to establish the correspondence of the remaining relational operators with their predicate-transformer counterparts.[43] Usually the latter are simpler.

5.9   Review: semantic structures

We now look back at what we have done, both in this chapter and earlier, to see how our different views fit together into a consistent whole. Figure 5.9.1 is the first step: it extends our earlier Fig. 5.5.1, and puts our spaces into context by giving the functions we have constructed between them.

The following are some of the principal features.

- *Probabilistic relational semantics for pGCL*

$Syn \xrightarrow{rp} PRS$ Our first encounter with $pGCL$, in Chap. 1, introduced its syntax (Syn) and the function rp which — although not named explicitly — was implicitly at least the topic of Sec. 1.3 in which the operational meaning of $pGCL$ programs was described in terms of a gambling game. That view was formalised in this chapter as the space $\mathbb{H}S$ (the probabilistic relational semantics PRS).

- *Expectation-transformer semantics for pGCL*

$Syn \xrightarrow{wp} PTS$ Later in Chap. 1 we defined an explicit function wp between $pGCL$ program texts and expectation transformers: it was summarised in Fig. 1.5.3.

Note however that we treated expectations there as formulae, whereas in this chapter we dealt directly with the functions of S that the formulae denote; and because of that more mathematical approach we were able to formalise the expectation-transformer space, as $\mathbb{T}S$ (or PTS in the figure).

The aim of the earlier Sec. 1.3.3 was therefore to lead up to the expectation-transformer view, arguing that it would correspond in a natural way with the operational gambling-game description of probabilistic programs.

---

[43]We deal with recursion in Sec. 6.12.

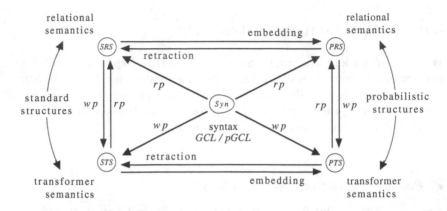

Syn (for *Syntax*) is the space of program texts, written in Dijkstra's original guarded-command language *GCL* but possibly using the new operator $_p\oplus$ (represented by the Syn-space's bulge towards the probabilistic side of the diagram).

SRS (for *Standard Relational Semantics*) is the space $S \to \mathbb{F}^+ S_\perp$ mentioned in Sec. 5.5.1, modelling demonic and possibly nonterminating standard programs.

PRS is the space $\mathbb{H}S$ introduced in Sec. 5.4, where states are taken to non-empty, convex, up- and Cauchy-closed sets of distributions.

STS is the standard (predicate-)transformer semantic space $\mathbb{P}S \leftarrow \mathbb{P}S$, originally proposed by Dijkstra for *GCL*.

PTS is the space $\mathbb{T}S$ we have constructed in this chapter, the expectation transformers that provide a semantics for probabilistic demonic programs. (We see in Chap. 8 that they describe *angelic* nondeterminism as well.)

Figure 5.9.1. SEMANTIC STRUCTURES

- *Complementary and consistent semantics*

$$Syn \xrightarrow{wp} PTS$$
$$= Syn \xrightarrow{rp} PRS \xrightarrow{wp} PTS$$

Thus the principal technical contribution of this chapter was to make the above correspondence precise via the construction of the function $PRS \xrightarrow{wp} PTS$, given in Sec. 5.5, which converts a relational program into an expectation transformer.

The equality above states that our function $PRS \xrightarrow{wp} PTS$ establishes the consistency of our two complementary views of probabilistic programs $pGCL$: the relational $Syn \xrightarrow{rp} PRS$ and the transformer $Syn \xrightarrow{wp} PTS$.

What consistency means for us, however, is that the function $\xrightarrow{wp}$ must not only map PRS to PTS — it must preserve the structure that the program operators determine *within* those spaces. Figure 5.9.2 shows a "zoomed-in" view of $PRS \xrightarrow{wp} PTS$ and it illustrates, for example, the two lemmas 5.8.1 and 5.8.2: if you take a pair of relational programs in $PRS * PRS$, then it does not matter whether you convert them separately to transformers (go up, in the figure) and then combine them with a transformer-style operator (go left), or instead combine them with the corresponding relational-style operator (go left) and then convert the result to a transformer (go up). The same transformer is determined, either way.

The great importance of this correspondence is as follows. If we have a physical probabilistic system r in mind, function $Syn \xrightarrow{rp} PRS$ is what we must think of when we try to capture a description *prog* of it in $pGCL$ — that is, we must ensure that $r = rp.prog$.

But the point of a formal description, once captured, is to allow us to discover properties of the original (real or proposed) physical system that we cannot see or deduce simply "by inspection" — that is precisely why we need a program logic. For example, what we might want to know about system r could be "what is the probability of its delivering a final state in some desirable subset S_b of the state space?" We write that "mathematically" as $wp.r.[S_b]$, using our wp as defined by Def. 5.5.2.

But we do not have to reason "mathematically": once we have understood Def. 5.5.2, we know we can work directly from the program text *prog*, instead of its mathematical representation r: that is, if predicate *post* describes subset S_b, then $wp.prog.[post]$ — calculated from Fig. 1.5.3 — gives us the same answer that $wp.r.[S_b]$ would have.[44]

---

[44] And note that we do not have to use the function set out in Def. 5.5.2 explicitly — indeed, it is fairly complex. We need only to know it exists and to be aware of its commuting properties.

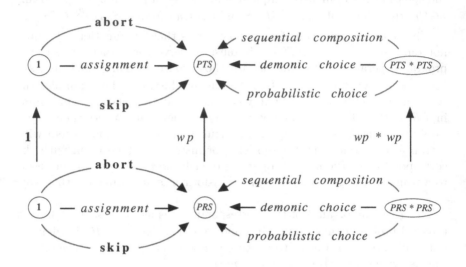

probabilistic   transformer   semantics

probabilistic   relational   semantics

On the right, the function $wp * wp$ takes pairs of relations to pairs of transformers by applying wp to each component separately. On the left, we indicate elements of *PRS* and *PTS* (*i.e. atomic* programs, those not made by the composition of others) by functions from arbitrary singleton sets into each space; the function **1** is the identity.

The diagram thus illustrates that the program operators and the wp function can be applied in either order without affecting the result, so that *PRS* and its image $wp.PRS$ within *PTS* correspond structurally.

Figure 5.9.2. COMPLEMENTARY AND CONSISTENT SEMANTICS

- *Healthiness conditions, and program algebra*

1. $(SRS \xrightarrow{wp})$ $\subseteq$ *conjunctive STS*   Once we have
2. $STS \xrightarrow{rp} SRS \xrightarrow{wp} STS$ $\supseteq$ *conjunctive STS*   decided on a
3.   hence $(SRS \xrightarrow{wp})$ $=$ *conjunctive STS*   programming
                                                                                                        notation and

and   its meaning(s)

4. $(PRS \xrightarrow{wp})$ $\subseteq$ *sublinear PTS*   we can, as
5. $PTS \xrightarrow{rp} PRS \xrightarrow{wp} PTS$ $\supseteq$ *sublinear PTS*   we have seen,
6.   hence $(PRS \xrightarrow{wp})$ $=$ *sublinear PTS*   construct a
                                                                                                        logic respecting
                                                                                                        those   mean-
ings. As noted above, that then becomes a tool for reasoning about the
programs at the source level. For us, this ability is given by *wp* acting
between expressions denoting predicates or expectations.

But we also have potentially a calculus of the programs themselves: two
program texts with the same meaning are themselves equal — that gives
us a program *algebra*.

For example, in standard *GCL* we know that demonic choice distributes
through sequential composition — that is, for any standard programs
$prog, prog_1, prog_2$ we have[45]

$$prog; (prog_1 \sqcap prog_2) \quad = \quad prog; prog_1 \sqcap prog; prog_2 . \qquad (5.15)$$

And with equalities like that we can reason about complex programs by
manipulating them into simpler forms. (Recall for example the program
equalities used in Chap. 4 (pp. 112 and 120) to discover data refinements;
and see Sec. B.1.)

Equality (5.15) is justified by — *i.e.* is proved from — the *conjunctivity*
healthiness condition of *GCL*, *i.e.* that for any postcondition S_b we have

$$wp.(prog; (prog_1 \sqcap prog_2)).S_b$$
$= \quad wp.prog.(wp.(prog_1 \sqcap prog_2).S_b)$
$= \quad wp.prog.(wp.prog_1.S_b \cap wp.prog_2.S_b)$

$=$ conjunctivity: $wp.prog$ distributes $\cap$
    $wp.prog.(wp.prog_1.S_b) \cap wp.prog.(wp.prog_2.S_b)$
$= \quad wp.(prog; prog_1).S_b \cap wp.(prog; prog_2).S_b$
$= \quad wp.(prog; prog_1 \sqcap prog; prog_2).S_b .$

And that leaves only conjunctivity itself to be proved.

One way to proceed is to assume that the only programs we are interested
in are those we can write in *GCL*, so that we deal exclusively with the *wp*-
image subset of *Syn* in *STS*. We prove by structural induction over the

---

[45]Note that this equality does *not* hold any longer in the probabilistic world, in
particular if *prog* is not standard.

program texts — *i.e.* syntactically — that for all programs *prog* in *GCL*, and all postconditions *post*, *post'*, we have

$$wp.prog.(post \wedge post') \quad = \quad wp.prog.post \ \wedge \ wp.prog.post' \ .$$

That is the approach Dijkstra used in his presentation [Dij76].[46]

Another approach, and the one we have taken here, is to argue that since *rp* takes every program in *Syn* to some relation in *SRS*, and because of the equality

$$Syn \xrightarrow{wp} STS \quad = \quad Syn \xrightarrow{rp} SRS \xrightarrow{wp} STS \qquad (5.16)$$

from the previous section (but in its standard version), we can concentrate on the image *wp.SRS* of the whole of *SRS*, a subset of *STS* guaranteed by (5.16) to contain *wp.Syn*. That is, we need only prove that for every $r \in SRS$ we have

$$wp.r.(S_b \cap S'_b) \quad = \quad wp.r.S_b \ \cap \ wp.r.S'_b \ .$$

Doing the proof is a routine exercise in relational semantics [Hes92, Chap. 6].

The probabilistic analogue of that proof is exactly what we carried out in Lem. 5.6.2, where we proved Line 4 above, *viz.*

$$(PRS \xrightarrow{wp}) \quad \subseteq \quad sublinear \ PTS \ ,$$

by which we mean that the *wp*-image of *PRS* is a subset of the sublinear transformers in *PTS* — *i.e.* for all *r* in $\mathbb{HS}$ (*i.e.* in *PRS*) the transformer *wp.r* is sublinear. Immediately following we showed how sublinearity reduced to conjunctivity over standard programs.

In neither case however did we prove that *all* transformers are sublinear (or conjunctive) — and indeed they are not. But at least all the transformers described by programs are such: thus we say that relying on sublinearity is *sound*.

But is it *complete* in the sense that there is no stronger property that all programs satisfy? That is an important question for our algebra, for the stronger the properties we can discover about our programs, the more algebraic laws we will have. In fact for standard programs conjunctivity is complete in this sense: if a transformer is positively conjunctive, then it is the *wp*-image of some program — and so with conjunctivity we have said it all.

---

[46] Joe Hurd has carried out a similar syntactic proof for *pGCL* using the higher-order logic tool *HOL* [HMM04].

The way that is proved is via the inclusion in Line 2, where we say that $STS \xrightarrow{rp} SRS \xrightarrow{wp} STS$ acts as the identity on at least the conjunctive subset of STS — that

if t is conjunctive, then $wp.(rp.t) = t$,

thus establishing immediately the completeness of conjunctivity: if t is conjunctive, then it is the wp-image of a relation. The proof is not so immediate as for soundness, but still is relatively routine [Hes92, Chap. 6]. The two inequalities together give us

$$(SRS \xrightarrow{wp}) \quad = \quad conjunctive\ STS$$

immediately, that positive conjunctivity *characterises* the image $wp.SRS$ of the relations SRS among the transformers STS.

Lem. 5.7.6 of this chapter is the probabilistic analogue of the second inclusion: that if t in PTS is sublinear, then it is the wp-image of some probabilistic relation r in PRS. And so with sublinearity, too, we have Line 6 above and hence a characterisation:

a transformer in PTS is in fact in the subset $wp.PRS$ if and only if it is sublinear.

The algebraic laws that result are those like

$$prog;\ (prog_1\ {}_p\oplus prog_2) \quad \sqsupseteq \quad prog;prog_1\ {}_p\oplus\ prog;prog_2 , \qquad (5.17)$$

which are proved in a similar way to our proof for (5.15), but result in an $\sqsupseteq$-inequality because sublinearity is a $\Leftarrow$-inequality rather than an equality as conjunctivity is.

In Chap. 8 we will revisit Fig. 5.9.1 and explore more of its properties; but in the next chapter we clear away some of the mathematics and look at pictures of what we have constructed. Then in Chap. 7 we use our theory to prove rigorously the loop rules we introduced in Chap. 2 and used in Chap. 3.

When we do return to semantic structures in Chap. 8, we will look beyond demonic choice and remove some of the restrictions we've enjoyed so far, treating in particular the extension to infinite state spaces, the arithmetic characterization of *standard programs* (semi-linearity) and deterministic programs (simple linearity); and we will explore the more general "angelic" transformers that do not correspond to relational programs (only *semi-sublinearity*).[47]

Chapter notes

The operational model on which the quantitative logic depends is based on the work of He *et al.* [HSM97]; in terms of the interaction between probability and nondeterminism, it shares a number of the characteristics with Markovian decision processes [FV96]. Many others have also studied the combination of probability and nondeterminism, including Vardi [Var85] and Lynch, Saias and Segala, the latter using probabilistic versions of *I/O* automata [LS89, LSS94, Seg95].

Our approach draws also on the theory of probabilistic powerdomains, whose application to program semantics was first investigated by Jones [Jon90]; Kwiatkowska and Baier [KB00] have also investigated probabilistic semantics from a powerdomain perspective. We do not know whether a recursive domain equation for probabilistic processes has yet been solved in the case where the *interval* domain is the range for model values (*e.g.* transitions, state properties *etc.*) Prakash Panangaden *et al.* [DGJP03] have however solved such equations for sub-probability measures.[48]

The proposal to use a quantitative logic based on expectations seems to be due originally to Kozen [Koz81]; Halpern and Pucella [HP02] have also studied the correspondence between a logic of expectations and convex-closed sets of probability distributions.

More generally, program logics can play a crucial role in model building, as they provide a rationale for defining equivalences in mathematical structures based on what properties can be reasonably expressed. Emerging from that rationale, for us, is a guiding principle for selecting the probabilistic *closure conditions* which significantly simplify the operational model. In theories of concurrency there is a similar feedback between operational and logical characterisations — besides Larsen and Skou's work [LS91] on bisimulation, Jonsson *et al.* [JHSY94] have studied refinement and equivalence of probabilistic processes in terms of testing and expected rewards; there are also many treatments of testing pre-orders, with the work of Christoff [Chr90] and of Wang and Larsen [YL92] being among the earliest.

---

[47]Since subclasses of transformers satisfy stronger healthiness conditions, they should obey more algebraic laws. For example, using simple linearity of *prog* we can strengthen (5.17) to an equality in the case that *prog* is deterministic. See Sec. B.1.

[48]We thank Michael Huth for pointing this out [Hut03].

6

The geometry
of probabilistic programs

6.1   Embedding distributions in Euclidean space

As we first remarked in Def. 5.4.3, the space $\overline{S}$ of distributions over state space S can be embedded in Euclidean N-space, where N is the cardinality of (finite) S. Since the relational programs $r \in \mathbb{H}S$ are functions $S \to \mathbb{C}S$, where $\mathbb{C}S$ is the set of probabilistically closed elements of $\mathbb{P}\overline{S}$, we know that for each initial state s we are dealing effectively with a probabilistically closed subset $r.s$ of $\mathbb{R}^N$.

It turns out that probabilistic closure can be interpreted geometrically — as can the operation wp of finding a greatest pre-expectation — and both have compelling visualisations that can suggest new results and, in some cases, how to prove them.

In this chapter we illustrate the visualisations by example; in most cases, we will remain in just three dimensions by supposing that we have a state space S containing only the states s_1, s_2, s_3.

A discrete sub-probability distribution Δ over $\{s_1, s_2, s_3\}$ — that is, an element of $\overline{\{s_1, s_2, s_3\}}$ — assigns a probability Δ_i to each s_i such that $\Delta_1 + \Delta_2 + \Delta_3 \leq 1$, and we can place it into three-dimensional Euclidean space as the point $(\Delta_1, \Delta_2, \Delta_3)$. Thus, if we use the normal names for the orthogonal Cartesian axes, we set $x := \Delta_1$, $y := \Delta_2$ and $z := \Delta_3$ and put the point at (x, y, z). Naturally all such points lie in the positive octant, because probabilities are non-negative; and because the coordinates sum to no more than one, the points lie on or below the plane $x + y + z = 1$ as well.

Thus our first geometrical analogue is one of the whole space $\overline{S}$ itself: it is a right tetrahedron with its apex at the origin. "Terminating" (summing-to-one) distributions lie on its base; sub-distributions, having a nonzero "**abort** probability," lie closer to its apex.

In higher dimensions we would say "right hyper-pyramid," and in two dimensions we are dealing with a right triangle.

6.2   Standard deterministic programs

Fix the initial state s, and consider the action of some standard and deterministic probabilistic program. In our impoverished setting, with only three final states to choose from, the possible actions of such programs are easy to imagine. Figure 6.2.1 shows "pictures" of the three sets of distributions produced by $s := s_1$, $s := s_2$ and $s := s_3$.[1]

---

[1]These three programs do not depend on their initial states, producing fixed distributions — thus for each program one picture says it all. Most programs do depend on the initial state, of course, and so usually a program will have a different picture for each initial state in which it can be run.

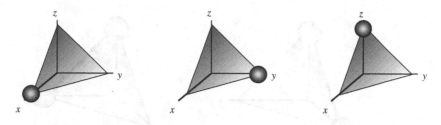

The three programs $s := s_1$, $s := s_2$ and $s := s_3$.

Figure 6.2.1. THREE SIMPLE PROGRAMS DEPICTED

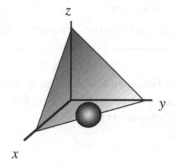

The program $s := s_1 \, {}_{0.5}\!\oplus s_2$.

Figure 6.3.1. PROBABILISTIC DETERMINISTIC PROGRAM DEPICTED

In general, such programs are represented by single point placed on one of the axes, at a distance one from the origin: there is just one point, because the program is deterministic; and the point is at distance one, on an axis, because it represents a point distribution.

6.3 Probabilistic deterministic programs

When the deterministic programs are probabilistic, they may move off the axes. (For simplicity, we consider only terminating programs for now.) Figure 6.3.1 shows a probabilistic deterministic program made by combining the first two shown earlier in Fig. 6.2.1.

In general, to combine two probabilistic deterministic programs *prog*, *prog′* with the operator $_p\oplus$, we draw a line between the two points representing

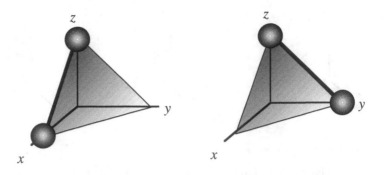

The programs $s := s_1 \sqcap s_3$ and $s := s_2 \sqcap s_3$.

Figure 6.4.1. STANDARD DEMONIC PROGRAMS DEPICTED

---

the two programs' distributions, and select a new point at a fraction $\bar{p}$ of the way along the line from *prog* to *prog'*. In Fig. 6.3.1 that puts the point midway between the two bottom vertices.

6.4   Demonic programs

If a program acts demonically from initial state s, then its picture for s will be a *clump* of points rather than just a single point.[2]

Because of our probabilistic-closure condition, though, not all sets can be clumps: they must for example be convex. Geometrically, a region is *convex* iff whenever it contains two points corresponding to *prog* and *prog'* it contains the whole line joining those points as well. And the same is true here — in our case it is modelling the fact that demonic choice between *prog* and *prog'* can be implemented by any probabilistic choice between them. Such choices, if they tend to select *prog* more often, will be represented by a point on the line from *prog* to *prog'*, but nearer "the *prog* end." Figure 6.4.1 shows demonic choices between several of our programs above; notice how their pictures are indeed "lines" rather than single points.

The effect of producing a line occurs even when one or both of the programs is not standard, as shown in Fig. 6.4.2.

---

[2]We say "clump" to emphasise the fact that not just any set of points will do: since the set is convex, it's more like a clod of mud than a cloud of dust. Clumps are exactly the elements of $\mathbb{C}S$.

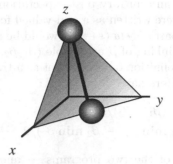

The program $s := (s_{1\ 0.5}\oplus s_2) \sqcap s_3$.

Figure 6.4.2. PROGRAM DEPICTED AS A LINE

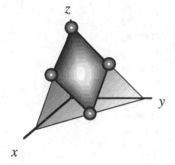

The program $s := (s_1 \sqcap s_3)_{0.5}\oplus (s_2 \sqcap s_3)$.

Figure 6.5.1. PROGRAM DEPICTED AS A DIAMOND

6.5 Refinement

Consider the program shown in Fig. 6.5.1: it is formed by selecting the mid-point (because we are using $_{0.5}\oplus$) of every line drawn between the two programs of Fig. 6.4.1. A bit of thought shows that a "diamond" shape results.

If we compare the two programs of Figs. 6.4.2 and 6.5.1, that is programs

$$s := (s_{1\ 0.5}\oplus s_2) \sqcap s_3 \quad \text{and} \quad s := (s_1 \sqcap s_3)_{0.5}\oplus (s_2 \sqcap s_3) ,$$

we note that the first is a refinement of the second, and we can see that in several ways. The first way — brute force — would be to use the definition of refinement directly: take an arbitrary post-expectation β, and calculate the pre-expectations. If β were written as a real-valued formula *beta* in the logic, the semantic counterpart of *beta* $\langle s \mapsto s_i \rangle$ would be just $\beta.s_i$ — which we will abbreviate as β_i, thinking of β as a tuple $(\beta_1, \beta_2, \beta_3)$ of the values it takes over S. Using the definition of assignment as a transformer, but at the semantic level, we therefore have

$$(\beta_1 + \beta_2)/2 \text{ min } \beta_3 \tag{6.1}$$
$$\text{and} \quad (\beta_1 \text{ min } \beta_3 + \beta_2 \text{ min } \beta_3)/2 \tag{6.2}$$

for the β-pre-expectations of the two programs — and it's then just a matter of arithmetic to show that $(6.1) \geq (6.2)$ for all $\beta_1, \beta_2, \beta_3 \geq 0$.

The second way to show refinement would be to use program algebra: we'd begin with Fig. 6.5.1, and use the laws of Sec. B.1 to calculate

$$s := (s_1 \sqcap s_3) \quad {}_{0.5}\oplus \quad s := (s_2 \sqcap s_3)$$

$$
\begin{array}{lllll}
= & s := s_1 & {}_{0.5}\oplus & s := s_2 & \text{Law 8 twice} \\
& \sqcap \quad s := s_1 & {}_{0.5}\oplus & s := s_3 & \\
& \sqcap \quad s := s_3 & {}_{0.5}\oplus & s := s_2 & \\
& \sqcap \quad s := s_3 & {}_{0.5}\oplus & s := s_3 & \\
\end{array}
$$

$$
\begin{array}{lllll}
= & s := s_1 & {}_{0.5}\oplus & s := s_2 & \text{Law 2} \\
& \sqcap \quad s := s_1 & {}_{0.5}\oplus & s := s_3 & \\
& \sqcap \quad s := s_3 & {}_{0.5}\oplus & s := s_2 & \\
& \sqcap \quad s := s_3 & & & \\
\end{array}
$$

$$
\begin{array}{lllll}
\sqsubseteq & s := s_1 & {}_{0.5}\oplus & s := s_2 & \text{Law 7} \\
& \sqcap \quad s := s_3 \, , & & & \\
\end{array}
$$

which, unsurprisingly, is more-or-less the same arithmetic we'd have done for the first approach. We have just worked "one level up," with programs directly rather than their pre-expectations.

The third way to see the refinement would be just to examine the two figures, and note that the line shown in Fig. 6.4.2 lies within the diamond of Fig. 6.5.1. That is, refinement is (in fact, remains) just set inclusion, if looked at the right way.[3]

---

[3]There is no need, for example, to have "approximate," that is probabilistic refinement to model our inability to measure real-life behaviour exactly. We merely make the specifications themselves as inexact as is necessary to accommodate our measurement tolerances.

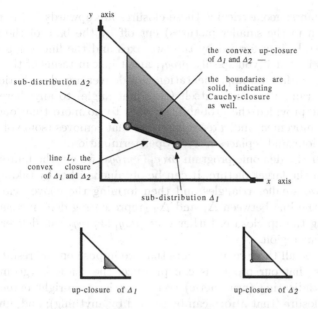

A program with some probability of nontermination, illustrating up closure.

Figure 6.6.1. A NONTERMINATING PROGRAM DEPICTED

6.6   Nontermination and sub-distributions

Nonterminating programs lead to proper sub-distributions, ones that sum to strictly less than one. In our three-element state space, the clumps representing the programs would occupy part of the interior of the tetrahedron, instead of remaining on its base.

For visualisation, in Fig. 6.6.1 we reduce the state space to just two elements so that our pictures lie in the x, y plane. Sub-distribution Δ_1 is the point $(1/2, 1/4)$, representing probability $1/2$ of terminating in state s_1, similarly $1/4$ for s_2 (and probability $1 - 1/2 - 1/4 = 1/4$ of failing to terminate at all, of looping or of "going to jail," in the terms used in Sec. 1.3); in $pGCL$ we could write it

$$prog_1 : \qquad s := (s_1 \, @ \, 1/2 \mid s_2 \, @ \, 1/4) \,.$$

Similarly, sub-distribution Δ_2 is the point $(1/5, 1/3)$, representing probability $1/5$ of terminating in state s_1 and $1/3$ in s_2; that one is

$$prog_2 : \qquad s := (s_1 \, @ \, 1/5 \mid s_2 \, @ \, 1/3) \,.$$

Although Δ_1 and Δ_2 are themselves represented by single points in the plane, the two programs producing them are represented by the *up clo-*

sure of those points; geometrically, those closures are upwards-facing right triangles (shown in the smaller pictures) cut off by the base of the two-dimensional tetrahedron formed by the two axes and the line $x + y = 1$. What it means is that if you specify $prog_1$, and think in terms of the sub-distribution Δ_1, in fact an implementation can deliver any distribution in the triangle — ranging from Δ_1 itself (the right angle) to anywhere up-and-to-the-right provided the probabilities sum to no more than one. In effect, up-right movement indicates refinement that removes (some of the) aborting behaviour and replaces it by proper termination.

If we consider the demonic program $prog_1 \sqcap prog_2$, we get the four-sided region shown in the larger picture. It can be obtained either by taking the union of the two smaller triangles and then forming the convex closure, or by drawing the line between Δ_1 and Δ_2 (representing demonic choice) and then taking the up closure. Either way, $prog_1 \sqcap prog_2$ can deliver any behaviour in that region.

We can now see all three requirements that we impose on the result sets of distributions that our programs can produce: the shape is (geometrically) convex, indicating (arithmetic) convexity; it is up-right extended, indicating up closure (that **abort** can be refined by anything); and, finally, it has a hard boundary, indicating Cauchy closure.[4]

6.7   Post-expectations, touching planes and pre-expectations

With all this apparatus, we are now ready to look geometrically at the calculation of weakest pre-expectations. (We move back to three dimensions.)

In Fig. 6.7.1 we show our program from Fig. 6.5.1, the diamond lying entirely in the base of the tetrahedron (lying there because it is terminating): it represents the action of the program

$$s := (s_1 \sqcap s_3) \; _{0.5}\oplus \; s := (s_2 \sqcap s_3) \; .$$

A post-expectation is represented by a family of planes all with the same normal (so that they are parallel).[5] Consider a post-expectation β with, as before, β_i being its value at s_i; the triple $(\beta_1, \beta_2, \beta_3)$ of non-negative reals determines a direction from the origin into the positive octant, and it is the

---

[4] As we showed earlier, in finite state spaces the Cauchy closure happens automatically — after all, figures obtained by drawing solid lines between points must have hard boundaries. In infinite state spaces, however, we must impose the condition explicitly. They are harder to visualise!

[5] In higher dimensions, they are hyperplanes.

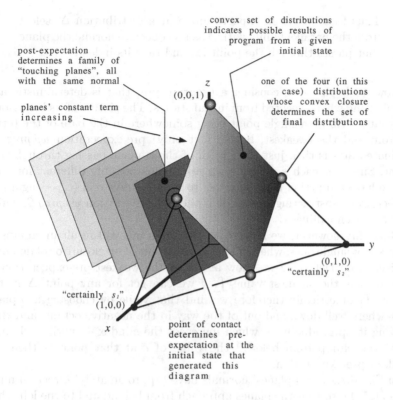

convex set of distributions
indicates possible results of
program from a given
initial state

post-expectation
determines a family of
"touching planes", all
with the same normal

planes' constant term
increasing

z

$(0,0,1)$

one of the four (in this
case) distributions
whose convex closure
determines the set of
final distributions

y

$(0,1,0)$
"certainly s_2"

"certainly s_1"
$(1,0,0)$

x

point of contact
determines pre-
expectation at the
initial state that
generated this
diagram

This example illustrates weakest pre-expectations over a three-element state
space $\{s_1, s_2, s_3\}$.

Figure 6.7.1. TOUCHING-PLANE GEOMETRY

common normal of the family of planes representing that post-expectation.
Any particular plane in the family has equation

$$\beta_1 x + \beta_2 y + \beta_3 z \ = \ c \,, \tag{6.3}$$

with the parameter c acting as an index into the family: increasing c
moves the corresponding c-plane up into the positive octant, maintaining
its orientation; decreasing c moves it down.

The principal operation carried out on expectations (considered as
random variables) is to take their expected value over a distribution:
if the distribution Δ is $(\Delta_1, \Delta_2, \Delta_3)$, then the expected value $\int_\Delta \beta$ is
$\beta_1 \Delta_1 + \beta_2 \Delta_2 + \beta_3 \Delta_3$ — and that is just the value obtained by evaluating
the left-hand side of (6.3) at the point (x, y, z) representing Δ. (Compare
Def. 5.2.2.)

Thus to evaluate an expectation β at a distribution Δ, select from the parallel family with β as its collective normal the plane that passes through the point Δ, and take its index c.

Now, to begin with, consider a program *prog* that is deterministic and terminating when executed from initial state s. The picture of its behaviour, for that s, will be a single point lying somewhere in the base of the tetrahedron; and the (weakest, but in fact only) pre-expectation *wp.prog.β.s* evaluated at s is then just the c term of the β-plane passing through that point. Since for each s we get such a c_s (thus possibly different for each s), we have a function from state (s) to non-negative real (c_s) — again an expectation: abstracting from s, it is the pre-expectation *wp.prog.β*, which we have been calling α.

Generally however, we know that our programs will result in clumps of points, not single ones, whether that is the result of up closure or of demonic choice. From Def. 5.5.2 we know also that the weakest pre-expectation in that case is the smallest value $\int_\Delta \beta$ we can get for any point Δ in the clump. Geometrically therefore we find that value by choosing a β-plane somewhere well down and out of the way in the negative octant, and then moving it up gradually — which increases the c index — until it "bumps into" the clump from below:[6] the value of c at that point is then our weakest pre-expectation.

In Fig. 6.7.1 the β-planes' normals have (approximately) direction numbers $(1, 5, 1)$, so that the planes approach from behind and to the left: they touch the clump at the "9-o'clock" position $(1/2, 0, 1/2)$ of the diamond, for which the c value — hence the pre-expectation — would be

$$1 * 1/2 \ + \ 5 * 0 \ + \ 1 * 1/2 \ \ = \ \ 1 .$$

6.8 Refinement seen geometrically

We have seen already that one program refines another if its clump lies within the clump of the other. Geometrically, if one clump lies within another, then the constant term for a plane touching the inner clump can be no less than the constant term for a parallel plane touching the outer clump: in effect, in moving from the negative octant, the plane must reach the outer clump first — continuing on, to reach the inner (more refined) one, can only increase the constant term.

---

[6]The plane might just graze the clump at a single point — like a book resting on a basketball — and so be tangential there; but it might also touch at a vertex (where the normal of the clump, and thus "being tangential," are both undefined); and it could even touch over a whole line or plane, as if the "basketball" needed inflation. In all cases the intersection must include no interior point of the clump [Kuh03, p17].

6.9   The geometry of the healthiness conditions

Figure 5.6.7 gave a number of conditions that *wp.prog* satisfies as a transformer; we now look at them geometrically.

6.9.1   Feasibility

Feasibility is the probabilistic version of Dijkstra's *Law of the Excluded Miracle* for standard programs, saying basically that "you can't get out (of a transformer) more than you put in." With our identification [false] $\equiv 0$, that means "if you put in false as a post-condition, you can't get more than false as a precondition" — and since [false] is the smallest pre-expectation, that's exactly what you get.

   Probabilistically, feasibility says that the pre-expectation cannot exceed the maximum value of the post-expectation. Since the precondition is given by the c term in the family of planes, we are in effect saying that the planes "cannot move up too far." That is ensured by the fact that the base of the tetrahedron caps all our clumps of distributions. That is, the "picture" of feasibility is the base of the tetrahedron.

6.9.2   Monotonicity

Surprisingly, this does not seem to have an easy geometric interpretation on its own — we view it as a special case of sublinearity.

6.9.3   Scaling

Scaling is an easy consequence of the fact that multiplying the postcondition β by a non-negative real does not change the direction (of normals) that it represents: thus the family of planes is unchanged, and contact occurs at the same position on the clump. The coefficients β_i have been scaled, however, and so the index c is multiplied by the same amount.

6.9.4   Sublinearity

This is the most sophisticated property, which implies all the others. We defer it until the next section.

6.9.5   Continuity

Continuity corresponds to Cauchy closure, and geometrically it means that there is at least one point in the clump that actually lies in the touching plane: that is, the clump is a "closed" set rather than an "open" one. What we can deduce from that, for example, is the Fact B.3.5 used in our derivation of the proof rules for loops (Chap. 7, and Sec. B.3 p. 331).

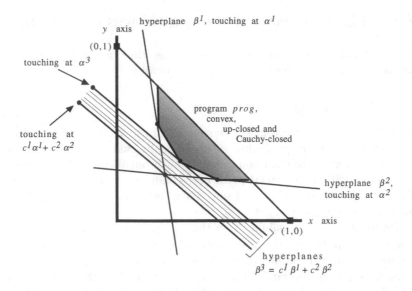

Figure 6.10.1. GEOMETRIC INTERPRETATION OF SUBLINEARITY

6.10   Sublinearity corresponds to convexity

We are left with sublinearity, for which we return to two dimensions as shown in Fig. 6.10.1. Suppose program *prog* produces the five-sided clump shown; we can see immediately that it is demonic (not the up closure of a single point) and possibly nonterminating (goes below the base of the right triangle).

Post-expectation β^1 corresponds to a family of planes whose normal is nearly horizontal; we show just one member, which is touching the clump high on the left.[7] The constant term for that plane is α^1.

Post-expectation β^2 corresponds to a family of planes whose normal is nearly vertical; it touches low on the right, with constant term α^2.

Post-expectation β^3 is formed as a linear combination of β^1 and β^2, as shown. Since the two families' normals are non-negative, as are the coefficients c^1, c^2 for the linear combination, the normal for β^3 will lie somewhere "between" the normals for β^1 and β^2, on the positive side: in the figure, we have it about midway between, pointing up and to the right at 45°.

We concentrate on just two members of the β^3 family of hyperplanes. The lower one is obtained by setting the constant term to the same linear combination of the constant terms α^1 and α^2 that we used to combine the normals β^1 and β^2 to get β^3. It will therefore pass through the intersection

---

[7]The normal is not shown in the picture.

of those other two planes, as shown — and in this case, it is "pushed away from" the clump representing the program.

The other member of the β^3 family is higher up, touching the clump, and has constant term α^3. Since it *is* the upper one, clearly we have

$$\alpha^3 \quad \geq \quad c^1\alpha^1 + c^2\alpha^2$$

— and that is our picture of sublinearity. (It is effectively a picture of Lem. 5.6.2, saying that all clumps have this "pushing-away" of planes behaviour, because of their convex- and up closure. The converse, Lem. 5.7.3, says that any pushing-away behaviour can be realised by some clump.)

In general, therefore, sublinearity is the arithmetic characterisation of our clumps' being convex and up-right closed.

We can see also that sublinearity is no more pessimistic than necessary. If in the figure all we knew about the clump was α^1, β^1, α^2 and β^2, and we wanted to find out which plane from the family α^3 would touch it, we'd have to pick the lower one: for all we know, the clump could occupy the whole space above the two planes defined by the given observations $\alpha^1, \beta^1, \alpha^2, \beta^2$.

6.11   Truncated subtraction

Finally, we must explain the $\ominus$ term in the formulation of sublinearity. It comes from the one part of the figure that is not up-right facing: the base of the tetrahedron, part of the boundary of any clump, faces *downwards*. Its normal will have direction numbers $(-1, -1, \cdots, -1)$, as many as are required for the dimension, and we can use it in linear combinations too. That accounts for the $\ominus\underline{c}$ in sublinearity. The truncation of the subtraction is just to remain within the non-negative domain.

6.12   A geometrical proof for recursion

Now we illustrate the use of the geometrical ideas of this chapter, working beyond the logic to outline a proof of the consistency of the relational- and transformer semantics of recursion. The proof must differ from our earlier consistency proofs (Sec. 5.8) because here the semantics is given by a least fixed-point, which impedes the "just unfold the definitions" style of reasoning we used before: for a fixed point, we would have to unfold forever.

Suppose we have a recursion ($\mathbf{mu}\ xxx \cdot \mathcal{C}$), where $\mathcal{C}$ is a program context in which the place-holder xxx can appear (1.16). Since $\mathcal{C}$ is itself built from program constructors (we are using structural induction over $pGCL$), we can assume that it has two, consistent, "higher-order" semantics — a relation-style $\mathbf{r}$, and a transformer-style $\mathbf{t}$ — satisfying

$$
\begin{aligned}
\mathbf{r} &\in \ \mathbb{H}S \to \mathbb{H}S \quad \text{and is continuous } (\dagger) \\
\mathbf{t} &\in \ \mathbb{T}S \to \mathbb{T}S \quad \text{and is continuous } (\ddagger)
\end{aligned}
$$

and $\mathbf{t} \circ wp \ = \ wp \circ \mathbf{r}$.   (higher-order consistency)

Note that although we have continuity of transformers in $\mathbb{T}S$ — it is a healthiness condition — we have not proved our assumption at ($\ddagger$) above that *transformer-transformers* such as $\mathbf{t}$ are continuous as well. That requires a separate (and routine) structural induction argument over $pGCL$, and as expected relies on the continuity of the transformers themselves. Similar remarks apply to $\mathbf{r}$ at ($\dagger$).

We need the continuity to express the fixed point as a limit.

The result for recursion is given in the following lemma.

Lemma 6.12.1 COMPLEMENTARY AND CONSISTENT SEMANTICS
                OF RECURSION

If $\mathbf{r}, \mathbf{t}$ are as above, then their least fixed-points are consistent:

$$
wp.(\mu.\mathbf{r}) \ \ = \ \ \mu.\mathbf{t} \ .
$$

Proof   Because $\mathbf{r}, \mathbf{t}$ are continuous, we can write their least fixed-points as $\mathbb{N}$-limits, *viz.*

$$
\begin{aligned}
\mu.\mathbf{r} \ &= \ (\cap n : \mathbb{N} \cdot r_n) \\
\text{and} \quad \mu.\mathbf{t} \ &= \ (\sqcup n : \mathbb{N} \cdot t_n)
\end{aligned}
$$

$$
\begin{aligned}
\text{where} \quad r_0.s \ &:= \ \overline{S} \quad \text{for all } s \in S \\
r_{n+1} \ &:= \ \mathbf{r}.r_n
\end{aligned}
$$

$$
\begin{aligned}
\text{and} \quad t_0.\beta \ &:= \ \underline{0} \quad \text{for all } \beta \in \mathbb{E}S \\
t_{n+1} \ &:= \ \mathbf{t}.t_n \ .
\end{aligned}
$$

We are using (†,‡) above and the fact that the least fixed-point of a continuous function in a *cpo* is the supremum of its *n*-fold iterates from the bottom element $\perp$ [Tar55].[8] The bottom of $\mathbb{H}S$ is the relation that for all initial states *s* returns the entire set of distributions $\overline{S}$, and the bottom of $\mathbb{T}S$ is the transformer that returns pre-expectation $\underline{0}$ for every post-expectation. They are of course the complementary semantics of **abort**, the $\sqsubseteq$-least program.[9]

Because $t_0 = wp.r_0$,[10] it is a straightforward mathematical induction over *n* to show from $\mathbf{t} \circ wp = wp \circ \mathbf{r}$ above that $t_n = wp.r_n$ for all $n \geq 0$. Thus our argument boils down to showing that *wp* itself is continuous, *i.e.* that

$$wp.(\cap n: \mathbb{N} \cdot r_n) \quad = \quad (\sqcup n: \mathbb{N} \cdot wp.r_n) \ . \tag{6.4}$$

+ We will show the $\sqsubseteq$ and $\sqsupseteq$ inequalities separately; in fact by monotonicity of *wp* the inequality $\sqsupseteq$ is immediate.

For $\sqsubseteq$ we use a geometric argument, relying on the ideas of Sec. 6.7. Fix a post-expectation β and an initial state *s*. On the left we define

$$c \quad := \quad wp.(\cap n: \mathbb{N} \cdot r_n).\beta.s \ ,$$

recalling from compactness that $(\cap n: \mathbb{N} \cdot r_n)$ cannot be empty, so that *c* (depending on β, s) is the constant coefficient of a hyperplane with normal β just touching the non-empty and closed set of distributions $(\cap n: \mathbb{N} \cdot r_n.s)$. We show that for any $\varepsilon > 0$ there is an n_ε such that

$$c - \varepsilon \quad < \quad wp.r_{n_\varepsilon}.\beta.s \ ,$$

which since ε is arbitrary will establish $c \leq (\sqcup n: \mathbb{N} \cdot wp.r_n.\beta.s)$.

Now consider the closed half-space $X \subseteq \overline{S}$ of distributions on-or-below the β-hyperplane that has coefficient $c - \varepsilon$, and assume for a contradiction

---

[8]In the fixed-point literature this would be called an ω-limit. If a function *f* is continuous over a *cpo*, then its least fixed-point $\mu.f$ is given by $(\sqcup n: \mathbb{N} \cdot f^n.\perp)$, where f^n is the *n*-fold functional composition of *f* and $\perp$ is the bottom element of the *cpo*. The main part of the argument is, using *f*'s treatment of continuity as described in Footnote 33 on p. 147, that

$$\begin{aligned} & f.(\sqcup n: \mathbb{N} \cdot f^n.\perp) \\ = \ & (\sqcup n: \mathbb{N} \cdot f^{n+1}.\perp) && \text{continuity of } f \\ = \ & (\sqcup n: \mathbb{N} \cdot f^n.\perp) \ . && f^0.\perp = \perp \end{aligned}$$

[9]Note that the refinement order for relations is $\sqsupseteq$ whereas for transformers it is our familiar $\sqsubseteq$. Thus $t_0 \sqsubseteq t_1 \sqsubseteq t_2 \cdots$ corresponds to $r_0 \sqsupseteq r_1 \sqsupseteq r_2 \cdots$ and in general we have

$$wp.r \sqsubseteq wp.r' \quad \text{iff} \quad r \sqsupseteq r' \ ,$$

where on the right by inclusion between relations we mean $(\forall s: S \cdot r.s \sqsupseteq r'.s)$.

[10]This is simply stating that the two semantics for **abort** are consistent.

that for all n we have $c-\varepsilon \geq wp.r_n.\beta.s$. Then — for all n — we have that the set $X \cap r_n.s$ is non-empty; and because those sets are compact, and nested as n increases, by the finite-intersection lemma also their intersection is non-empty. That means that the set $(\cap n\colon \mathbb{N} \bullet r_n.s)$ extends into X as well, whence $c \leq c - \varepsilon < c$, which is the contradiction establishing that for some n we have $c - \varepsilon \not\geq wp.r_n.\beta.s$. We take that n as our n_ε.

Thus, since we can choose ε as small as we like, we have for any β, s that

$$wp.(\cap n\colon \mathbb{N} \bullet r_n).\beta.s \quad = \quad c \quad \leq \quad (\sqcup n\colon \mathbb{N} \bullet wp.r_n).\beta.s \ , \qquad [11]$$

i.e. that $wp.(\cap n\colon \mathbb{N} \bullet r_n) \sqsubseteq (\sqcup n\colon \mathbb{N} \bullet wp.r_n)$ as required.

With the opposite inequality (+) above, we now have the desired (6.4).

$\square$

Chapter notes

The relation between convex sets (of distributions, in our case) and optimisation problems has been understood for a long time — Farkas' lemma for example first emerged in 1895, with the earliest published proof appearing in 1899 [Far99]. Other mature disciplines relying on linear programming techniques include Markov decision processes [FV96], used for many years to model systems which evolve within uncertain contexts. They model the interaction of probability with actions controlled by so-called *maximising-* or *minimising* "agents," and they share many of the characteristics of probabilistic programs.

In probabilistic verification specifically there are even more recent examples of linear programming applied within model-checking frameworks such as those used by Bianco and de Alfaro [BdA95].

An interesting geometric perspective on games, and many historical references, is provided by Kuhn [Kuh03].

---

[11]The equality $(\sqcup n\colon \mathbb{N} \bullet wp.r_n.\beta.s) = (\sqcup n\colon \mathbb{N} \bullet wp.r_n).\beta.s$ is direct from the definition of $\sqcup$ for functions.

7

Proved rules for probabilistic loops

7.1   Introduction

In Chap. 2 we introduced tools for dealing with loops: invariants and variants. Neither of those is part of the definition of $pGCL$, just as their "traditional" standard versions are not part of the definition of GCL: rather the rules that loops, their invariants and their variants obey must be proved from the semantics of the language containing them.

The probabilistic "loop rules" introduced earlier were Def. 2.2.1, Lem. 2.4.1 and Lem. 2.7.1. The first defined *probabilistic invariants*; the second showed how to *combine* such invariants with termination probabilities; and the third treated the use of *probabilistic variants* to demonstrate almost-certain termination. In this chapter we give the soundness proofs for all of them.

In the standard case, soundness is shown via the weakest "liberal" precondition wlp, where $pre \Rrightarrow wlp.prog.post$ means

> from any state satisfying precondition *pre*, the sequential program *prog* is guaranteed to terminate in a state satisfying *post*
> — *if it is guaranteed to terminate at all.*

The last phrase is what distinguishes wp and wlp. (Compare (1.1) in Sec. 1.1.) It is not true however that $wlp.prog.post$ is just the predicate

$$wp.prog.\text{true} \quad \Rightarrow \quad wp.prog.post \quad ^1$$

— in fact the wlp of a program adds genuinely new information. Neither wp nor wlp can be derived from the other: program **abort** $\sqcap$ **skip** is wp-equivalent to **abort**, but wlp-equivalent to **skip**.[2] What we do know, though, is that

$$wlp.prog.post \;\wedge\; wp.prog.\text{true} \quad \equiv \quad wp.prog.post \, . \qquad (7.1)$$

That is, if the program establishes *post* whenever it does terminate, and indeed it is *guaranteed* to terminate, then it is guaranteed to establish *post*.

With the above apparatus, the treatment of standard loops is fairly straightforward [Dij76]. With some knowledge of fixed-point theory, it is not hard to show that if the loop body preserves some standard invariant *Inv*, then *Inv* is a "weak" invariant for the loop as a whole — *i.e.* predicate *Inv* implies the weakest liberal precondition for the loop to (re-)establish

---

[1] Although we allow ourselves now to write "$\Rrightarrow$" between Booleans, assuming an implicit embedding as explained in Footnote 4 on p. 39, there is still an important difference between "$\Rightarrow$" and "$\Rrightarrow$". The former constructs a new formula from its operands; the latter, if used here (incorrectly), would assert that $wp.prog.\text{true}$ implied $wp.prog.post$ in all states.

[2] If wp could be derived from wlp, or *vice versa*, then equivalence in one semantics would imply equivalence in the other: by transitivity of equivalence, in the case above that would make **skip** and **abort** equivalent — which they are not (in either semantics).

Inv if it terminates — and then (7.1) does the rest. The termination part *wp.prog*.true is usually handled by a separate, variant argument.

In the probabilistic case, however, things are not so simple.[3] Weak invariance of *I* can be defined just as before, with virtually the same fixed-point argument, once one has defined the probabilistic "greatest liberal pre-expectation" *wlp* [MM01b]; and there is a probabilistic analogue of (7.1) that links it to probabilistic *wp*, in which the normal conjunction ∧ is replaced by the *probabilistic conjunction* & we saw first in Sec. 1.6. But operator & — unlike ∧ — is not idempotent: for one-bounded expectation *R* we might have $R \& R \Rrightarrow R$, but equality occurs only when *R* is zero or one. The result is that the probabilistic version

$$wlp.prog.post \ \& \ wp.prog.\text{true} \ \Rrightarrow \ wp.prog.post \qquad (7.2)$$

of (7.1) is too weak in general — its left-hand side is too low.

To get the most general Case 3 of our earlier Lem. 2.4.1, therefore, we will need some fairly intricate arithmetic — and that is the first contribution of this chapter.

The other matter we deal with is probabilistic variants, and again they need a slightly more sophisticated treatment than in the standard case.

Finally, we introduce and discuss "probabilistic loop guards," which — although they require a further extension of our loop rules — can lead to more succinct programs and simpler reasoning in some cases.

Notationally we revive the "syntactic style" of Part I — for this chapter only — using Roman letters *R* or words *Inv* for expressions denoting expectations (rather than the Greek letters α, β used earlier in Part II for the expectation functions themselves); this allows us to make a more direct link with Chap. 2, of which this material is a continuation.

One important difference, however, is that in this chapter the expectations' values are bounded above by one throughout — *i.e.* we operate in the real interval [0, 1] exclusively.[4] Note that one-boundedness makes both $\mathbb{E}S$ and $\mathbb{T}S$ into a *cpo*.

---

[3]Recall Footnote 11 on p. 42.

[4]Using [0, 1] is of equivalent power to the use of "full" but bounded expectations, because *feasibility* ensures that *wp.prog* preserves one-boundedness and *scaling* allows us to manipulate the bound to translate from one- to general bounding and back again. (See *e.g.* Sec. 7.4.2.)

We use one-bounding here because it simplifies the link with *wlp*; and it gives us a "top" element in our partial order of expectations, which allows us to define greatest fixed-points.

7.2   Partial loop correctness

The standard *weakest liberal precondition* of a program describes its *partial* correctness, identifying those initial states from which the program either establishes a given postcondition or fails to terminate [Hoa69, Dij76]. This contrasts with the more conventional *weakest precondition* we have been using so far, which requires termination as well and thus describes *total* correctness. We write *wlp.prog.R* for the *greatest liberal pre-expectation* of program *prog* with respect to post-expectation R.

The *wlp* semantics differs from the *wp* in these two respects:

1. The nowhere-terminating program is defined *wlp*.**abort**.$R := $ [true] for all postconditions R. (Compare *wp*.**abort**.$R := $ [false].)

2. The weakest-liberal-precondition semantics of a recursive program is given by a *greatest-* (rather than least-) fixed point.

For the remainder of *pGCL* the definitions are analogues of those in Fig. 1.5.3, with *wp* replaced by *wlp* on the right in the structurally recursive cases.

An important point is that the *wlp*-generated transformers do not have the same algebra as those generated by *wp*. Inspection of Case 1 above shows that **abort** is not scaling; and so in general we cannot assume sublinearity when using *wlp*. The liberal expectation semantics in its own right is treated elsewhere in more depth [MM01b]; here we will simply use the algebraic properties, summarised at Sec. B.3, which relate it to *wp*.

When considering loops, a special case of recursion, Case 2 above tells us that the *wlp* semantics is as follows:

Definition 7.2.1 WEAKEST LIBERAL PRECONDITION FOR LOOP
    For iteration *loop* given as **do** $G \to$ *body* **od**, and any post-expectation R, we define

$$wlp.loop.R   :=   (\nu Q \cdot [G] * wlp.body.Q + [\overline{G}] * R) .   {}^{5}$$

□

Notice that *wlp* is applied "hereditarily" to the loop body.

---

^{5}We use ν to indicate taking the greatest fixed-point. The definition of *wlp* for a loop is based on the simpler form (1.19) of loops' *wp*-semantics, from p. 21, rather than appealing directly to the original recursion over programs (1.18).

The advantage of (1.18) is a simpler theory, because iteration is then no more than a special case of recursion, and so can be ignored when proving general results (as can any other syntactic sugar); the advantage of (1.19) however (and similarly Def. 7.2.1) is practical, as we are about to see.

Their equivalence is shown using (7.12) in Sec. 7.7.2.

From Def. 7.2.1 we derive immediately the usual rule for partial correctness of loops, based on the preservation of an invariant.

Lemma 7.2.2 Let expectation I be a *wlp-invariant* of *loop*, thus satisfying

$$[G] * I \quad \Rightarrow \quad wlp.body.I . \tag{7.3}$$

Then in fact

$$I \quad \Rightarrow \quad wlp.loop.([\overline{G}] * I) . \tag{7.4}$$

Proof   We substitute I for Q in the right-hand side of Def. 7.2.1, setting $R := [\overline{G}] * I$, and find

$$
\begin{array}{ll}
& [G] * wlp.body.I \; + \; [\overline{G}] * ([\overline{G}] * I) \\
\Leftarrow & [G] * [G] * I \; + \; [\overline{G}] * [\overline{G}] * I & \text{assumption} \\
\equiv & I , & * \text{ idempotent for standard expectations}
\end{array}
$$

obtaining the result immediately from the elementary property of greatest fixed-points in a *cpo* that $f.x \geq x$ implies $\nu.f \geq x$.[6]   □

It is worth noting that the assumption (7.3) of Lem. 7.2.2 is weaker than the one used in the standard rule for partial correctness of loops, where one conventionally finds *wp* instead:

$$[G] * I \quad \Rightarrow \quad wp.body.I . \tag{7.5}$$

The difference is real, even in the standard case, but only if we are genuinely interested in partial correctness. With Lem. 7.2.2 we can show for example that we have

$$wlp.(\textbf{do } x \neq 0 \rightarrow \textbf{abort od}).[x = 0] \quad \equiv \quad [\text{true}] , \tag{7.6}$$

choosing $I := [\text{true}]$ to do so: because *where the loop is guaranteed to terminate* it indeed establishes $x = 0$. We could not prove that using only (7.5).

The reason (7.5) is nevertheless used in the standard case is that it avoids introducing the extra concept of *wlp*: that is, if $I \Rightarrow wp.loop.[\text{true}]$ then we must have $[G] * I \Rightarrow wp.body.[\text{true}]$ in any case, in which circumstances (7.3) and (7.5) are equivalent.

---

[6]We write $\nu.f$ for the GREATEST FIXED-POINT of function f.

For probabilistic programs the above analysis does not apply,[7] and as shown below use of the stronger (7.5) is required for soundness in some properly probabilistic cases (*e.g.* Program (7.7) following Thm. 7.3.3).

7.3   Total loop correctness

In the standard case Fact B.3.1 is used to combine partial loop correctness with a termination argument, to give total loop correctness. Here we rely on its probabilistic analogue Fact B.3.2 — with it, and Lem. 7.2.2, we have immediately a rule for total correctness of probabilistic loops.

Lemma 7.3.1 Let invariant I satisfy $[G] * I \Rrightarrow wlp.body.I$. Then

$$I \& T \quad \Rrightarrow \quad wp.loop.([\overline{G}] * I) \, ,$$

where T is the termination probability $wp.loop.[\text{true}]$ of the loop.
   Proof

$$
\begin{array}{lll}
& wp.loop.([\overline{G}] * I) & \\
\equiv & wp.loop.(([\overline{G}] * I) \& [\text{true}]) & \text{definition \&} \\
\Lleftarrow & wlp.loop.([\overline{G}] * I) \ \& \ wp.loop.[\text{true}] & \text{Fact B.3.2} \\
\Lleftarrow & I \& T \, . & \text{assumption, Lem. 7.2.2}
\end{array}
$$

$\square$

Lemma 7.3.1 suffices for many situations, in particular those in which either I or T is standard since in that case $I \& T \equiv I * T$, provided that

---

[7]The reasoning fails at the point of concluding $[G] * I \Rrightarrow wp.body.I$ from

$$[G] * I \Rrightarrow wp.body.[\text{true}] \quad \text{and} \quad [G] * I \Rrightarrow wlp.body.I \, .$$

Applying Fact B.3.2 to those two inequalities gives only

$$
\begin{array}{lll}
& wp.body.I & \\
\equiv & wp.body.([\text{true}] \& I) & \\
\Lleftarrow & wlp.body.I \& wp.body.[\text{true}] & \text{Fact B.3.2} \\
\Lleftarrow & ([G] * I) \& ([G] * I) & \\
\Lleftarrow & [G] * (I \& I) & [G] \text{ standard} \\
\text{!! } \Rrightarrow & [G] * I \, , & \text{only } \Rrightarrow \text{ because \& not idempotent}
\end{array}
$$

where the last inequality is going the wrong way.
   For example, having the facts

$$
\begin{array}{llll}
& 1/2 & \equiv & wlp.(x:=0 \, _{1/2}\oplus \textbf{abort}).[x = 1] \quad (\text{Take the } \textbf{abort} \text{ side.}) \\
\text{and} & 1/2 & \equiv & wp.(x:=0 \, _{1/2}\oplus \textbf{abort}).[\text{true}] \quad (\text{Take the } x:=0 \text{ side.})
\end{array}
$$

does not give us $1/2 \equiv wp.(x:=0 \, _{1/2}\oplus \textbf{abort}).[x = 1]$. And indeed it is not true: that program fragment is *not* guaranteed to establish $x = 1$ half the time.
   What saves us from the unsound reasoning is that $1/2 \& 1/2$ is zero (not $1/2$ again).

I is one-bounded as we are assuming in this chapter.[8] Sec. 7.4 below uses the above lemma to prove the first two cases of our earlier Lem. 2.4.1.

When both I and T are probabilistic, however, the precondition $I \& T$ given us by Lem. 7.3.1 can be too low (*i.e.* pessimistic, though still correct). And we can show that replacing $\&$ by the idempotent $\sqcap$, *i.e.* using $I \sqcap T$ on the left-hand side, is not the solution.[9]

Thus we must improve Lem. 7.3.1 in a different way, which we do below. For simplicity we assume initially that *body* is pre-deterministic, so that we can appeal to its linearity (2.18).

The strategy is to develop an alternative invariant I' that is nowhere less than the I we are given, and usually greater than it; when we eventually form the precondition $I' \& T$ we get our original I back again.

First we show exact *wp*-invariance of T itself.

Lemma 7.3.2 For all iterations *loop* as above we have

$$[G] * T \quad \equiv \quad [G] * wp.body.T .$$

Proof   We calculate

$$
\begin{aligned}
&\quad [G] * T \\
\equiv &\quad [G] * wp.loop.[\text{true}] \\
\equiv &\quad \qquad\qquad\qquad\qquad\qquad\qquad\qquad \textbf{do}\cdots\textbf{od}\text{-algebra (2.2)}\\
&\quad [G] * ([G] * wp.body.(wp.loop.1) + [\overline{G}] * 1) \\
\equiv &\quad [G] * wp.body.T . \qquad\qquad\qquad\qquad\qquad\qquad [G] \text{ standard}
\end{aligned}
$$

$\square$

---

[8]The termination T is one-bounded by definition, as it is a probability; alternatively, we have $T \equiv wp.loop.[\text{true}] \Rrightarrow \max[\text{true}] \equiv 1$ by feasibility.

[9]Take invariant $I := [n = 0]/2 + [n = 1]$ in the program *loop*, defined

$$
\begin{aligned}
\textbf{do } &n = 0 \;\rightarrow\; n := -1\,{}_{\frac{1}{2}}{\oplus}\,n := +1 \\
[\!]\quad &n > 0 \;\rightarrow\; \textbf{skip} \\
\textbf{od } &,
\end{aligned}
$$

where we are using Dijkstra's more general "multiple-guard" loop; in this case it is read

$$\textbf{do } n \geq 0 \rightarrow (n := -1\,{}_{\frac{1}{2}}{\oplus}\,n := +1) \text{ if } n{=}0 \text{ else } \textbf{skip } \textbf{od} ,$$

so that G is $n \geq 0$. Thus we have

$$
\begin{aligned}
T &\equiv [n < 0] + [n = 0]/2 \\
I \sqcap T &\equiv [n = 0]/2 \\
I * [\overline{G}] &\equiv 0 ,
\end{aligned}
$$

but in fact $I \sqcap T \equiv [n = 0]/2 \not\Rrightarrow 0 \equiv wp.loop.0 \equiv wp.loop.(I * [\overline{G}])$.

There would be little point in trying $*$ instead of $\&$, since $*$ isn't idempotent either.

We now prove our main theorem for total correctness of deterministic loops (but see (‡) below); note that we are assuming a wp-invariance property (stronger than the wlp-invariance assumption of Lem. 7.3.1).

Theorem 7.3.3 INVARIANT-IMPLIES-TERMINATION LOOP RULE   If I is a wp-invariant of *loop* with deterministic *body*, and $I \Rrightarrow T$, then

$$I \quad \Rrightarrow \quad wp.loop.(\lceil \overline{G} \rceil * I) \ .$$

Proof   We show first that wp-invariance of I implies wlp-invariance of

$$I' \quad := \quad I + 1 - T \ .$$

Note we are using $I \Rrightarrow T$ to ensure one-boundedness of I'. (That is in fact the only place we appeal to the inequality.)[10] We reason

$$
\begin{array}{lll}
& wlp.body.I' & \\
\equiv & wlp.body.(I + 1 - T) & \text{definition } I' \\
\equiv & & \text{Fact B.3.3 twice; } body \text{ deterministic} \\
& wp.body.I + wlp.body.1 - wp.body.T & \\
\equiv & wp.body.I + 1 - wp.body.T & \text{Fact B.3.4} \\
\Leftarrow & \lceil G \rceil * (wp.body.I + 1 - wp.body.T) & \\
\equiv & \lceil G \rceil * wp.body.I + \lceil G \rceil - \lceil G \rceil * wp.body.T & \\
\equiv & \lceil G \rceil * wp.body.I + \lceil G \rceil - \lceil G \rceil * T & \text{Lem. 7.3.2} \\
\Leftarrow & \lceil G \rceil * \lceil G \rceil * I + \lceil G \rceil - \lceil G \rceil * T & \text{assumed } wp\text{-invariance of } I \\
\equiv & \lceil G \rceil * (I + 1 - T) & \lceil G \rceil \text{ standard} \\
\equiv & \lceil G \rceil * I' \ . & \text{definition } I'
\end{array}
$$

From Lem. 7.3.1 we then conclude immediately

$$I \quad \equiv \quad I' \& T \quad \Rrightarrow \quad wp.loop.(\lceil \overline{G} \rceil * I') \quad \equiv \quad wp.loop.(\lceil \overline{G} \rceil * I) \ ,$$

since for the final equality we have

$$
\begin{array}{ll}
& \lceil \overline{G} \rceil * I' \\
\equiv & \lceil \overline{G} \rceil * (I + 1 - T) \\
\equiv & \lceil \overline{G} \rceil * I + \lceil \overline{G} \rceil - \lceil \overline{G} \rceil * T \\
\equiv & \lceil \overline{G} \rceil * I \ . \hspace{4cm} \overline{G} \Rrightarrow T\text{: see (†)}
\end{array}
$$

$\square$

† In the last step we used that when G is false, termination is certain because the loop exits immediately. The compact idiom is an advantage of using one-bounded expectations: when G is false the value of $\lceil \overline{G} \rceil$ is one; since we are one-bounded, that means T is exactly one also, and can be removed from the multiplication above.

‡ Thm. 7.3.3 is extended to the nondeterministic case by Thm. B.2.2, in which an appeal is made to the geometric interpretation of continuity we saw in Sec. 6.9.5.

---

[10]We have $0 \Rrightarrow I'$ directly from $T \Rrightarrow 1$.

It is not hard to show that Thm. B.2.2 in turn implies Lem. 7.3.1: thus they are of equal power: Thm. B.2.2 merely "automates" the construction of the more complex invariant I', and hides its use behind the scenes.

The following example shows the *wp*- (rather than *wlp*-) invariance of the invariant I to be necessary for soundness of Thm. 7.3.3 in general. (Recall from Sec. 7.2 that it is not necessary in the standard case.) Let *loop* be

$$\textbf{do } b \rightarrow$$
$$b := \textsf{false} \ _{1/2}\oplus \textbf{ abort} \qquad\qquad (7.7)$$
$$\textbf{od} \ ,$$

for Boolean b, and note that we have for termination

$$T \quad\equiv\quad [\neg b] + [b]/2 \ .$$

Define $I := 1/2$, so that $I \Rightarrow T$ as required by Thm. 7.3.3, and reason

$$\qquad wlp.body.I$$
$$\equiv \quad wlp.(b := \textsf{false} \ _{1/2}\oplus \textbf{abort}).(1/2)$$
$$\equiv \quad (1/2)(wlp.(b := \textsf{false}).(1/2)) \ + \ (1/2)(wlp.\textbf{abort}.(1/2))$$
$$\equiv \quad (1/2)(1/2) + (1/2)(1)$$
$$\equiv \quad 3/4$$
$$\Leftarrow \quad [b] * 1/2$$
$$\equiv \quad [b] * I$$

which shows *wlp*-invariance of I; that would be the other requirement of the theorem if we relaxed *wp*-invariance. But

$$\qquad wp.loop.(\lceil \overline{G} \rceil * I)$$
$$\equiv \quad wp.loop.([\neg b]/2)$$
$$\equiv \qquad\qquad\qquad\qquad\qquad\qquad\qquad \text{unfold the loop}^{11}$$
$$\qquad wp.(\textbf{if } b \textbf{ then } b := \textsf{false} \ _{1/2}\oplus \textbf{abort fi}).([\neg b]/2)$$

$$\equiv \quad [b] * ((1/2)(1/2) + (1/2)(0)) \ + \ [\neg b] * [\neg b]/2$$
$$\equiv \quad [b]/4 + [\neg b]/2 \ ,$$

showing the conclusion of Thm. 7.3.3 to be false in this case: the precondition $[b]/4 + [\neg b]/2$ is not at least I, since when b holds for example the former is $1/4$ and the latter remains $1/2$.

7.4   Full proof of the loop rule Lem. 2.4.1

With our results above, we can now briefly summarise the steps needed to prove our loop rule of Sec. 2.4. (In this subsection we allow our expectations to exceed one.) The lemma read as follows:

---

[11] We are using that $b := \textsf{false}; loop$ is just $b := \textsf{false}$ and that $\textbf{abort}; loop = \textbf{abort}$.

Lemma 2.4.1 (repeated from p. 43) Let expectation T be the termination probability of *loop*, so that

$$T \quad := \quad wp.(\textbf{do } G \to prog \textbf{ od}).1 \; ,$$

and let I be a probabilistic invariant for it. We consider three cases:

1. If $I \equiv [Inv]$ for some standard Inv, define $preE := T * [Inv]$.
2. If $[Term] \Rrightarrow T$ for some standard $Term$, so that $Term$ contains only states where termination is almost certain, define $preE := I * [Term]$. (Thus if T is itself standard, we can again define $preE := I * T$.)
3. If $I \Rrightarrow T$, then define $preE := I$.

Then in each case we have that $preE$ is a sufficient pre-expectation for the loop to terminate while maintaining the invariant:

$$preE \quad \Rrightarrow \quad wp.loop.([\overline{G}] * I) \; .$$

We now revisit each of the three cases, in turn.

7.4.1   Case 1: standard invariant

Let I be *wlp*-invariant, and suppose $I \equiv [Inv]$ for some standard Inv. Then we have

$$
\begin{array}{lll}
& preE & \\
:= & T * [Inv] & \\
\equiv & T \,\&\, [Inv] & \text{$[Inv]$ standard, $T \Rrightarrow 1$} \\
\equiv & T \,\&\, I & \text{$[Inv] \equiv I$} \\
\Rrightarrow & wp.loop.([\overline{G}] * I) \; . & \text{Lem. 7.3.1}
\end{array}
$$

Note that in this case we require only weak invariance of I.

7.4.2   Case 2: standard sub-termination

Let I be *wp*-invariant, and suppose $[Term] \Rrightarrow T$ for some standard $Term$, and assume that I is not everywhere zero. Because (in this subsection) I is not necessarily one-bounded, we must scale it down before applying our lemma. From scaling we know that if I is invariant then so is $I/\max I$, *i.e.* that

$$
\begin{array}{lll}
& wp.body.(I/\max I) & \\
\dagger \equiv & (wp.body.I)/\max I & \text{note } wlp \text{ is not appropriate here} \\
\Leftarrow & ([G] * I)/\max I & I \text{ is invariant} \\
\equiv & [G] * (I/\max I) \; , &
\end{array}
$$

and so we proceed as follows:

$$preE$$
$$:= \quad I * [Term]$$
$$\equiv \qquad\qquad\qquad\qquad\qquad\qquad I \text{ bounded, though not necessarily by one}$$
$$\max I * (I/\max I) * [Term]$$
$$\equiv \qquad\qquad\qquad\qquad\qquad\qquad I/\max I \Rrightarrow 1; \text{ and } [Term] \text{ is standard}$$
$$\max I * ((I/\max I) \,\&\, [Term])$$
$$\Rrightarrow \quad \max I * ((I/\max I) \,\&\, T) \qquad\qquad\qquad \text{assumption; \& monotonic}$$
$$\Rrightarrow \qquad\qquad\qquad\qquad (I/\max I) \text{ is invariant, shown above; Lem. 7.3.1}$$
$$\max I * wp.loop.(\lceil\overline{G}\rceil * I/\max I)$$
$$\Rrightarrow \quad wp.loop.(\lceil\overline{G}\rceil * I) \,. \qquad\qquad\qquad\qquad\qquad \text{scaling for } loop$$

If on the other hand I is everywhere zero, then $preE \equiv 0$ also and the case holds trivially.

Note that in this case we required strong invariance of I.[12] That is because we have not investigated the definition and properties of wlp outside the $[0,1]$ domain used elsewhere in this chapter: if we had attempted to use wlp then at (†) the expressions $wlp.body.I$ would have occurred, not necessarily well defined for $I \nRrightarrow 1$. Thus, without knowing more about such an "extended" wlp, we must take the safe course here.

If however I is one-bounded to start with, then weak invariance suffices.

7.4.3   Case 3: invariant-below-termination

This is Thm. 7.3.3 directly, since $I \Rrightarrow T$ implies I is one-bounded. Recall however that we can also use the more general (2.12), which treats the case when I exceeds one.

Strong invariance is required in this case whether one is exceeded or not, because it is used in Thm. 7.3.3.

7.5   Probabilistic variant arguments

Here we give the full proof of our variant rule, first mentioned as Lem. 2.7.1:

Lemma 7.5.1 VARIANT RULE FOR LOOPS
Let V be an integer-valued expression in the program variables, defined at least over some subset Inv of the state space. Suppose further that for iteration $loop$

1. there are fixed integer constants L (low) and H (high) such that

$$G \wedge Inv \quad \Rrightarrow \quad L \leq V < H \,,$$

---

[12]By STRONG INVARIANCE we mean wp-invariance.

and

2. the standard expectation $[Inv]$ is a strong invariant for *loop* and

3. for some fixed probability $\varepsilon > 0$ and for all integers N we have

$$\varepsilon\,[G \,\wedge\, Inv \,\wedge\, (V = N)] \quad \Rightarrow \quad wp.body.[V < N] \ .$$

Then termination is certain from any state in which Inv holds: we have $[Inv] \Rrightarrow T$, where T is the termination condition of *loop*.

Proof   We show first that Assumption 2 allows Assumption 3 to be strengthened as follows:

$$
\begin{array}{ll}
& wp.body.[Inv \,\wedge\, (V < N)] \\
\equiv & wp.body.([Inv]\ \&\ [V < N]) & \text{standard predicates} \\
\Lleftarrow & wp.body.[Inv]\ \&\ wp.body.[V < N] & \text{sub-conjunctivity} \\
\Lleftarrow & [G \wedge Inv]\ \&\ \varepsilon\,[G \,\wedge\, Inv \,\wedge\, (V = N)] & \text{Assumptions 2,3} \\
\equiv & \varepsilon\,[G \,\wedge\, Inv \,\wedge\, (V = N)]\ . & [G],[Inv] \text{ standard}
\end{array}
$$

Thus we can add $(Inv\wedge)$ to the right-hand side of Assumption 3.

Now we carry out an induction, showing that for all $n \geq 0$ we have

$$\varepsilon^n\,[Inv \,\wedge\, (V < L+n)] \quad \Rightarrow \quad T\ , \tag{7.8}$$

i.e. that the probability of termination from Inv when $V < L+n$ is no less than ε^n.

For the base case we reason from Assumption 1 that

$$\varepsilon^0\,[Inv \,\wedge\, (V < L)] \quad \Rrightarrow \quad [\overline{G}] \quad \Rrightarrow \quad T\ .$$

For the step case we reason

$$\varepsilon^{n+1}\,[Inv \,\wedge\, (V < L+n+1)]$$

$$
\begin{array}{lll}
\equiv & \begin{array}{l} [G]\ *\ \varepsilon^{n+1}\,[Inv \,\wedge\, (V < L+n+1)] \\ \max\quad [\overline{G}]\ *\ \varepsilon^{n+1}\,[Inv \,\wedge\, (V < L+n+1)] \end{array} & [G] \text{ standard} \\
\\
\Rrightarrow & [G] * \varepsilon^{n+1}\,[Inv \,\wedge\, (V < L+n+1)]\ \ \max\ \ T & [\overline{G}] \Rrightarrow T \\
\\
\equiv & \begin{array}{l} \varepsilon^{n+1}\,[G \,\wedge\, Inv \,\wedge\, (V < L+n)] \\ \max\quad \varepsilon^{n+1}\,[G \,\wedge\, Inv \,\wedge\, (V = L+n)] \\ \max\quad T \end{array} & \text{arithmetic} \\
\\
\Rrightarrow & & \text{inductive hypothesis} \\
& \varepsilon * T \ \ \max\ \ \varepsilon^{n+1}\,[G \,\wedge\, Inv \,\wedge\, (V = L+n)]\ \ \max\ \ T \\
\\
\Rrightarrow & & \varepsilon * T \Rrightarrow T; \text{ Assumption 3 strengthened} \\
& \varepsilon^n * wp.body.[Inv \,\wedge\, (V < L+n)]\ \max\ T
\end{array}
$$

$$\equiv \quad wp.body.(\varepsilon^n\,[Inv\,\wedge\,(V < L + n)])\ \mathsf{max}\ T \hspace{3em} \text{scaling}$$
$$\Rightarrow \quad wp.body.T\ \mathsf{max}\ T \hspace{3em} \text{inductive hypothesis again}$$
$$\equiv \quad T\ . \hspace{3em} wp.body.T \Rightarrow T$$

With (7.8) now established for all $n \geq 0$, from Assumption 1 we can conclude

$$\varepsilon^{H-L}\,[Inv]$$
$$\equiv \quad \varepsilon^{H-L}\,[G \wedge Inv]\ \mathsf{max}\ \varepsilon^{H-L}\,[\overline{G} \wedge Inv]$$
$$\equiv \quad \varepsilon^{H-L}\,[G \wedge Inv]\ \mathsf{max}\ T \hspace{3em} [\overline{G}] \Rightarrow T$$
$$\Rightarrow \quad \varepsilon^{H-L}\,[Inv \wedge (L \leq V < H)]\ \mathsf{max}\ T \hspace{3em} \text{Assumption 1}$$
$$\Rightarrow \quad \varepsilon^{H-L}\,[Inv \wedge (V < L + (H - L))]\ \mathsf{max}\ T$$
$$\equiv \quad T\ \mathsf{max}\ T \hspace{3em} \text{(7.8) above}$$
$$\equiv \quad T\ .$$

That, with Assumption 2 and $\varepsilon^{H-L} \neq 0$, gives us $[Inv] \Rightarrow T$ directly from Lem. 2.6.1. $\hfill \Box$

7.6   Finitary completeness of variants

We now show that, if the state space is finite, the technique set out in Lemmas 2.7.1 and 7.5.1 is complete for proving almost-certain termination. We construct a variant explicitly: for any state it is the least n such that the iteration has nonzero probability of terminating in no more than n steps from that state. The following lemma establishes its existence and properties.

Lemma 7.6.1 COMPLETENESS OF VARIANT RULE   Consider our usual iteration *loop*. Note that $\lfloor T \rfloor \equiv \lfloor wp.loop.1 \rfloor$ is (the characteristic function of) that subset of the state space from which termination of *loop* is almost-certain.[13]

We have that there is an integer variant-function V of the state such that whenever $[G] * \lfloor T \rfloor = 1$ (termination is certain but has not yet occurred) the probability of V's strict decrease in the very next iteration is nonzero; more precisely, we construct V such that

$$[G] * \lfloor T \rfloor * [V = N] \quad \Rightarrow \quad \lceil wp.body.[V < N] \rceil$$

for all N.[14]

---

[13]We use $\lfloor \cdot \rfloor$ for the mathematical FLOOR function, so that $\lfloor x \rfloor$ is the greatest integer no more than the real x.

[14]To show that from any state satisfying the standard Q the program *prog* has nonzero probability of establishing the standard R, we simply prove $[Q] \Rightarrow \lceil wp.prog.[R] \rceil$. (Recall that we write $\lceil \cdot \rceil$ for the CEILING of x.)

Proof   Define the N-indexed expectations

$$T_0 \quad := \quad \lceil \overline{G} \rceil$$
$$T_{n+1} \quad := \quad \lceil \overline{G} \rceil \ \mathsf{max} \ wp.body.T_n \ ,$$

so that T_N is the probability of termination within N iterations. The variant is then given by

$$V \quad := \quad (\min n \mid T_n > 0) \ , \quad {}^{15} \tag{7.9}$$

which is well defined in states where T is not zero (and thus in particular where it is one); define V arbitrarily otherwise. Then $V = N$ in any state in $\lfloor T \rfloor$ means that, from that state, there is a nonzero probability of termination within N iterations.

We now show that whenever $[G] * \lfloor T \rfloor * [V = N]$ holds (*i.e.* is one) the probability of establishing $V < N$ on the very next iteration is nonzero. When $N = 0$ the result is trivial (antecedent false); for $N > 0$ we reason

	$\lceil wp.body.[V < N] \rceil$	
$\Leftarrow$	$\lceil wp.body.T_{N-1} \rceil$	$T_{N-1} \Rightarrow [V < N]$: see (†) below
$\Leftarrow$	$\lceil [G] * wp.body.T_{N-1} \rceil$	
$\equiv$	$\lceil [G] * T_N \rceil$	definition T_N
$\equiv$	$[G] * \lceil T_N \rceil$	G standard
$\Leftarrow$	$[G] * \lfloor T \rfloor * [V = N] \ ,$	$\lfloor T \rfloor * [V = N] \Rightarrow \lceil T_N \rceil$

establishing the desired inequality.   □

†   For the second step we reasoned as follows. The inequality is trivially true if T_{N-1} is zero. If it is nonzero then the least n such that T_n is nonzero must be $N-1$ or less — and that means, from (7.9), that V itself must be $N-1$ or less. Thus, overall, we have that if T_{N-1} is nonzero then $[V < N]$ must be one.

We now use the lemma to show that if we assume finiteness of the state space the expression V constructed above satisfies the conditions of Lem. 2.7.1, so establishing completeness.

Theorem 7.6.2 VARIANT RULE IS SOUND AND COMPLETE
The termination rule of Lemmas 2.7.1 and 7.5.1 is sound and complete for certain termination over a finite state space.
Proof   Soundness was established by Lem. 7.5.1 directly, even when the state space is infinite.

For completeness, note that if the state space is finite then any expression — and in particular the expression (7.9) constructed in Lem. 7.6.1 — is

---

[15]This comprehension gives a quantifier min, a bound variable n with implicit type N, and a predicate "$T_n > 0$" which constrains the values n can assume. The expression over which the min is taken is omitted, and so defaults to the bound variable itself. As usual, the parentheses give the scope explicitly.

trivially bounded above and below. Similarly the probability of its decrease is bounded away from zero (being a finite infimum of positive quantities); thus we choose ε for Lem. 2.7.1 to be the minimum, taken over the states in the finite set $\lfloor T \rfloor$, of $\lceil wp.body\,[V < N] \rceil$ with N set to the value of the variant in that state.

All that remains therefore is to show that $\lfloor T \rfloor$ is an invariant of *loop*, so that we can take $Inv := \lfloor T \rfloor$ in our lemma. For that we have

$$
\begin{aligned}
& [G] * \lfloor T \rfloor \\
\equiv \quad & [G] * \lfloor wp.loop.1 \rfloor \\
\equiv \quad & \lfloor [G] * wp.loop.1 \rfloor && G \text{ standard} \\
\Rightarrow \quad & \lfloor wp.body.(wp.loop.1) \rfloor && \text{definition } wp.loop \\
\Rightarrow \quad & wp.body.\lfloor wp.loop.1 \rfloor && \text{Fact B.3.6} \\
\equiv \quad & wp.body.\lfloor T \rfloor \,,
\end{aligned}
$$

as required.   □

7.7   Do-it-yourself semantics: Probabilistic loop-guards

The theorems and proofs so far in this chapter have completed our semantics: we now have all the logical and mathematical structures we will use for reasoning about probabilistic sequential programs.[16] In this final section we give a running example of language *extension,* to illustrate one of the nice things you can do with a semantics when you have it: it is then easy to modify a programming language while maintaining consistency.

In the terminology of systems design, we can say that having an underlying semantics helps to avoid the accidental introduction of "feature interactions," where extensions can combine with extant behaviours in surprising and unpleasant ways. In our case, a feature interaction would be a new language construct that looked fine "on its own," but which in a larger program resulted in code that could be given two incompatible interpretations. Motivating our example will be an earlier code fragment which we will now try to treat more concisely.

In Sec. 2.11 we discussed the geometric-distribution program; referring to Fig. 2.11.1 there, we recall its use of an extra Boolean variable b to organ-

---

[16]In Chap. 8, the last of this part, we have an exploration of the more exotic "angelic" transformers where (following *e.g.* Back and von Wright [BvW98]) we undertake a systematic classification of the transformer space to isolate its various operationally motivated subsets and determine the algebraic sublinearity-style characterisation of each. (See *e.g.* p. 240.)

In Part III we examine more general systems, via probabilistic (more generally, quantitative) temporal logic and its associated μ-calculus.

ise the "with-probability-1/2-each-time" termination of an n-incrementing loop. Common sense suggests we should be able to write that more directly as

$$
\begin{aligned}
&n:= 1; \\
&\textbf{do } 1/2 \to \\
&\quad n:= n+1 \\
&\textbf{od },
\end{aligned}
\qquad (7.10)
$$

without having to introduce the "technical" b.[17] That is, we simply say that the loop's chance of continuing is $1/2$ on each iteration: the guard *itself* is that probability.

This "do-it-yourself semantics" section will use Program (7.10) as an example to take us through the steps needed to introduce a new construct: motivation (syntax and informal meaning); consistency (its semantics expressed in the same framework as the existing language); support for reasoning (selecting and appealing to previous theorems and meta-theorems about other constructs to suggest what properties should be proved for the new one) and finally utility (examples using the extension, to see whether reasoning with it is really worth the effort).

7.7.1   Syntax and informal meaning

In general we extend the syntax of iterations so that we write an expression in the interval $[0,1]$, rather than a Boolean in $\{\text{true}, \text{false}\}$, for the loop guard.[18]

Definition 7.7.1 PROBABILISTICALLY GUARDED ITERATION: SYNTAX
A *probabilistically guarded iteration* is written

$$\textbf{do } p \to \ body \ \textbf{od },$$

where probability p is a real-valued expression over state variables such that $0 \leq p \leq 1$.[19]

Informally, an (invisible) choice $_p\oplus$ is carried out before each potential iteration: if the left branch is selected, then the loop is entered and the

---

[17]Program (7.10) initialises n to 1 in order to maintain n's meaning on exit, that it has counted the number of probabilistic choices that were made. In fact the initialisation $n:= 0$ in Fig. 2.11.1 is another artefact of the use of b: in that program, either we must set b probabilistically in two separate places (which is bad coding style, since the two identical statements should be written only once), or we must accept that entry to the loop is guaranteed on the first iteration.

We return to this program at Fig. 7.7.9.

[18]The semantics for such guards was implicit in Jones' work [Jon90, Chap. 7], but without demonic choice.

[19]For consistency with our upper-case convention for expressions we should write "P" (as for Boolean loop-guards G); but we continue with lower-case by analogy with $_p\oplus$.

body is executed, after which the process is repeated; otherwise the loop terminates immediately.

For compatibility with existing notation, we allow ourselves to assume implicit embedding brackets around a Boolean-valued loop guard: *i.e.* by "**do** $G \to \cdots$" we now mean "**do** $[G] \to \cdots$" when G is Boolean.   □

Access to $[0, 1]$-valued *expressions* (rather than only to constants) gives us considerable generality: depending on the value of p in the current state, entry to the loop can be "denied" (if p is zero), "possible" (if $0 < p < 1$) or "assured" (if p is one). For example, because $0 \leq n \leq N$ is invariant,[20] we can see that the iteration

$$n := N;$$
$$\textbf{do } n/N \to$$
$$n := n - 1$$
$$\textbf{od}$$

is certain to allow loop-entry initially, when n is N, and is certain to deny it if n ever reaches zero. In between, a "sliding scale" operates, depending on the value of n.

However our compatibility desideratum in Def. 7.7.1 — that "**do** $G \to \cdots$" (extant construct) and "**do** $[G] \to \cdots$" (new construct) should be the same — provides the first danger of a destructive feature interaction. Are we sure that they really *are* the same?

This kind of question can lead to endless (and fruitless) philosophical discussion — but only in the absence of a mathematical semantics. With the semantics, there is *no* discussion: either one can prove the equality of the transformers [21]

$$wp.(\textbf{do } G \to body \textbf{ od}) \qquad \text{and} \qquad wp.(\textbf{do } [G] \to body \textbf{ od}) \,,$$

for all Boolean expressions G and program fragments *body*, or one cannot. If they are proved equal, then we accept that Def. 7.7.1 is a reasonable starting point; if not, we reject it and look for a better definition.

For that proof, however, we need semantics.

---

[20] As we see later, that follows from $(0 \leq n \leq N) \land (n/N \neq 0) \Rrightarrow 0 \leq n{-}1 \leq N$.

[21] Naturally we must use the existing semantics (p. 21) on the left and the new semantics (to come) on the right, since those are what we are hoping will give equal transformers.

Note that it is not (quite) enough to appeal to the definitional equality (p. 19) of the (invisible) probabilistic choice $_0\oplus$, say, and the conditional " **if false else** ". Since the informal semantics above of "**do** $0 \to \cdots$" is self-referential ("...the process is repeated"), the argument would be circular. Fixed-points (Sec. 7.7.2 below) will cut the knot, and that is of course why they are used so much in program semantics.

7.7.2   Semantics in our existing model

Following the lead established by our treatment of conventional iteration, we define our new construct's transformer semantics as follows:

Definition 7.7.2 ITERATION SEMANTICS FOR PROBABILISTIC GUARDS
For iteration *loop* written **do** $p \rightarrow$ *body* **od** we define

$$wp.loop.R \quad := \quad (\mu Q \cdot wp.body.Q \ _p\oplus \ R) \quad ^{22}$$
$$\text{and} \quad wlp.loop.R \quad := \quad (\nu Q \cdot wlp.body.Q \ _p\oplus \ R) \, .$$

□

Note that we are appealing to our earlier experience, in that we define both *wp-* and *wlp* forms of the semantics right at the start: we expect to need the latter for construction of "invariant-style" proof rules.

As a confidence-building measure, we can now compare the new definition with our earlier ones by specialising p to a standard guard $[G]$, in which case we have (Def. 7.7.2 right-hand sides, and Sec. 1.5.2 conditional) the expectations

$$(\mu Q \cdot wp.body.Q \text{ if } G \text{ else } R) \tag{7.11}$$
$$\text{and} \quad (\nu Q \cdot wlp.body.Q \text{ if } G \text{ else } R)$$

for the *wp-* and *wlp* semantics respectively. Recalling Footnote 33 on p. 21 and Def. 7.2.1 shows that our new definitions do appropriately generalise the old and, incidentally, resolve the "feature-interaction" issue of the previous section: the self-referential "knot" is cut by the observation that if the functions (of Q) defined at (1.19) and (7.11) are equal, then so are their fixed points.[23]

---

[22]Recall from Footnote 45 on p. 63 that we can use probabilistic choice $_p\oplus$ in ordinary arithmetic; here, by lifting it to the functional level, we are using it between expectations.

[23]The link between (1.18) and the simpler formulation at (1.19), a detail that is usually glossed over, is made by the general fixed-point identity (given their existence) that

$$\mu.F.x \quad = \quad \mu.(f.x) \qquad \text{whenever } F.g.x = f.x.(g.x) \text{ for all } g, x, \tag{7.12}$$

where $F \in (X \leftarrow X) \rightarrow (X \leftarrow X)$, $f \in X \rightarrow (X \leftarrow X)$, $g \in X \leftarrow X$ and $x \in X$.
In our case X is the space $\mathbb{E}S$, and

$$F.g \quad \text{is} \quad wp.((body;g) \text{ if } G \text{ else skip})$$
$$\text{while} \quad f.x.y \quad \text{is} \quad wp.body.y \text{ if } G \text{ else } wp.\text{skip}.x \, .$$

Essentially (7.12) says mathematically that F ("the program that calls g") makes use of its transformer argument g ("the recursive call") only by applying it to x ("the overall postcondition"), and in no other way; in programming terms, that is saying that F makes its recursive call to g only when the overall postcondition x is what immediately follows the return from g.

Thus, as pointed out *e.g.* by Nelson [Nel89], identity (7.12) applies to iterations precisely because they are tail-recursive, and is what allowed Dijkstra to give a succinct *wp* formulation for them [Dij76] (and prevented him from treating full recursion).   ...

7.7.3   Invariants for partial correctness

A *theorem* "in" a logic is any specific formula that can be proved from its axioms and inference rules,[24] and we think of such formulae as "truths" about any model the logic describes. In our programming logic, for example, the statement

$$[\text{true}] \quad \Rrightarrow \quad wp.(\text{do } 1/2 \to \text{skip od}).[\text{true}]$$

is a theorem — that the loop shown is almost certain to terminate — and with our semantic definition Def. 7.7.2 we can prove that theorem by reasoning

$$wp.(\text{do } 1/2 \to \text{skip od}).[\text{true}]$$

$\equiv$ unfold using Def. 7.7.2

$$wp.(\text{skip}; \text{ do } 1/2 \to \text{skip od}).[\text{true}] \;\; {}_{1/2}\oplus \;\; [\text{true}]$$

$\equiv$ $1/2 * wp.(\text{do } 1/2 \to \text{skip od}).[\text{true}] \quad + \quad 1/2 \,,$ $wp.\text{skip}$

after which we apply arithmetic along the lines of our earlier example (2.6) of recursive termination.

The truth this expresses about our model is that a series of independent probability-1/2 choices cannot ignore either of its outcomes indefinitely (except with probability zero).

In contrast, *meta-theorems* are theorems "about" (rather than "in" or "of") a logic, *e.g.* "if a formula like *this* is a theorem of the logic, then so is a formula like *that*." They can increase the practical power of a logic considerably, but they do not extend its theory.[25] Our principal examples have been the meta-theorems about invariants and variants of iterations, and in this section we treat (an extension of) the standard meta-theorem "if it is a theorem that *Inv* is preserved by the loop body, then it is also a theorem that *Inv* is true on termination if it was true initially."

In the following, let *loop* be **do** $p \to body$ **od** as usual. With the same greatest-fixed-point technique we applied at Lem. 7.2.2, *i.e.* replacing Q by

---

...[23]As a result, using (7.12) we end up comparing the fixed points

                   $(\mu Q \bullet \; wp.body.Q \text{ if } G \text{ else } wp.\text{skip}.R)$ from (1.19),
     and $(\mu Q \bullet \; wp.body.Q \text{ if } G \text{ else } R)$ from (7.11),

which — being based on equal functions of Q — are themselves equal.

[24]We have not set out exactly what our formal language is, because it is not our purpose to be so precise in this presentation. The way in which that formalisation would be carried out however can be seen by comparing the (formal) Hoare logic [Hoa69] for standard programs (but in its total-correctness form) with the equivalent Dijkstra-style presentation [Dij76] we have adapted for our use here. Although the latter is less logically formal, it is no less precise.

[25]Literally, the THEORY of a logic is the set of its theorems.

I, we have immediately that if

$$I \quad \Rrightarrow \quad wlp.body.I \ _p\oplus \ R \ , \qquad\qquad (7.13)$$

for any one-bounded expectations I, R, then in fact

$$I \quad \Rrightarrow \quad wlp.loop.R \ . \quad [26]$$

In the interests of making connections with our earlier standard-guard rule, however, we use some simple arithmetic to "build-in" extra information that might help in practice: that in states where p is one we can ignore R; and that in states where termination occurs p cannot have been one. We achieve this by a three-way case-split, and the result is that for one-bounded I, R we have

Lemma 7.7.3 Let expectation I be a *wlp-invariant* of *loop* for post-expectation R, by which we mean it satisfies (7.13), or equivalently

$$
\begin{aligned}
[p=1] * I &\quad \Rrightarrow \quad wlp.body.I & (7.15) \\
[0 < p < 1] * I &\quad \Rrightarrow \quad wlp.body.I \ _p\oplus \ R & (7.16) \\
\text{and} \qquad [p=0] * I &\quad \Rrightarrow \quad R \ . & (7.17)
\end{aligned}
$$

$$\text{Then in fact} \qquad I \quad \Rrightarrow \quad wlp.loop.R \ . \qquad (7.18)$$

$\square$

The three-way formulation (7.15–7.17) more obviously generalises our earlier Lem. 7.2.2 than (7.13) appears to on its own (although they are equivalent, as we pointed out). For when p is some standard $[G]$, we can ignore the "extra" Clause (7.16) because its left-hand side is always zero (compare "antecedent false" in Boolean logic); then in (7.15) the term $[p=1]$ becomes just $[G]$ itself; and in (7.17) the term $[p=0]$ becomes $[\overline{G}]$.

On the other hand, in simple probabilistic cases where p is nowhere zero or one (*e.g.* it is a constant like $1/2$), the clauses (7.15) and (7.17) and the term $[0 < p < 1]$ in (7.16) can be ignored, leaving just (7.13).

---

[26] It's worth noting that the invariant technique is COMPLETE in the sense that for any iteration *loop* and post-expectation R there is an invariant I which can be used as above: that is, for any pre-expectation Q with

$$Q \quad \Rrightarrow \quad wlp.loop.R \ , \qquad\qquad (7.14)$$

there is an invariant I satisfying both (7.13) and the implication $Q \Rrightarrow I$. In fact it is *wlp.loop.R* itself, which satisfies (7.13) by definition (as a fixed point in Def. 7.7.2), and satisfies $Q \Rrightarrow I$ from (7.14) trivially.

That means that any valid assertion of the form (7.14) can in principle be proved using the invariant technique.

Recalling the standard-guard version raises another point, however: why must we say "is invariant *for post-expectation R*" in Lem. 7.7.3, when earlier we simply said "is invariant"? Usually, we appeal to transitivity of implication $\Rightarrow$ to eliminate R from the antecedent of the lemma (Clauses 7.16–7.17): we did that in Lem. 7.2.2, and it is done in the loop rule for standard programs.

That is, in Lem. 7.2.2, we prove $I \Rightarrow wlp.loop.([\overline{G}] * I)$, and post-expectation R is not mentioned at all. Instead we rely subsequently on transitivity and $[\overline{G}] * I \Rightarrow R$ to get us the rest of the way, to the overall conclusion $I \Rightarrow wlp.loop.R$.

A similar approach here would suggest choosing the post-expectation to be $[p \neq 1] * I$, which would eliminate Clause (7.17), and would change Clauses 7.15 and 7.16 to the single

$$[p \neq 0] * I \quad \Rightarrow \quad wlp.body.I \; {}_p\!\oplus\; I \,, \tag{7.19}$$

since the "$[p \neq 1] *$" can be dropped on the right-hand side of ${}_p\!\oplus$. By transitivity the modified rule would then be suitable for all R for which there was some I satisfying both Clause (7.19) and the implication $[p \neq 1] * I \Rightarrow R$. Why don't we do this?

The reason is that (7.19) turns out to be too restrictive: there are some combinations of $p, body, R$ which are valid but for which nevertheless no such I can be found.[27] Thus we must leave Lem. 7.7.3 as it is.

---

[27]Here is an example: with what probability does the iteration *toggle*, defined

$$\textbf{do } 1/2 \to b := \overline{b} \textbf{ od} \,,$$

establish Boolean b finally? An informal calculation gives $1/2 + 1/2^3 + \cdots = 2/3$ if b holds initially, and $1/2^2 + 1/2^4 + \cdots = 1/3$ if it does not, suggesting that

$$2\,[b]\,/3 + [\overline{b}]\,/3 \quad \equiv \quad \frac{1 + [b]}{3} \quad \Rightarrow \quad wlp.toggle.[b] \tag{7.20}$$

— and although that is easily verified formally by an appeal to (7.13), it cannot be proved by our modified rule. First we would need an I satisfying (7.19), *viz.*

$$I \quad \Rightarrow \quad wlp.(b := \overline{b}).I \; {}_{1/2}\!\oplus\; I \,, \tag{7.21}$$

to give us $I \Rightarrow wlp.loop.I$. Then we would need to finish off by establishing (7.20) with transitivity — *i.e.* outside the loop rule — simply by noting that

$$\begin{array}{lll} (1 + [b])/3 & \Rightarrow & I \quad \text{(weak-enough pre-expectation)} \\ \text{and} \quad I & \Rightarrow & [b] \,. \quad \text{(strong-enough post-expectation)} \end{array} \tag{7.22}$$

But in fact (7.22) alone tells us there can be no such I, because if there were we would have by transitivity that

$$\frac{1 + [b]}{3} \quad \Rightarrow \quad [b] \,,$$

which is not true. (Consider the case $\overline{b}$.) Note that (7.21) therefore plays no role in the impossibility argument.

$\cdots$

7.7.4   *Partial- and total correctness*

Our second meta-theorem parallels Sec. 7.3, where we put together a partial correctness result (as in the previous section) with a proof of termination. We have

Lemma 7.7.4 Let invariant I satisfy the conditions 7.15–7.17 of Lem. 7.7.3, or equivalently (7.13). Then

$$I \,\&\, T \quad \Rrightarrow \quad wp.loop.R \;,$$

where T is the termination probability $wp.loop.[\mathsf{true}]$ of the loop.

   Proof   As for Lem. 7.3.1.   □

A practical difficulty occurs, just as before, with the case where both I and T are probabilistic — the pre-expectation $I \,\&\, T$ can be too low to be useful. To fix it, we take the same approach; the proof, however, must be slightly adapted.

Theorem 7.7.5 Invariant-implies-termination loop rule
                          for probabilistic guards

If *loop* has a deterministic *body*, and one-bounded expectation I is a probabilistic *wp*-invariant for R, *i.e.* it satisfies

$$I \quad \Rrightarrow \quad wp.body.I \;_p\oplus\; R \;, \tag{7.23}$$

then $I \Rrightarrow wp.loop.R$ provided $I \Rrightarrow T$.

   Proof   We show first that if I satisfies (7.23) then I' satisfies (7.13), where as before (Thm. 7.3.3) we define

$$I' \quad := \quad I + 1 - T \;.$$

We reason

	$wlp.body.I' \;_p\oplus\; R$	*rhs* of (7.13)
$\equiv$	$wlp.body.(I + 1 - T) \;_p\oplus\; R$	definition I'
$\equiv$		Fact B.3.3 twice; *body* deterministic
	$(wp.body.I + wlp.body.1 - wp.body.T) \;_p\oplus\; R$	
$\equiv$	$(wp.body.I + 1 - wp.body.T) \;_p\oplus\; R$	Fact B.3.4
$\equiv$	$(wp.body.I \;_p\oplus\; R) \;+\; p \;-\; p * wp.body.T$	definition $_p\oplus$
$\Lleftarrow$	$I \;+\; p \;-\; p * wp.body.T$	(7.23)
$\equiv$	$I \;+\; p \;-\; (T - \overline{p})$	$T \equiv wp.body.T \;_p\oplus\; 1$

---

[27] Thus the suggested modification, although sound, would make the rule *incomplete* in the sense of Footnote 26 above: there would be some theorems of the form (7.14) that we could not use our modified rule to prove — and we must therefore keep R in the antecedent.

$$\equiv \qquad I + 1 - T$$
$$\equiv \qquad I' . \qquad\qquad\qquad\qquad\qquad \text{definition } I'$$

From Lem. 7.7.4 we then continue

$$I \;\equiv\; I' \,\&\, T \;\Rrightarrow\; wp.loop.R ,$$

as required. □

The reason the above proof of Thm. 7.7.5 is slightly more direct than for our earlier Thm. 7.3.3 is just that we were forced to retain the overall post-expectation R in its formulation — perhaps a compensating (but once-only) advantage.

Upgrading the special cases Lem. 2.4.1 of invariant-plus-variant reasoning to our new construct is also straightforward; we re-state the result in the notation of this section. Note that for this lemma we do not assume that I, R are one-bounded.

Lemma 7.7.6 TOTAL CORRECTNESS FOR PROBABILISTIC LOOPS
Let expectation T be the termination probability of *loop*, so that

$$T \;:=\; wp.(\mathbf{do}\ p \to prog\ \mathbf{od}).[\text{true}] ,$$

and let expectation I be *wp*-invariant for it with respect to post-expectation R. We consider three cases:

1. If $[Inv] \equiv I$ for some standard *Inv*, define $Q := T * [Inv]$. (In fact weak invariance suffices for this case.)

2. If $[Term] \Rrightarrow T$ for some standard *Term*, define $Q := I * [Term]$.

3. If $\varepsilon * I \Rrightarrow T$ for some fixed $\varepsilon > 0$, define $Q := I$.

Then in each case we have

$$Q \;\Rrightarrow\; wp.loop.R .$$

Proof   As for Sec. 7.4. □

7.7.5 Extended variants

For the almost-certain termination argument for the new construct, we will appeal for inspiration both to the variant technique of Sections 2.7 and 7.5 and to the Zero-One Law of Sec. 2.6 — for neither will do on its own: the Zero-One Law does not apply to standard iterations (except trivially); and for the terminating loop

$$\mathbf{do}\ 1/2 \to \mathbf{skip}\ \mathbf{od} \qquad\qquad (7.24)$$

there is obviously no variant in the ordinary sense (since there are no variables).

We proceed as follows. Informally, suppose we have a constant $\delta < 1$ such that whenever $p > \delta$ some variant V is strictly decreased with probability at least some fixed $\varepsilon > 0$. Then the variant's at-least-ε decrease in the $\delta < p \leq 1$ region (of the state space) forces the iteration to visit the complementary $0 \leq p \leq \delta$ repeatedly; yet each time it does so, its probability of immediate termination is $\overline{p}$, which is at least $\overline{\delta}$. Thus the probability of *eventual* termination is always at least $\overline{\delta}$, bounded away from zero: and so, from the Zero-One Law, termination is almost-certain.

This approach specialises to our earlier rule Lem. 2.7.1 (proved as Lem. 7.5.1) by taking δ to be zero; and it handles **do** $1/2 \rightarrow$ **skip od** by taking $\delta := 1/2$, say, since then the antecedent $1/2 > 1/2$ for decrease of the (in fact non-existent) variant is false. We now give a proof of soundness.

Lemma 7.7.7 VARIANT RULE FOR PROBABILISTIC-GUARD LOOPS   Let V be an integer-valued expression in the program variables, defined at least over some subset Inv of the state space. Suppose further that for iteration *loop* defined **do** $p \rightarrow body$ **od** we have that

1. there are fixed integer constants L, H such that

$$p \neq 0 \wedge Inv \quad \Rightarrow \quad L \leq V < H \;, \qquad (7.25)$$

    that

2. the predicate Inv satisfies

$$p \neq 0 \wedge Inv \quad \Rightarrow \quad wp.body.Inv \qquad (7.26)$$

    and that

3. for some fixed $\varepsilon > 0$ and $\delta < 1$, and for all integers N, we have

$$\varepsilon \, [p > \delta \wedge Inv \wedge (V = N)] \quad \Rightarrow \quad wp.body.[V < N] \;. \quad {}^{28} \;(7.27)$$

Then termination of *loop* is almost-certain from any state in which the standard Inv holds.

Proof   Introducing a fresh Boolean variable b, we rewrite the loop in the conventional form

$$
\begin{aligned}
&b := \mathsf{true} \,{}_p\!\oplus \mathsf{false}; \\
&\mathbf{do}\ b \rightarrow \\
&\qquad body; \\
&\qquad b := \mathsf{true} \,{}_p\!\oplus \mathsf{false} \\
&\mathbf{od}\ ,
\end{aligned}
\qquad (7.28)
$$

and then we appeal to Lem. 7.5.1.

---

[28] Note that the variant is not required to "stay put" when $p \not> \delta$ — it can increase then, and it doesn't matter. Refer Footnote 44 on p. 214 below.

Let V be the variant we are given; from it we form a two-layered variant $V' := ([b], V)$ by combining it with b. We argue as follows that V' satisfies the conditions of Lem. 7.5.1 for Program (7.28).

- Condition 1 is satisfied for both layers, for $[b]$ by definition (it has only two values), and for V by assumption (7.25) because if b is true then we must have $p \neq 0$.

- Condition 2 is satisfied because if $b \wedge Inv$ holds, then we also have $p \neq 0 \wedge Inv$; thus we have $wp.(body;\ b:=\ \mathsf{true}\ _p\oplus \mathsf{false}).Inv$ because we have $wp.body.Inv$ from (7.26), and because b is fresh so that Inv is unaffected by $b:= \mathsf{true}\ _p\oplus \mathsf{false}$.

- Condition 3 holds for $\varepsilon' := \varepsilon$ min $\overline{\delta}$. There are two cases, based on the value of p at the beginning of the loop body:

  - If $p > \delta$ then by assumption (7.27) we know that $body$ decreases the inner variant V with probability at least ε. And because b holds, the subsequent assignment to it cannot increase the outer variant $[b]$, since $[\mathsf{true}]$ is its maximum value — thus in this case V' decreases with probability at least ε.
  - If $p \leq \delta$ then, because b holds and by assumption, the explicit assignment to b decreases $[b]$ with probability $\overline{p}$, which is at least $\overline{\delta}$. Since $[b]$ is the outer layer of V', it does not matter what happens to V in this case — and so V' is decreased with probability at least $\overline{\delta}$.

The conditions of Lem. 7.5.1 are therefore satisfied, and we have almost-certain termination of (7.28), and equivalently of $loop$, as required.[29]   □

7.7.6   Completeness of the variant technique for finite state spaces

The remaining question is whether our more elaborate rule Lem. 7.7.7 is still as powerful as Lem. 7.5.1 was shown to be in Sec. 7.6 for finite state spaces. We prove here that it is.

Lemma 7.7.8 COMPLETENESS OF VARIANT RULE FOR
                  PROBABILISTIC GUARDS OVER FINITE STATE SPACES

As before, let $loop$ be defined **do** $p \to body$ **od**, and let $Term$ be the states from which its termination is almost certain.

---

[29]Equivalences and inequalities like these between programs can be formalised along the lines of the argument at Footnote 32 on p. 102 which related the two versions of the random walker.

Then, provided the state space is finite, there is a standard invariant *Inv* with *Term* $\Rightarrow$ *Inv*, and an integer variant-function V, satisfying the conditions of Lem. 7.7.7.

Proof   Set δ less than one so that $\delta < p$ iff $p = 1$, which is possible because the state space is finite. (If p is not everywhere one, take *e.g.* the largest non-one value of p; otherwise choose δ arbitrarily.)

Define $Term := \lfloor wp.loop.[\text{true}] \rfloor$, those states from which termination is almost certain.

We consider the related iteration *loop'* defined **do** $(p = 1) \to body$ **od**, and its *Term'*: because the guard p of our original *loop* is no less probable than the guard $p = 1$ of *loop'* — that is, because $p \Leftarrow [p = 1]$ — we have $Term \Rightarrow Term'$. In other words, if *loop* terminates almost-certainly from a given state, then so does *loop'*.[29 again, 30]

From Lem. 7.6.1 for *loop'*, there is a variant V satisfying the conditions of Lem. 7.5.1 for some probability $\varepsilon > 0$, guard $p = 1$ and invariant *Term'*. We claim that the same V and ε, together with the δ chosen above, satisfy the conditions of Lem. 7.7.7 for *loop*, with guard p and invariant *Term*:

- Condition 1 is satisfied trivially, because the state space is finite.

- Condition 2 is satisfied because

$$
\begin{array}{lll}
 & [p \neq 0] * Term & \\
\equiv & [p \neq 0] * \lfloor wp.loop.1 \rfloor & \text{definition } Term \\
\equiv & [p \neq 0] * \lfloor wp.body.(wp.loop.1)\,{}_p\!\oplus 1 \rfloor & \text{definition } wp.loop \\
\Rightarrow & \lfloor wp.body.(wp.loop.1) \rfloor & \text{arithmetic} \\
\Rightarrow & wp.body.\lfloor wp.loop.1 \rfloor & \text{Fact B.3.6} \\
\equiv & wp.body.Term\, . & \text{definition } Term
\end{array}
$$

- We compare the antecedents of Condition 3 in Lem. 7.7.7 and in our appeal to Lem. 7.5.1. We have

$$
\begin{array}{lll}
 & \text{antecedent of Condition 3 in Lem. 7.7.7} & \\
\text{that is} & \varepsilon\,[p > \delta \,\wedge\, Term \,\wedge\, (V = N)] & \\
\equiv & \varepsilon\,[p = 1 \,\wedge\, Term \,\wedge\, (V = N)] & \text{choice of } \delta \\
\Rightarrow & \varepsilon\,[p = 1 \,\wedge\, Term' \,\wedge\, (V = N)]\, , & Term \Rightarrow Term'
\end{array}
$$

which is the antecedent of Condition 3 in Lem. 7.5.1 with invariant *Term'*. Thus we have the consequent $wp.body.[V < N]$ as required.

That shows the existence of $Inv, V, \varepsilon, \delta$ under the conditions given. $\square$

---

[30]Of course, with its early termination the iteration *loop'* might produce "incorrect" final states that *loop* would not — but, when discussing termination only, we do not care.

7.7.7   Utility of the extension

Utility of expression is by now established and is, after all, why we considered the new construct in the first place; but that alone is nowhere near enough.[31] By this section's title we mean utility for *reasoning*.

We reconsider the geometric-distribution program of Fig. 2.11.1, revised here as Fig. 7.7.9; and we see from the figure that the simple calculations about its behaviour are no more difficult than before. More interesting however is the following.

With the notation $_p\oplus$ we have been taking for granted some probabilistic-choice mechanism parametrised by p; in practice however it might be that we have access only to "probabilistic bits," *i.e.* coin-flips $_{1/2}\oplus$. An obvious approach to building $_p\oplus$ from $_{1/2}\oplus$ — but at first sight an impractical one — is to construct probabilistically a real number $0 \leq x \leq 1$ as a binary expansion $0.x_1x_2\cdots$ where each x_n is chosen with $_{1/2}\oplus$; once that is done, we implement the probabilistic $_p\oplus$ as the conditional " **if** $x \leq p$ **else** ".

That can be made practical by observing that we don't need to construct all of x — almost certainly some prefix of it will be sufficient to determine whether $x \leq p$. With that in mind, we note the connection with the geometric-distribution program — the sequence of $_{1/2}\oplus$ choices — and we now exploit it.[32]

We have seen that Program (7.29) produces a final distribution $n \mapsto 1/2^n$ for $n \geq 1$; and it is a property of binary expansions $0.p_1p_2\cdots$ of p in $[0,1]$ that

$$p = \sum_{n\geq 1} p_n/2^n . \quad {}^{33}$$

But that summation is just the expected value of p_n after running Program (7.29) — that is, we have that

> the probability p is the expected value of the nth bit p_n of its own binary expansion, considered as a function of n, in the final state n produced by Program (7.29).

---

[31]Another widely known example of this issue — that "can we express ourselves easily?" is not enough — is the "demonically non-deterministic expressions," investigated by many researchers (including us). They are suggested here by our habit of writing *e.g.* "$x := 1 \sqcap 2$" as an abbreviation for "$x := 1 \sqcap x := 2$" (1.17). Giving "$1 \sqcap 2$" an independent meaning might lead to many advantages of notation.

The semantic issues (the equivalent of our Sections 7.7.2–7.7.6) turn out to be formidably difficult, however [Mor97]. Had they not, we might have considered "probabilistic expressions" as independent entities — and the iteration "**do** $1/2 \rightarrow \cdots$" would have become the special case of "**do** $G \rightarrow \cdots$" in which G was "true $_{1/2}\oplus$ false".

[32]This example was suggested by Joe Hurd's analysis [Hur02].

[33]For this to hold even when $p = 1$ we must assume one is represented as $0.11\cdots$, but as we are arguing intuitively here it hardly matters. In our formal reasoning below we use simple arithmetic, and do not appeal to binary expansions.

$$
\begin{array}{ll}
\textit{Generate} & n\!:=\ 1; \\
\textit{geometric} & \textbf{do } 1/2 \to \\
\textit{distribution.} & \qquad n\!:=\ n+1 \\
& \textbf{od}
\end{array}
\qquad (7.29)
$$

Termination with probability one is an immediate consequence of Lem. 7.7.7 with $\delta := 1/2$.

To establish the final distribution, we use post-expectation $[n = N]$. The probability of establishing $n = N$ is (we guess) $[n \leq N]/2^{N-n+1}$; and it is easily verified that (7.23) is satisfied, *viz.*

$$
wp.(n\!:=\ n+1).([n \leq N]/2^{N-n+1}) \ _{1/2}\!\oplus \ \ [n = N]
$$
$$
\equiv \quad [n+1 \leq N]/2^{N-n} \ _{1/2}\!\oplus \ \ [n = N]
$$

$$
\equiv \qquad\qquad\qquad\qquad\qquad \text{if } [n = N] \neq 0 \text{ then } N - n + 1 = 1
$$
$$
\qquad [n < N]/2^{N-n+1} \ + \ \ [n = N]/2^{N-n+1}
$$

$$
\equiv \qquad ([n < N] + [n = N])/2^{N-n+1}
$$
$$
\equiv \qquad [n \leq N]/2^{N-n+1} \ .
$$

The final calculation $wp.(n\!:=\ 1).([n \leq N]/2^{N-n+1})$ gives us probability $[1 \leq N]/2^N$ of achieving $n = N$.

For the expected number of steps to termination we use instead the post-expectation n, and we guess as before that its expected final value is $n + K$ for some constant K. That suggests invariant $n + K$ itself, and to determine K we calculate:

$$
wp.(n\!:=\ n+1).(n + K) \ _{1/2}\!\oplus \ n
$$
$$
\equiv \quad n+1+K \ _{1/2}\!\oplus \ n
$$
$$
\equiv \quad n \ + \ (1+K)/2
$$
$$
\equiv \quad n + K \ , \qquad\qquad\qquad\qquad\qquad\qquad\qquad \text{set } K := 1
$$

which gives an expected number of $_{1/2}\!\oplus$-evaluations of $wp.(n\!:=\ 1).(n+K) \equiv 2$.

Note that in both cases we have achieved exact invariance (2.20), which justifies our saying "the expected value *is*" rather than "is at least."

Figure 7.7.9. A GEOMETRIC DISTRIBUTION

But recall that p_n is 0/1-valued — *i.e.* it is standard — so that we can then say

> equivalently, Program (7.29) sets the index n so that $p_n=1$ with probability p.

That is the key observation — and to put it into practice, we perform a data refinement (informally), fixing p and replacing n by a real number x determined by the coupling invariant $x = 0.p_np_{n+1}\cdots$; *i.e.* variable x is the value of the suffix starting from the nth position of p's binary expansion. That gives the program (working from (7.29) informally)

$$
\begin{array}{rl}
\text{initial } n=1 \text{ means "all of } p\text{"} \rightarrow & x := p; \\
& \textbf{do } 1/2 \rightarrow \\
n := n+1 \text{ means "shift left one bit"} \rightarrow & \quad x := 2x - \lceil x \geq 1/2 \rceil \\
& \textbf{od} \\
\text{postcondition } p_n=1 \rightarrow & \{x \geq 1/2\}\ ,
\end{array}
$$

$$(7.30)$$

where we have written the assertion $\{x \geq 1/2\}$ as a comment to remind us of the iteration's "target" postcondition. [34]

Useful as the intuitive arguments above are, luckily we need not rely on them: the formal proof is now very straightforward. Our hope is that the expected value of $[x \geq 1/2]$ is just the current value of x; we choose x itself as our invariant, therefore, and we check (7.23) via

$$
\left.
\begin{array}{rl}
& wp.(x := 2x - \lceil x \geq 1/2 \rceil).x \quad {}_{1/2}\oplus \quad \lceil x \geq 1/2 \rceil \\
\equiv & 2x - \lceil x \geq 1/2 \rceil \quad {}_{1/2}\oplus \quad \lceil x \geq 1/2 \rceil \\
\equiv & (2x - \lceil x \geq 1/2 \rceil + \lceil x \geq 1/2 \rceil)/2 \\
\equiv & x\ .
\end{array}
\right\}
$$

$$(7.31)$$

Applying the initialisation $wp.(x := p)$ to that invariant x gives us probability just p of establishing $x \geq 1/2$, as required, so that $_p\oplus$ can indeed be implemented as in Fig. 7.7.10 (where we have adjusted the body slightly so that it uses more conventional syntax).[35]

---

[34]More correctly we would write $\{\,[x \geq 1/2]\,\}$; but to reduce clutter we omit any embedding brackets $[\cdots]$ immediately enclosed by assertion brackets $\{\cdots\}$.

[35]Note that although we appear to be operating over the reals, we do not need to consider distributions over an uncountable state space: since p should be a *computable* real number (and hence x must be as well), there are only countably many reals to consider. $\cdots$

$$
this \ _p\oplus \ that \qquad =^{37} \qquad
\begin{aligned}
&x:= \ p; \\
&\textbf{do } 1/2 \to \\
&\qquad x:= \ 2x; \\
&\qquad \textbf{if } x \geq 1 \textbf{ then } x:= \ x-1 \textbf{ fi} \\
&\textbf{od}; \\
\\
&this \quad \textbf{if } x \geq 1/2 \textbf{ else} \quad that \ .
\end{aligned}
$$

The left-hand probabilistic choice $_p\oplus$ is implemented on the right as a sequence of coin flips followed by a final Boolean test.

The only formal reasoning needed to check that the loop contributes correctly to this behaviour is the four-line calculation at (7.31).

The expected number of flips is just two, and is independent of p.

Figure 7.7.10. Coin-flipping implementation of $_p\oplus$

A final issue is that this program is not guaranteed absolutely to terminate.[36] In fact it is not possible to build an absolutely terminating conversion of $_{1/2}\oplus$ into $_p\oplus$ unless p is a "dyadic" rational having denominator some power of two. However, as we saw in Fig. 7.7.9, for any probability p the *expected* number of probabilistic bits required to implement $_p\oplus$ — *i.e.* the expected executions of $_{1/2}\oplus$ — is just two, whether p is dyadic or not.

7.7.8 Invariant-finding heuristics

Our final example of the new construct illustrates the case where the probabilistic loop guard can depend on the state. It also brings together in one place a number of the reasoning techniques we have encountered in

---

[35]Hurd also points out that in practice p might be limited to rationals only, so that it can be represented in the program as a pair of integers (N, D). A corresponding coupling invariant that represents x as (n, D) then allows the data refinement that gives the alternative implementation

$$
this \ _{\frac{N}{D}}\oplus \ that \quad = \quad
\begin{aligned}
&n:= \ N; \\
&\textbf{do } 1/2 \to \\
&\qquad n:= \ 2n; \\
&\qquad \textbf{if } n \geq D \textbf{ then } n:= \ n-D \textbf{ fi} \\
&\textbf{od}; \\
\\
&this \quad \textbf{if } 2n \geq D \textbf{ else} \quad that \ ,
\end{aligned}
$$

and a similarly short proof.

[36]This is also discussed by Hurd [Hur02].

[37]We assume for the equality (Fig. 7.7.10) that x is a local variable on the right. See Footnote 6 on p. 269 for an example of this program in action.

There are two cowboys A and B who fight a duel by taking alternate shots at each other: on each shot Cowboy A hits B with probability a; and similarly Cowboy B hits A with probability b. If A goes first, what is his survival probability?

We describe the situation with a program

$$p := a; \quad \textbf{do}\ \overline{p} \rightarrow p := \tilde{p}\ \textbf{od}\ \ \{p = a\}\ , \tag{7.32}$$

in which for convenience we introduce a "conjugate" operator with the property that $\tilde{a} = b$ and $\tilde{b} = a$. Again we write the postcondition explicitly, to remind us of our goal. (And we omit the embedding brackets.)

Cowboy A begins; with probability a the loop is *not* entered, because he has shot B and wins. If the loop is entered — he has missed, with probability $\overline{a}$ — then p becomes b and it is Cowboy B's turn.
The iterations continue until one cowboy shoots the other; if A wins then p will equal a on termination.

Figure 7.7.11. THE DUELLING COWBOYS

---

the two chapters about loops (this one, and the earlier Chap. 2), and it gives a simple context in which to revisit the issue of how invariants can be found.[38]
The program is given in Fig. 7.7.11.[39]

- *Reason within a standard invariant*

Although variable p is obviously real-valued, we notice immediately that in fact it can take only two values because

$$p \in \{a, b\} \tag{7.33}$$

is a standard invariant of the loop, established by the initialisation $p := a$ and trivially preserved by the body $p := \tilde{p}$. Thus by our earlier Lemmas 1.7.1 and 2.10.2 we can assume the truth of (7.33) wherever we need it below.

- *Guess a probabilistic invariant*

An operationally motivated technique for "capturing" a probabilistic invariant in quantitative logic is to estimate, even informally, a set of states within which the loop is guaranteed to remain — a standard invariant as above — whether or not it is sufficiently small to establish the postcondition; the smaller that estimated set is, however, the more useful it will be.

---

[38]We discussed it earlier, in Sec. 2.2, for Boolean-guarded loops.
[39]It is based on a well-known puzzle, and was suggested to us by T.S. Hoang.

Then for each state s_i, say, in that set separately — *and ignoring whether or not the state can actually be reached by execution of the loop from the given initialisation* [40] — we ask "what would be the greatest guaranteed expected value E_i of the post-expectation if the loop were started from this state?"

If the post-expectation is standard, then the question is correspondingly rephrased "what would be the greatest guaranteed probability E_i of establishing the postcondition if the loop were started from this state?"

In either case, a possibility for the invariant is then $(\sum_i E_i * [s = s_i])$; and checking it formally will of course show whether it is correct.

In this example, given (7.33), there are only two states — where p is a or it is b — and so the guessed invariant would be of the form

$$w_a\,[p = a]\ +\ w_b\,[p = b] \tag{7.34}$$

for two constants w_a, w_b ("win a" and "win b") that we don't yet know.

- *Check the invariant*

Since we don't know the constants, our check will be schematic — an excellent way of discovering the values by calculation. Referring to (7.23), we will discover values for w_a, w_b that make it true. We have

$$\qquad\qquad\qquad\qquad\qquad\qquad\text{(7.23) right-hand side, for this case}$$

$$wp.(p := \tilde{p}).(w_a\,[p = a]\ +\ w_b\,[p = b])\ \ _{\overline{p}}\oplus\ \ [p = a]$$

$$\equiv\quad (w_a\,[\tilde{p} = a]\ +\ w_b\,[\tilde{p} = b])\ _{\overline{p}}\oplus\ [p = a] \qquad\qquad \text{assignment}$$

$$\equiv\qquad\qquad\qquad\qquad\qquad\qquad \text{reason within invariant (7.33)}\,^{[41]}$$

$$\qquad (w_a\,[p = b]\ +\ w_b\,[p = a])\ _{\overline{p}}\oplus\ [p = a]$$

$$\equiv\quad (p + \overline{p} * w_b)\,[p = a]\ +\ (\overline{p} * w_a)\,[p = b] \qquad\qquad \text{arithmetic}$$

$$\equiv\quad (a + \overline{a} * w_b)\,[p = a]\ +\ (\overline{b} * w_a)\,[p = b] \qquad\qquad \text{arithmetic}$$

$$\equiv\quad w_a\,[p = a]\ +\ w_b\,[p = b] \qquad\quad \text{the required left-hand side, provided}\ldots$$

$\ldots$ we set w_a, w_b so that they satisfy the equations

$$\begin{aligned} w_a &=\ a + \overline{a} * w_b \\ w_b &=\ \overline{b} * w_a\,. \end{aligned} \tag{7.35}$$

If for example Cowboy A hits two-thirds of the time, but Cowboy B only half the time, we get $w_a = 4/5$ and $w_b = 2/5$ as solutions of (7.35).

---

[40] Recall Footnote 20 on p. 48.

[41] Even without (7.33) we would have at least $\Leftarrow$ for this step — but we want an exact invariant if we can get one.

- *Is the answer sane?*

In simple cases like this one, direct informal calculation in the style of elementary probability theory should confirm our answer. We would reason

> The chance w_a that A is the survivor is given by
>
> $$w_a \;=\; a + \overline{a} * \overline{b} * w_a \,,$$
>
> since either he wins immediately (left operand) or both he and Cowboy B miss, in which case they start again.

This is consistent with the two equations (7.35). The reasoning above that led to those equations — some five elementary steps in the program logic — is of course checking as well that the program (7.32) is an accurate description of the informally stated problem.

In both cases however, formal and informal, we have so far ignored termination.[42]

---

[42]We have also been wondering all along what would happen if a and b were equal — since in that case our "clever" encoding of the postcondition would not be able to distinguish between the two winners.

We chose not to deal with it pre-emptively, however, because it seemed more interesting to wait until the formal reasoning threw the issue up as a necessity: thus we see now that $a \neq b$ is needed to argue that we can treat the $p = a$ and $p = b$ cases separately, in the last step of our reasoning, to extract the two constraints on w_a, w_b. When $a = b$, all we get is

$$a + \overline{a}w_b + \overline{b}w_a \;=\; w_a + w_b \,, \tag{7.36}$$

whose solutions are not unique.[43]

Rather than complicate the program in the obvious way (with an extra variable, say), instead we imagine that the two values a, b are reals "of distinct colours," and so can never be equal. The remaining arithmetic simply ignores the colour.

Formally we'd make the type of p a disjoint union $[0, 1] + [0, 1]$, with a from one side and b from the other.

[43]At the risk of overdoing this example, we remark that the prospect of non-unique solutions for an exact invariant could be worrying: from (2.21) we recall that we should have "the exact expected value of any expression $postE$ that agrees with I everywhere on $\overline{G}$" — and there cannot be more than one "exact expected value."

With a closer look at (7.36) when $a = b$, we see however

$$a + \overline{a}w_b + \overline{a}w_a \;=\; w_a + w_b \,,$$

whence $w_a + w_b = 1$ uniquely for the sum, provided a (and thus b) is nonzero. The guessed invariant (7.34) is then

$$w_a\,[p = a] \;+\; w_b\,[p = b] \;\;\equiv\;\; (w_a + w_b)\,[p = a = b] \;\;\equiv\;\; [p = a = b] \,,$$

and so all is well: the overall precondition $wp.(p := a).[p = a = b]$ is just one.

If $a = b = 0$ then termination is the main issue anyway, as we see in a moment.

- *Termination*

In most cases the termination argument is trivial: when $0 < a, b$ the Zero-One Law applies, because loop-exit occurs with probability at least $a \min b$ on each iteration. Formally we would apply Lem. 7.7.7 with $\delta := \overline{a} \max \overline{b}$.

In the degenerate case where $a = b = 0$, the cowboys are stuck in the loop forever and we cannot prove termination.

That leaves the case where only one of a, b is nonzero, which at first glance seems slightly problematic — the loop has a chance of termination only on "every other" iteration.

However we have proved our variant rule Lem. 7.7.7 to be complete (Lem. 7.7.8), and so this case cannot be beyond us. (The effective state space, by (7.33), has only two elements and so certainly is finite.)

In fact we take

$$\delta \quad := \quad \overline{a} \min \overline{b} \, ,$$
$$\varepsilon \quad := \quad 1$$

and variant function $\lfloor \overline{p} \rfloor$,

reasoning informally that

— when $\overline{p}$ is one its floor $\lfloor \overline{p} \rfloor$ is certainly decreased from one to zero (*i.e.* with probability $\varepsilon = 1$) by execution of the loop body $p := \tilde{p}$;

— yet when $\overline{p}$ is not one, it is by construction no more than δ, so the behaviour of the variant is irrelevant.[44]

Thus we have almost certain termination unless $a = b = 0$.[45]

That concludes our discussion of the construct

do $p \rightarrow$ *body* **od** ,

which we now accept as a fully fledged member of *pGCL*.

7.8 Summary

Our main goal in this chapter was to justify the rules we introduced in Chap. 2: they are now proved as Thm. 7.3.3/B.2.2 for total correctness of iterations when the termination condition is known, and Thm. 7.6.2 for

---

[44]Recall that in Lem. 7.7.7 the variant may increase when the loop-entry probability is δ or below — in this program, it increases from zero back to one.

[45]Minimalists might like to note that the two arguments can be combined by taking $\delta := \overline{a} \min \overline{b}$, as above, but variant function $[\overline{p} > \delta]$. This in effect argues for termination irrespective of whether just one or both of a, b are nonzero — we rely only on the more accurate cowboy, who cannot always miss unless they both do.

almost-certain termination, *i.e.* with probability one. And with our earlier examples of Sections 2.5, 2.8 and Chap. 3 we have shown that probabilistic reasoning for partial correctness — on this scale at least — is not much more complex than standard reasoning.

A secondary goal was to introduce the alternative "probabilistic guard" iteration, in many ways more direct for reasoning than the traditional Boolean-based form. It was delayed until this point however because the latter is more obviously related to standard iterations, both in its appearance and in its proof rules.

For total correctness it seems harder to achieve simplification using grossly pessimistic variants (a familiar technique in the standard case). Our experience so far suggests that it is often necessary to use accurate bounds on the number of iterations remaining, and that can require intricate calculation.

Beyond the inductive approach of Sec. 2.9, we do not have general rules for determining the termination condition when it is not equal to one; at this stage it seems those situations have to be handled by using the *wp* semantics to extract a recurrence relation to which standard probabilistic methods can then be applied. A promising approach however is to use (probabilistic) data refinement to extract not a recurrence relation but a simple(r) program, involving only the variant captured by a single variable. That program's termination condition is equal to the original, but could perhaps be taken straight from the literature, where one would thus have access to a large collection of termination "paradigms." [46]

A longer-term approach to probabilistic termination is to build a temporal logic over the expectation transformers in the style of Part III, generalising the construction of modal logic over standard transformers. The resulting properties are then very like those of Ben-Ari, Pnueli and Manna [BAPM83] and allow termination conditions to be determined for quite complicated programs using structured arguments in the style, for example, of UNITY [CM88].

---

[46]For example, a complexity bound of $O(N^2)$ has been established that way for the Herman's Ring algorithm of Sec. 2.8, using the random walk as a paradigm [MM04a].

Chapter notes

Proving termination is crucial for most applications — Jones [Jon90] gave an expectation-based *wp*-style semantics and rules for analysing partial- and total correctness for probabilistic deterministic programs, and Hurd's *HOL*-based theory [Hur02] mechanises a theory for termination with probability one. Pnueli and Zuck [AZP03] have exploited fairness to propose a method for *parametrised model-checking* which suggests the possibility of fully automated termination proofs for classes of parametrised systems of arbitrary size.

Hart, Sharir and Pnueli proved completeness theorems for both finite- and infinite state programs [HSP83], and Hart and Sharir [HS86] have sound- and complete temporal logics for reasoning about termination with probability one.

In a refinement-oriented framework for program development the proof of termination can often be deferred until the last stages of development — this can be seen for example in Abrial's development of the *IEEE Firewire Protocol* using the *B*-method [CMA02] which can be verified using the simple variant techniques proposed here [MMH03]. Fidge and Shankland have also treated *Firewire* [FS03].

The boundedness of expectations, of course, has turned out to be crucial to our general approach; however there are situations where unboundedness is desirable, for instance when analysing the expected time to termination. Hurd *et al.* have a mechanised version of *pGCL* [HMM04] that includes a theory of infinite expectations, and McIver's approach [McI02] using derived *pGCL* operators is sound even for nonterminating programs.

8

Infinite state spaces, angelic choice and the transformer hierarchy

8.1 Introduction

This chapter rounds off our exploration of the transformer spaces, which has been the main topic of Part II. We remove the restriction to finite state spaces, and we classify the transformers in two ways that we show to be equivalent: in terms of their construction via choice operators (deterministic, probabilistic, demonic and angelic) on the one hand; and in terms of their algebraic healthiness properties (linear, sublinear, semi-linear, semi-sublinear...) on the other.

Sec. 8.2 sketches the steps necessary to accommodate infinite state space (with the details given in Sec. B.4).

For the classification, in Sections 8.3–8.5 we investigate systematically the structure of the whole expectation-transformer space — not just the conjunctive-like part of it on which we have concentrated so far — and we reveal a hierarchy of program classes, with the more complex being built from the simpler ones. (Back and von Wright [BvW90] have done the same for standard programs.) We associate with each class of programs a characteristic "healthiness condition" in the expectation logic (extending Sections 1.6 and 5.6), showing each one to be both necessary and sufficient for membership of its associated class: in most cases, the conditions are arithmetic variations on the sublinearity we have seen already.

We recall that standard predicate transformers can be presented as nested classes of increasing sophistication: the *deterministic* programs have a single, predictable final state for any initial state, and are characterised by disjunctivity and conjunctivity in the (Boolean) programming logic [Dij76];[1] *demonic* programs may contain "worst-case" nondeterministic choices, and are characterised by conjunctivity alone; and demonic/*angelic* programs may contain "best-case" nondeterministic choice as well. Those last have been shown [BvW90] to be characterised simply by monotonicity.

Our expectation transformers can be similarly presented. In this chapter we remind ourselves that deterministic (but probabilistic) programs have a single, predictable *distribution* of output behaviours;[1] and we discover that their associated expectation transformers are characterised by "linearity." Going on, we construct demonic programs by $\sqcap$-choices between deterministic ones; and, as we know from Chap. 5, their expectation transformers are characterised by "sublinearity." Beyond that, demonic/angelic programs are made by $\sqcup$-choices between demonic ones, with their expectation transformers characterised by what we will call "semi-sublinearity." In Sections 8.3, 8.4 and 8.5 we build these three classes, of deterministic, demonic and demonic/angelic programs.

Finally, in Sec. 8.6 we identify the algebraic property of "semi-linearity" that characterises the standard programs (*i.e.* those from Dijkstra's original *GCL*) when embedded within the richer probabilistic space. Naturally, that property can be applied orthogonally to the other three.

Figures 8.7.1 (the structure of transformer spaces) and 8.7.2 (algebraic properties used to characterise them) on pp. 240ff summarise our classification.

---

[1]In this informal discussion we are assuming termination. More generally, pre-deterministic programs are wholly predictable only from initial states at which termination is guaranteed.

For the remainder of Part II we return to the mathematical notation in which lower-case Greek letters α, β are random variables, and upper-case Greek letters Δ, Δ' are distributions. Upper-case Roman letters P, Q are used for predicates over the state space S, and (equivalently) for subsets of it; and calligraphic letters $\mathcal{A}, \mathcal{P}$ are used for sets of expectations or predicates. Finally, for states $s, s' \ldots$ in S we write $\{s, s' \ldots\}$ uniformly for the set containing those states, for the predicate holding only for those states and for the characteristic function taking the value one on those states and zero elsewhere.

8.2 Infinite state spaces

Our principal results about $\mathbb{H}S$ and $\mathbb{T}S$ were presented in Chap. 5, where we showed that the *regular* subset $\mathbb{T}_r S$ of $\mathbb{T}S$ — the image under wp of $\mathbb{H}S$ — is characterised by sublinearity. The proof of that relied on finiteness of the state space (especially in its appeal to the "geometric" lemmas of Sec. B.5), and so did the subsequent proof that sublinearity implies continuity (and that therefore all regular transformers are continuous). We now indicate how to extend those results to infinite state spaces.

Taking the second result first, we deal with the failure of the continuity proof by *imposing* continuity: we extend the characterisation of $\mathbb{T}_r S$ so that it includes continuity explicitly — thus what we will prove is that,

> even for infinite state spaces, a transformer is in $\mathbb{T}_r S$ iff it is sublinear and continuous.[2]

In our earlier treatment of iteration, we have occasionally used examples over infinite state spaces (anticipating this section), and have stressed the importance of each expectation's being bounded above. In a finite state space that happens automatically; but in an infinite state space it too must be imposed.[3]

One generalisation we do not need to make, however, is from discrete distributions over finite S to measures over an infinite S. That is because our programs' only access to probability is via a binary operator $_p\oplus$, and

---

[2]Note that we do not change the *definition* Def. 5.5.3 of $\mathbb{T}_r S$ — it remains the *wp*-image, the embedding $wp.(\mathbb{H}S)$, of the probabilistic relational programs. It is its algebraic characterisation that we update.

[3]Recall Footnote 39 on p. 25 and Footnote 50 on p. 66.

they therefore can generate only discrete result distributions, even when S is infinite.[4]

We now outline the approach; details are given in Sec. B.4.

To re-establish our characterisation of $\mathbb{T}_r S$, we introduce the notion of "finitary" expectations each of which is nonzero only on some finite subset P of S. Usually, finitary expectations can be treated as (full) expectations over that finite subset P, where they therefore have the properties we have established already. If we can show that those properties are preserved by taking limits — in this case of larger and larger finite subsets P within S — then the properties will hold in the infinite case as well, since any expectation α can be written as a $\sqcup$-limit of its finitary restrictions $\alpha{\downarrow}P$.[5]

We show that the definition Def. 5.7.1 of rp can be given in terms of finitary expectations only, and we prove directly from the updated definition that all transformers in $\mathbb{T}_r S$ are continuous even when S is infinite; in doing so we rely on the fact that result sets of distributions in $\mathbb{H}S$ are Cauchy closed. (Our earlier proof Lem. 5.6.6 did not appeal to that explic-

---

[4]The short explanation for this is that our semantics guarantees it trivially, since it is over discrete distributions and is closed under all the conventional program-construction operators. Nevertheless that might seem surprising, at first, in view *e.g.* of the following argument that attempts to exploit our abstraction techniques to carry out a step-by-step replacement of demonic choices by ever-finer probabilistic ones.

Let the state space be the unit interval $[0, 1]$, and construct an N-indexed sequence of programs *prog* in which Program $prog_0$ chooses demonically from $[0, 1]$, Program $prog_1$ chooses *probabilistically* between $[0, 1/2]$ and $[1/2, 1]$ and then demonically within the chosen sub-interval; and, in general, Program $prog_{n+1}$ divides $[0, 1]$ into twice as many intervals as $prog_n$ did, still using at each stage however only a finite number of discrete probabilistic choices to select an (ever smaller) interval within which a final demonic choice is made. The higher n becomes, the more uniform an outcome we achieve, and the less we are affected by the demon.

Indeed we have $prog_0 \sqsubseteq prog_1 \cdots$ *etc.*, and we might expect in the limit to reach the program $prog_\infty$ that makes a uniform probabilistic choice from $[0, 1]$, with all the demonic choice "refined away."

The problem with the argument is that we cannot construct even the first program; although it contains no probabilistic choice at all, it does contain unbounded *demonic* choice, which makes it non-continuous and — for us, therefore — not constructible. (Recall for example that the abbreviation introduced on p. 21 in Sec. 1.5.2 specifically mentions a finiteness restriction that would exclude the statement $x{:} \in [0, 1]$.) We cannot construct the subsequent programs, either, for the same reason.

In fact we can see with continuity directly that the program "choose uniformly from $[0, 1]$" is problematic: every postcondition comprising only finitely-many points gives pre-expectation zero (the probability of achieving it); yet the least upper bound of those postconditions is the whole interval $[0, 1]$, for which the probability should be one. (As mentioned in Footnote 7 on p. 297 to come, this effect is related to the fact that $[0, 1]$ is uncountable.)

Our treatment of continuous distributions is addressed briefly elsewhere [MM01b], where for the problem of continuity we rely on techniques developed by Edalat [Eda95].

[5]We write $\alpha{\downarrow}P$ for the expectation equal to α on P but zero elsewhere.

itly.) Then we must re-prove the connection in the other direction as well: that for any continuous transformer t and state s the set of distributions $rp.t.s$ is Cauchy closed.

The earlier proofs then go through much as before, except that we rely more heavily on topological notions (principally compactness) to reduce infinite expressions to finite ones where required.

The result is to show that our characterisation of wp-images of programs in $\mathbb{H}S$, the already-established healthiness conditions of Fig. 5.6.7, are applicable to the infinite case: the main (and "invisible") difference is that continuity is no longer a consequence of the others and, instead, must be imposed as a separate condition in its own right. However the first three conditions in Fig. 5.6.7 are still consequences of sublinearity as shown earlier and, in the special case where the scalars are $\{0, 1\}$-valued, all are generalisations of properties of standard predicate transformers. Sublinearity thus remains the appropriate generalisation of conjunctivity, and feasibility, for example, continues to generalise strictness.

The results of Sections 8.3–8.5, to come, remain valid within this infinite extension. For simplicity however we present those results as for finite state spaces.[6]

8.3   Deterministic programs

8.3.1   ...are single-valued

Standard deterministic programs deliver a single output state for any fixed input; but probabilistic deterministic programs deliver a single output *distribution* for any fixed input. (Recall Figs. 6.2.1 and 6.3.1 from p. 167.) Thus even though separate runs of a deterministic, probabilistic program may deliver different outputs from the same initial state, over a group of runs a tabulation of results will reveal a probability distribution of those outputs; and because the program is deterministic, a second group of runs from the same initial state will reveal the same distribution.[7]

Since standard deterministic programs are disjunctive as well as conjunctive — and those properties are ($\neg$)-duals — we conjecture that one of the characteristic properties of probabilistic deterministic programs will be *super*- as well as sub-additivity, since those properties are ($-$)-duals: that is, deterministic programs are "simply" additive, at least. Our main result in this section is therefore to show that with scaling, $\ominus$-subdistribution and

---

[6]We will use footnotes to highlight issues that arise for the infinite case.

[7]...to within statistical confidence measures dependent on the number of runs.

continuity, the additional property of additivity indeed characterises determinism. (These characterising properties are together called "linearity" and are summarised, with others, in Fig. 8.7.2 on p. 241 below.)

We begin by fixing the idea of a deterministic operational program — it is one that is characterised by having a single output distribution for each initial state.[8]

Definition 8.3.1 DETERMINISTIC PROGRAM   A program r in $\mathbb{H}S$ is *deterministic* iff for each state s there is a single distribution Δ_s so that

$$r.s \ := \ \{\Delta : \overline{S} \mid \Delta_s \sqsubseteq \Delta\} \ .$$

$\square$

Notice how nonterminating programs (*i.e.* pre-deterministic) are also covered by the definition, and that up closure and Cauchy closure are built-in. (Recall Fig. 6.6.1 from p. 171.)

8.3.2   ... are linear *as transformers*

For our algebraic characterisation of deterministic transformers, we define the property "linearity."

Definition 8.3.2 LINEAR TRANSFORMERS   An expectation transformer t is said to be *linear* iff it is $\ominus$-subdistributive, scaling and "additive," where by *additive* we mean that for all expectations α, α' in $\mathbb{E}S$ we have

$$t.(\alpha + \alpha').s \ = \ t.\alpha.s \ + \ t.\alpha'.s \ . \ \ ^9$$

$\square$

The proof that deterministic programs map to linear expectation transformers under *wp* is straightforward.

Lemma 8.3.3 If r in $\mathbb{H}S$ is deterministic then *wp.r* is linear.

Proof   Lem. 5.6.2 ensures that *wp.r* is sublinear at least; then its additivity follows directly from Def. 8.3.1 and the additivity of $\int_\Delta$ (equivalently the additivity of expectation in elementary probability [GW86]).[10] $\square$

More involved is the converse, that linearity implies determinism. Note (Def. 1.6.2) that feasibility is implied by $\ominus$-subdistributivity and scaling, and hence by linearity.

---

[8]This generalises our earlier Def. 5.1.3 in that we are now defining deterministic programs *within* the larger demonic space.

[9]We are extending our earlier definition at (2.18) on p. 66.

[10]When S is infinite, Lem. B.4.7 shows that continuity follows from the Cauchy closure of *r.s* for all s.

Lemma 8.3.4 If a transformer t in $\mathbb{T}S$ is linear, then it is the wp-image of a deterministic relational program.

Proof For expectation transformer t in $\mathbb{T}S$ we define the deterministic program r in $\mathbb{H}S$ as follows:

$$r.s \quad := \quad \{\Delta : \overline{S} \mid \Delta_s \sqsubseteq \Delta\}$$
$$\text{where} \quad \Delta_s.s' \quad := \quad t.\{s'\}.s \text{ , for } s \text{ in } S,$$

writing $\{s'\}$ for the "point" expectation that is one at s' and zero elsewhere.

Observe that r is well defined: distribution Δ_s is indeed an element of $\overline{S}$ because

$$
\begin{aligned}
 &(\textstyle\sum_{s':S} \Delta_s.s') && \\
= \;&(\textstyle\sum_{s':S} t.\{s'\}.s) && \text{definition } \Delta_s \\
= \;&t.(\textstyle\sum_{s':S}\{s'\}).s && t \text{ additive}^{11} \\
= \;&t.\underline{1}.s && \text{definition } \underline{1} \\
\leq \;&1 \;. && t \text{ linear, implies feasible}
\end{aligned}
$$

We now show that in fact $t = wp.r$. Observe first that any expectation α in $\mathbb{E}S$ is a sum $(\sum_{s:S} (\alpha.s) * \{s\})$ of characteristic functions. We reason

$$
\begin{aligned}
 &t.\alpha.s && \\
= \;&t.(\textstyle\sum_{s':S} \alpha.s' * \{s'\}).s && \\
= \;&(\textstyle\sum_{s':S} \alpha.s' * t.\{s'\}.s) && t \text{ linear} \\
= \;&(\textstyle\sum_{s':S} \alpha.s' * \Delta_s.s') && \text{definition of } \Delta_s \\
= \;&\textstyle\int_{\Delta_s} \alpha && \text{definition of } \int \\
= \;&wp.r.\alpha.s \;. \quad ^{12} && \text{definition } wp \text{ and } r
\end{aligned}
$$

$\square$

We call the space of linear — equivalently deterministic — transformers $\mathbb{T}_\circ S$, and have our theorem:

Theorem 8.3.5 LINEARITY CHARACTERISES DETERMINISM

A transformer is linear iff it is the wp-image of a deterministic relational program.

Proof   Lemmas 8.3.3 and 8.3.4. $\square$

Fig. 8.3.6 illustrates the result.

---

[11] For infinite S we appeal here to the continuity of t as well, since we have the equality

$$(\sum_{s':S} \cdots) \quad = \quad (\sqcup P : \mathbb{F}S \bullet (\sum_{s':P} \cdots)) \;.$$

[12] If S is infinite, we use finitary α in this proof and sum over its support rather than over all of S; that proves $t.\alpha = wp.r.\alpha$, but only for finitary α. Since both t and $wp.r$ are continuous, however, the equality then extends to all α.

Let *prog* be the deterministic program

$$s := \text{heads} \; {}_{\frac{1}{2}} \oplus \text{tails}$$

over the state space {heads, tails}.

We have $wp.prog.\{\text{heads}\}.s = 1/2$ for all initial values of s, and similarly $wp.prog.\{\text{tails}\}.s = 1/2$. Now (punning sets and characteristic functions) we have {heads} + {tails} = {heads, tails}, and

$$wp.prog.\{\text{heads, tails}\}.s \quad = \quad 1 \quad = \quad 1/2 + 1/2 \,,$$

thus illustrating additivity.

Figure 8.3.6. DETERMINISTIC PROGRAMS ARE ADDITIVE

8.3.3 ... and are disjunctive if standard

We end this section by looking at the specialisation of our characterisation to standard transformers, where (as we know) deterministic programs are disjunctive as well as conjunctive [Dij76]. We will show in Sec. 8.6 that if t in $\mathbb{T}S$ maps characteristic functions to characteristic functions then it corresponds to a standard *predicate* transformer in the obvious way: conjunction and disjunction are then encoded respectively as $\sqcap$ and $\sqcup$ in the restricted subset of standard expectations. Hence our hope is that if t is also linear then we can deduce disjunctivity (of the corresponding predicate transformer), which by the above remarks corresponds to distributivity of $\sqcup$ applied to standard arguments. To see that, we assume that predicates P, Q are disjoint, and reason as follows:

$$\begin{array}{lll}
 & t.(P \sqcup Q) & \\
\equiv & t.(P + Q) & P, Q \text{ disjoint and standard} \\
\equiv & t.P + t.Q & t \text{ linear} \\
\equiv & t.P \sqcup t.Q \,. & \text{see below}
\end{array}$$

In the last step we use that t is feasible (implied by linearity) and that $t.P + t.Q$ is standard to infer that $t.P.s$ and $t.Q.s$ cannot both be one for any s in S (for otherwise $t.(P \sqcup Q).s$ would be two).

Disjunctivity of t now follows in general, since with conjunctivity (from (1.24) on p. 31) it is implied by disjunctivity of disjoint post-conditions.

8.4 Demonic programs

8.4.1 ... are constructed with $\sqcap$

Demonic programs are constructed by closing the deterministic programs $\mathbb{T}_\circ S$ under demonic choice.

We show here that they are exactly the regular (equivalently, sublinear) transformers $\mathbb{T}_r S$ described in Sec. 5.5, and in this chapter we call them $\mathbb{T}_{\sqcap} S$ to emphasise their construction via $\sqcap$. Figure 8.4.2 shows a "typical" demonic program: first natural number n is chosen demonically from $1..N$; then a uniform probabilistic choice is made from the numbers $1..n$. The final distribution over a set of runs is not (wholly) predictable, because of the nondeterminism: in separate sets one could find quite different distributions. But each distribution will have the property that higher frequencies are associated with lower final states, if the number of runs is sufficiently large; and it can be shown that the program is in fact capable of producing *all* such "anti-monotonic" distributions.

8.4.2   ... and are sublinear *as transformers*

To establish our characterisation of $\mathbb{T}_{\sqcap} S$, we note first from simple arithmetic that if transformers t_i are additive, for i in some index set I, then their (demonic) infimum $(\sqcap i \colon I \cdot t_i)$ is *sub*-additive (by subdistribution of greatest lower bounds through addition), and that scaling and $\ominus$-subdistribution are preserved as well. Thus the elements of the (even infinitary) $\sqcap$-closure of $\mathbb{T}_\circ S$ are sublinear.[13]

---

[13]In the finite case, they are continuous too, since that is implied by sublinearity (Lem. 5.6.6 on p. 147). Thus even an infinitary demonic choice over a *finite* state space preserves continuity.

The infinite case is considerably more complex, however, because infinitary demonic choice in an infinite state space does not necessarily preserve continuity. And we cannot simply prove that in fact continuous demonic programs *can* be reached by finitary infima... because it is not true even in the standard case: for example the program

$$n \colon \leq n \qquad\qquad (8.1)$$

over the natural numbers is continuous (as a predicate transformer); but because the size of the set of possible final states is not bounded (as the initial state is varied), there is no finite set of deterministic standard programs whose infimum is (8.1).

Instead we construct a given demonic program as the supremum of infima, each infimum being taken over a finite set of deterministic programs: each finitary infimum is continuous because its components are; and even infinitary suprema preserve (up-)continuity. For the case above we have

$$n \colon \leq n \quad = \quad (\sqcup N \colon \mathbb{N} \cdot \{n \leq N\} n \colon \leq n) \,,$$

where the assertion $\{n \leq N\}$ on the right causes abortion for initial states exceeding N. Each N-term is thus the infimum of no more than $1 * 2 * \cdots * (N{+}1)$ pre-deterministic programs.

A similar strategy can be carried out for expectation transformers: the details are given in Sec. B.4.5.

$$n: \in \{1..N\}; \qquad\qquad \leftarrow \text{choose demonically}$$
$$n: = (\lVert\, i: \{1..n\} \bullet i \,@\, 1/n) \qquad \leftarrow \text{choose uniformly}$$

This demonic program produces any "anti-monotonic" distribution over $1..N$ that assigns higher probabilities to lower outcomes.

Figure 8.4.2. DEMONIC PROGRAM OVER FINITE STATE SPACE

Let *prog* be the program

$$s: = \mathsf{heads} \quad \sqcap \quad s: = \mathsf{heads} \,{}_{\frac{2}{3}}\!\oplus \mathsf{tails}$$

over the state space $\{\mathsf{heads}, \mathsf{tails}\}$.

This program contains both probabilistic and demonic choice; it chooses between heads and tails, assigning a probability of *at least* 2/3 to the former. Now we have $wp.prog.\{\mathsf{heads}\}.s = 2/3$ because the nondeterminism is demonic (not angelic) — although heads can be chosen always, we can be *sure* it will be chosen only two-thirds of the time. Similarly $wp.prog.\{\mathsf{tails}\}.s = 0$ because tails might never be chosen at all. Now

$$wp.prog.\{\mathsf{heads}, \mathsf{tails}\}.s \quad = \quad 1 \quad > \quad 2/3 + 0\,,$$

thus illustrating sublinearity and the failure of additivity.

Figure 8.4.3. DEMONIC PROGRAMS ARE SUBLINEAR BUT NOT ADDITIVE

For the converse, we note that if t is sublinear then from Lem. 5.7.6 we have $wp.(rp.t) = t$;[14] thus take for the deterministic "components" of t the programs in $\mathbb{T}_\circ\, S$ generated (Def. 8.3.1) by choosing Δ_s in all possible ways from $rp.t.s$ for each s. The result is then a consequence of the correspondence between $\cup$ in $\mathbb{H}S$ and $\sqcap$ in $\mathbb{T}S$.

Thus any sublinear transformer may be written as the (possibly infinitary) demonic choice $\sqcap$ between members of $\mathbb{T}_\circ\, S$, and we have proved this theorem:

Theorem 8.4.1 SUBLINEARITY CHARACTERISES $\sqcap$-CLOSURE
The $\sqcap$-closure of the deterministic transformers $\mathbb{T}_\circ\, S$ is exactly the sublinear transformers $\mathbb{T}_\sqcap\, S$. □

Fig. 8.4.3 gives an illustration of sublinearity.

---

[14]This is our principal use of the Galois connection: note that sublinearity of t implies it is not miraculous, and that in turn ensures definedness of rp. Section B.4.3 extends that to the infinite case; but recall Footnote 13 on the previous page.

8.5 Angelic programs

8.5.1 ... are constructed with ⊔

Our final step is to move from $\mathbb{T}_\sqcap S$ to the more general space formed by allowing angelic choice as well. In the standard case, one drops conjunctivity; here we will drop sub-additivity.

Our analysis of the general space $\mathbb{T}S$ is motivated by the treatment of standard predicate transformers: there, monotonicity is considered of the first importance and moreover it is the only remaining healthiness condition once strictness and conjunctivity are given up (both having been shown to be too strong for specifications). Also well known is that any monotonic standard transformer is the disjunction of conjunctive ones [BvW90]; and the aim for this section is an analogy for the expectation transformers — to find a subset of the monotonic expectation transformers suitable for modelling both implementable programs and specifications, and to discover how the healthiness conditions are affected. As usual we fall back on our experience of the standard model, where the conjunction of specified properties corresponds to a disjunction (or angelic choice) of predicate transformers. We use *maximum* to model angelic choice, and we then consider the set of expectation transformers that are closed under all three forms of choice: probabilistic, demonic and angelic.[15]

Our generalisation of angelic choice in the probabilistic domain is thus as follows:

Definition 8.5.1 ANGELIC CHOICE   For expectation transformers t, t' in $\mathbb{T}S$, expectation α in $\mathbb{E}S$ and state s in S we define the angelic choice between t and t' to be

$$(t \sqcup t').\alpha.s \; := \; t.\alpha.s \; \sqcup \; t'.\alpha.s \; . \quad \text{[16]}$$

□

We denote the closure of $\mathbb{T}_\sqcap S$ under $\sqcup$ by $\mathbb{T}_\sqcup\sqcap S$, and now look for an arithmetic characterisation of it. We observe that the new property will be a weakening of sublinearity (since $\mathbb{T}_\sqcup\sqcap S$ contains all sublinear transformers). But since the operator $\sqcup$ preserves the scaling and $\ominus$-subdistribution properties from Fig. 5.6.7 (p. 148), we are encouraged to concentrate on "semi-sublinearity," defined next.

---

[15]Still this does not produce *all* monotonic transformers — see for example Fig. 8.7.1 on p. 240 to come — but it is sufficient for most specification tasks.

[16]For conciseness we write "⊔" on both sides: on the left, it is an operator we are defining; on the right, it is *maximum*.

8.5.2 ... and are semi-sublinear as transformers

Semi-sublinearity is a weakening of sublinearity in which only one, rather than two, scaled terms appear.

Definition 8.5.2 SEMI-SUBLINEAR TRANSFORMERS

Expectation transformer t in $\mathbb{T}S$ is said to be *semi-sublinear* iff it satisfies the properties of scaling and $\ominus$-subdistribution.   □

We know that semi-sublinearity and monotonicity imply feasibility, scaling and continuity. However, as we shall see, semi-sublinearity is also all we need to recover the whole of $\mathbb{T}_{\sqcup\sqcap} S$.

We show first that all members of $\mathbb{T}_{\sqcup\sqcap} S$ are semi-sublinear.

Lemma 8.5.3 If t is in $\mathbb{T}_{\sqcup\sqcap} S$ then it is semi-sublinear.

Proof   We need only show that semi-sublinearity is preserved by (possibly infinite) applications of $\sqcup$, and that is immediate from arithmetic and Def. 8.5.2.[17]   □

Our major goal however is to show that $\mathbb{T}_{\sqcup\sqcap} S$ comprises *exactly* the semi-sublinear transformers, and for that we need the converse of Lem. 8.5.3: that all semi-sublinear transformers in $\mathbb{T}S$ are members of $\mathbb{T}_{\sqcup\sqcap} S$.

Motivated by Back and von Wright's construction [BvW90, Lemma 8] for standard programs, given a semi-sublinear t we consider all post-expectations α and express t as the angelic choice over an α-indexed family t_α of sublinear transformers in $\mathbb{T}_\sqcap S$. Each transformer t_α will be the $\sqsubseteq$-weakest transformer that itself is monotonic and semi-sublinear and agrees with t at α:[18] because t_α is the weakest such, it will be everywhere no more than t; and because each t_α agrees with t at α, their supremum will be at least (thus in fact equal to) t. The surprise will be that the t_α's turn out to be fully (rather than just semi-) sublinear, making them just the members of $\mathbb{T}_\sqcap S$ we are looking for.

---

[17]Recall also that up-continuity is preserved by suprema.

[18]The impact of requiring semi-sublinearity in t_α is that the otherwise obvious choice

$$t_\alpha.\beta \;\; := \;\; t.\alpha \text{ if } (\alpha \Rrightarrow \beta) \text{ else } 0 \tag{8.2}$$

is not allowed — by scaling, for example, we see that $t_\alpha.(2\alpha)$ must be $2(t.\alpha)$ rather than the $t.\alpha$ that (8.2) would give. Those properties of t_α motivate Def. 8.5.4 that makes it monotonic and semi-sublinear by construction, leaving the job of Lem. 8.5.7 mainly to establish its sub-additivity.

The appropriate definition of t_α is the following:

Definition 8.5.4 α-COMPONENT   For t a semi-sublinear transformer in $\mathbb{T}S$ and expectation α in $\mathbb{E}S$, let the α-*component* t_α of t be defined by

$$t_\alpha.\beta.s \ := \ (\sqcup c, c' \colon \mathbb{R}_\geq \mid c\alpha - \underline{c'} \Rrightarrow \beta \cdot c(t.\alpha.s) - c') , \qquad 19$$

for β in $\mathbb{E}S$ and s in S. $\qquad\qquad\qquad\qquad\qquad\qquad\qquad\qquad\qquad\Box$

To see that t_α is well defined, by which we mean that when applied to β in $\mathbb{E}S$, the result is non-negative and bounded, we appeal first to Lem. 8.5.6 below to see that $t_\alpha.\beta$ must be bounded (because t is feasible); to see that the result is non-negative we need only put c, c' to be $0, 0$ in Def. 8.5.4 and then notice that the definition maximises.

We now prove the three crucial properties of these α-components. The first property is that t_α is no less than t at α.

Lemma 8.5.5 For all semi-sublinear transformers t and expectations α we have $t.\alpha \Rrightarrow t_\alpha.\alpha$.
    Proof   Take $c, c' := 1, 0$ in Def. 8.5.4. $\qquad\qquad\qquad\qquad\qquad\Box$

The second property is that t_α is no more than t in general.

Lemma 8.5.6 For all semi-sublinear transformers t and expectations α we have $t_\alpha \sqsubseteq t$.
    Proof   Take arbitrary expectation β in $\mathbb{E}S$ and state s. For all c, c' in $\mathbb{R}_\geq$,

$$c * \alpha - \underline{c'} \Rrightarrow \beta$$
$\qquad$ implies $\quad c * \alpha \ominus \underline{c'} \Rrightarrow \beta$ $\qquad\qquad\qquad\qquad\qquad\qquad$ $\underline{0} \Rrightarrow \beta$
$\qquad$ implies $\quad t.(c * \alpha \ominus \underline{c'}).s \leq t.\beta.s$ $\qquad\qquad\qquad\qquad$ t monotonic
$\qquad$ implies $\quad c * (t.\alpha.s) \ominus c' \leq t.\beta.s$ $\qquad\qquad\qquad$ t semi-sublinear
$\qquad$ implies $\quad c * (t.\alpha.s) - c' \leq t.\beta.s$. $\qquad\qquad\qquad$ definition $\ominus$

Since the inequality holds for arbitrary c, c' it holds for the supremum taken over them in Def. 8.5.4. $\qquad\qquad\qquad\qquad\qquad\qquad\qquad\qquad\qquad\Box$

Finally we show sublinearity of the α-components.

Lemma 8.5.7 For semi-sublinear t in $\mathbb{T}S$ and α in $\mathbb{E}S$ the component t_α is sublinear.
    Proof   The scaling of t_α is obvious from its definition. The $\ominus$-subdistributivity is straightforward also: for expectation β and non-negative b we have

$$t_\alpha.(\beta \ominus \underline{b}).s$$
$=\qquad (\sqcup c, c' \colon \mathbb{R}_\geq \mid c * \alpha - \underline{c'} \Rrightarrow \beta \ominus \underline{b} \cdot c * (t.\alpha.s) - c')$ $\qquad$ Def. 8.5.4

---

[19] Recall the syntax for comprehensions as set out in Footnote 23 on p. 140: here we take all non-negative c, c' that satisfy $c\alpha - \underline{c'} \Rrightarrow \beta$, make all terms $c(t.\alpha.s) - c'$ resulting from those, and then take the supremum.

$\geq$ $\qquad\qquad\qquad\qquad\qquad\qquad\qquad$ take $c' := c'' + b$
$$(\sqcup c, c'' \colon \mathbb{R}_{\geq} \mid c * \alpha - \underline{(c'' + b)} \Rrightarrow \beta \ominus \underline{b} \cdot c * (t.\alpha.s) - (c'' + b))$$

$\geq \qquad (\sqcup c, c'' \colon \mathbb{R}_{\geq} \mid c * \alpha - \underline{c''} \Rrightarrow \beta \cdot (c * (t.\alpha.s) - c'') - b) \qquad$ arithmetic
$= \qquad (\sqcup c, c'' \colon \mathbb{R}_{\geq} \mid c * \alpha - \underline{c''} \Rrightarrow \beta \cdot (c * (t.\alpha.s) - c'')) - b \quad$ distribute $(-)$
$= \qquad t_\alpha.\beta.s - b \qquad\qquad\qquad\qquad\qquad\qquad\qquad\qquad$ Def. 8.5.4

and hence, since $t_\alpha.(\beta \ominus \underline{b})$ is everywhere non-negative, we can deduce that indeed

$$t_\alpha.(\beta \ominus \underline{b}).s \;\geq\; (t_\alpha.\beta \ominus \underline{b}).s \;.$$

The main point however is sub-additivity, for which we continue the argument as follows:

$\qquad t_\alpha.(\beta_1 + \beta_2).s$
$= \qquad (\sqcup c, c' \colon \mathbb{R}_{\geq} \mid c * \alpha - \underline{c'} \Rrightarrow \beta_1 + \beta_2 \cdot c * (t.\alpha.s) - c') \qquad$ Def. 8.5.4

$= \qquad (\sqcup c_1, c_2, c_1', c_2' \colon \mathbb{R}_{\geq} \qquad\qquad\qquad\qquad$ take $c, c' := c_1 + c_2, c_1' + c_2'$
$\qquad\quad \mid (c_1 + c_2)\alpha - \underline{(c_1' + c_2')} \Rrightarrow \beta_1 + \beta_2$
$\qquad\quad \cdot (c_1 + c_2)(t.\alpha.s) - (c_1' + c_2'))$

$\geq \qquad (\sqcup c_1, c_2, c_1', c_2' \colon \mathbb{R}_{\geq} \qquad\qquad\qquad\qquad\qquad\qquad$ arithmetic
$\qquad\quad \mid c_1 * \alpha - \underline{c_1'} \Rrightarrow \beta_1 \wedge c_2 * \alpha - \underline{c_2'} \Rrightarrow \beta_2$
$\qquad\quad \cdot (c_1 * (t.\alpha.s) - c_1') + (c_2 * (t.\alpha.s) - c_2'))$

$= \qquad\qquad\qquad\qquad$ c_1, c_1' and c_2, c_2' vary independently
$\qquad\quad (\sqcup c_1, c_1' \colon \mathbb{R}_{\geq} \mid c_1 * \alpha - \underline{c_1'} \Rrightarrow \beta_1 \cdot c_1 * (t.\alpha.s) - c_1')$
$\qquad + (\sqcup c_2, c_2' \colon \mathbb{R}_{\geq} \mid c_2 * \alpha - \underline{c_2'} \Rrightarrow \beta_2 \cdot c_2 * (t.\alpha.s) - c_2')$

$= \qquad t_\beta.\beta_1.s \;+\; t_\alpha.\beta_2.s \;. \qquad\qquad\qquad\qquad\qquad$ Def. 8.5.4

$\square$

That gives our main theorem for this section:[20]

Theorem 8.5.8 SEMI-SUBLINEARITY CHARACTERISES $\sqcup$-CLOSURE
The $\sqcup$-closure of the demonic transformers $\mathbb{T}_\sqcap S$ is exactly the set of semi-sublinear transformers $\mathbb{T}_\boxdot S$.

---

[20] For the infinite case, see Sec. B.4.6.

Let *prog'* be the program

$$s:= \text{heads} \ {}_{\frac{1}{3}}\oplus \text{tails} \quad \sqcap \quad s:= \text{tails} \ ,$$

and take *prog''* as in Fig. 8.4.3 earlier. Programs *prog'* and *prog''* are in $\mathbb{T}_\sqcap S$; let *prog* in $\mathbb{T}_\boxdot S$ be the angelic choice *prog'* $\sqcup$ *prog''* between them.

Program *prog* (confusingly) chooses *angelically* whether to use a heads-biased or tails-biased coin, and then applies the chosen bias demonically.

Now we have $wp.prog.\{\text{heads}\}.s = 2/3 \sqcup 0 = 2/3$ because the nondeterminism is angelic between the biases, and once the heads bias is chosen, the worst it can do is to select heads only two-thirds of the time. Similarly $wp.prog.\{\text{tails}\}.s = 2/3$.

Yet

$$wp.prog.\{\text{heads}, \text{tails}\}.s \quad = \quad 1 \quad < \quad 2/3 + 2/3 \ ,$$

thus illustrating the failure of sublinearity.

Figure 8.5.9. ANGELIC PROGRAMS ARE NOT SUBLINEAR

Proof   For semi-sublinear expectation transformer t and expectation β in $\mathbb{E}S$ we have

$$
\begin{aligned}
&t.\beta \\
\Rightarrow \quad &t_\beta.\beta && \text{Lem. 8.5.5} \\
\Rightarrow \quad &(\sqcup\alpha\colon \mathbb{E}S \cdot t_\alpha).\beta \\
\Rightarrow \quad &t.\beta \ . && \text{Lem. 8.5.6}
\end{aligned}
$$

Thus we have $t = (\sqcup\alpha\colon \mathbb{E}S \cdot t_\alpha)$, and conclude from Lem. 8.5.7 that indeed t is the supremum of sublinear transformers.

The converse implication is given by Lem. 8.5.3. □

Figures 8.5.9 and 8.5.10 illustrate the behaviour of transformers in $\mathbb{T}_\boxdot S$.

8.6   Standard programs

8.6.1   ... are constructed without $_p\oplus$

The standard transformers that arise from Dijkstra's *GCL, i.e.* without the $_p\oplus$ that we have added, are as we have seen embedded within the expectation transformer space — they take standard post-expectations to standard pre-expectations. But what do they do to proper expectations? In the spirit of the preceding sections, we now identify the general algebraic property that tells us whether any given expectation transformer is actually just a predicate transformer in disguise.

Let *prog* be the program

$$s:=\ \textbf{heads}\ {}_{\frac{1}{3}}\oplus\textbf{abort}\quad\sqcup\quad s:=\ \textbf{tails}\ {}_{\frac{2}{3}}\oplus\textbf{abort}\ .$$

This program chooses angelically between setting s to heads or tails, with some (varying) chance of nontermination.

Now we have $wp.prog.\{\textsf{tails}\}.s = 0 \sqcup 2/3 = 2/3$ for all s, and similar reasoning shows that in fact $wp.prog.(x * \{\textsf{tails}\}).s = 2x/3$ for any non-negative real x, which demonstrates scaling. An illustration of $\ominus$-subdistribution is therefore that

$$
\begin{array}{lll}
& wp.prog.(\{\textsf{tails}\} \ominus 1/2).s & \\
= & wp.prog.(1/2 * \{\textsf{tails}\}).s & \text{arithmetic of } \ominus \\
= & 1/2 * 2/3 & \text{above} \\
= & 1/3 & \\
\geq & 2/3 \ominus 1/2 & \\
= & wp.prog.\{\textsf{tails}\}.s \ominus 1/2\ . &
\end{array}
$$

Figure 8.5.10. ANGELIC PROGRAMS ARE SEMI-SUBLINEAR

We concern ourselves only with continuous (hence monotonic) and feasible elements of $\mathbb{P}S \leftarrow \mathbb{P}S$, defining a function that injects them into the transformer space $\mathbb{T}S$. The range of that function then determines the "standard subset" of the probabilistic model, and thus the characteristic properties of standard behaviour.

We begin by defining the embedding:

Definition 8.6.1 STANDARD RELATIONS   We write $\mathbb{S}S$ for the set of continuous and feasible elements of $\mathbb{P}S \leftarrow \mathbb{P}S$.   □

Note that with $\mathbb{S}S$ we are imposing conditions analogous to those that defined the relational standard model $S \leftrightarrow S_\perp$ of Sec. 5.5; but because we are dealing with predicate transformers, rather than relations, we do not assume conjunctivity.

When the state space S is finite, continuity is equivalent to monotonicity.

Definition 8.6.2 STANDARD EMBEDDING   For standard predicate transformer $t: \mathbb{S}S$, its *embedding* t^+ into the probabilistic space $\mathbb{T}S$ is defined

$$t^+.\alpha.s\ :=\ (\sqcup P{:}\mathbb{F}S\mid s\in t.P\bullet\ \sqcap_P\alpha)$$

for α in $\mathbb{E}S$ and s in S, where $\sqcap_P\alpha$ denotes the minimum value of α over P.[21]

We write $\mathbb{T}_sS$ for the $(\cdot)^+$-image of $\mathbb{S}S$ in $\mathbb{T}S$, so that

$$\mathbb{T}_sS\ :=\ \{t{:}\mathbb{S}S\bullet t^+\}\ ,$$

---

[21] If the predicate transformer t were not continuous, then the appropriate definition would be $(\sqcup P{:}\mathbb{P}S\mid s\in t.P\bullet\ \sqcap_P\alpha)$; but continuity of t ensures that its behaviour on $\mathbb{P}S$ is determined by its behaviour on $\mathbb{F}S$, and the two definitions are thus equivalent.

and where there is no risk of confusion we call the elements of $\mathbb{T}_s S$ the *standard* elements of $\mathbb{T}S$. □

In effect the embedding determines the smallest set P that contains all final states resulting from (the program represented by) t acting on initial state s — if t corresponded to a standard *relational* program, then that P (depending on each s) would determine it.[22]

Note that feasibility of t means that $s \in t.\emptyset$ is false; thus we need never consider the empty infimum $\sqcap_\emptyset \alpha$ in Def. 8.6.2. However the empty supremum is zero by definition (occurring when $s \in t.P$ for no P).

An elementary property the embedding should satisfy is that embedded standard programs retain their original behaviour when applied to standard postconditions:

Lemma 8.6.3 For any t in $\mathbb{S}S$ we have

$$ t^+.[P] \quad = \quad [t.P] \, , $$

for all subsets P of the state space S.[23]

Proof   Take arbitrary state s in S; then

$$
\begin{aligned}
& t^+.[P].s \\
=\ & (\sqcup Q\!:\!\mathbb{P}S \mid s \in t.Q \cdot \sqcap_Q [P]) && \text{Def. 8.6.2} \\
=\ & (\sqcup Q\!:\!\mathbb{P}S \mid s \in t.Q \cdot 1 \text{ if } (Q \subseteq P) \text{ else } 0) \\
=\ & 1 \text{ if } (\exists Q\!:\!\mathbb{P}S \mid s \in t.Q \wedge Q \subseteq P) \text{ else } 0 \\
=\ & 1 \text{ if } (s \in t.P) \text{ else } 0 && t \text{ monotonic} \\
=\ & [t.P].s \, .
\end{aligned}
$$

□

Note that this gives us that $(\cdot)^+$ is indeed an embedding: if two predicate transformers differ, then they must disagree on standard arguments, whence from above also their embeddings must disagree.

8.6.2   ... and are semi-linear *as transformers*

In the remainder of this section we develop a characterisation of $\mathbb{T}_s S$ within $\mathbb{T}S$. In fact the principal characteristic of $\mathbb{T}_s S$ is "semi-linearity," a property similar in form to sublinearity: examination of the proofs of Sec. 1.6 will show that semi-linearity implies feasibility and scaling. It does not imply continuity (nor even monotonicity), which therefore we include explicitly as part of the characterisation of $\mathbb{T}_s S$.

---

[22]From the explanation given earlier of the syntax of comprehensions (Footnote 19 on p. 229), we see that it is the *smallest* P because we are maximising $\sqcap_P \alpha$, and the smaller the P the greater the value of $\sqcap_P \alpha$.

[23]Although we write the embedding brackets [·] explicitly here, since this equation is "about" embedding, we often omit them in the sequel to reduce clutter.

Definition 8.6.4 SEMI-LINEAR TRANSFORMERS   Expectation transformer $t \colon \mathbb{T}S$ is said to be *semi-linear* iff

$$c(t.\alpha) \ominus \underline{c}' \quad \equiv \quad t.(c\alpha \ominus \underline{c}')$$

for all $\alpha \colon \mathbb{E}S$ and $c, c' \colon \mathbb{R}_{\geq}$. □

Giving necessary conditions for membership of $\mathbb{T}_s S$ is now straightforward:

Lemma 8.6.5 For all $t \colon \mathbb{S}S$, the embedding t^+ is semi-linear and continuous.

Proof   Semi-linearity follows immediately from distributivity properties of $\sqcap$ and $\sqcup$. For continuity we take a directed subset $\mathcal{A}$ of $\mathbb{E}S$, and reason

$$
\begin{array}{lll}
& t^+.(\sqcup\mathcal{A}).s & \\
= & (\sqcup P \colon \mathbb{F}S \mid s \in t.P \bullet \sqcap_P (\sqcup\mathcal{A})) & \text{Def. 8.6.2} \\
= & (\sqcup P \colon \mathbb{F}S \mid s \in t.P \bullet (\sqcup\alpha \colon \mathcal{A} \bullet \sqcap_P \alpha)) & \text{P finite; $\mathcal{A}$ directed}[24] \\
= & (\sqcup\alpha \colon \mathcal{A} \bullet (\sqcup P \colon \mathbb{F}S \mid s \in t.P \bullet \sqcap_P \alpha)) & \\
= & (\sqcup\alpha \colon \mathcal{A} \bullet t^+.\alpha.s) \ . & \text{Def. 8.6.2}
\end{array}
$$

□

For sufficiency — that $\mathbb{T}_s S$ contains all continuous and semi-linear elements of $\mathbb{T}S$ — we introduce a function from $\mathbb{T}S$ back to $\mathbb{S}S$ that acts as the inverse of $(\cdot)^+$ on $\mathbb{T}_s S$:

Definition 8.6.6 STANDARD RETRACTION   For expectation transformer $t \colon \mathbb{T}S$, define its *retraction* t^- in $\mathbb{S}S$ as

$$(s \in t^-.P) \quad := \quad (t.[P].s = 1) \ ,$$

for s in S, and P a subset of S. □

If for some t in $\mathbb{T}S$ the retraction t^- is not continuous and feasible, then we say that $(\cdot)^-$ is undefined at t. It is easy to see however that t^- is feasible when t itself is feasible.[25]

We now note an important property of semi-linear transformers in $\mathbb{T}S$.

---

[24]This step exploits the continuity of $\sqcap_P(\cdot)$ for finite P, and depends on the directedness of $\mathcal{A}$. It can be shown via the "arbitrary c" technique used in the proof of Lem. 5.6.6 that continuity follows from sublinearity: indeed it is a special case of it.

[25]For non-looping programs the retraction can be thought of as replacing all occurrences of $_p\oplus$ by $\sqcap$. If the program contains loops, however, then there is a difference: the retraction of

$$\textbf{do } b \ \rightarrow \ b := \textbf{true } _{1/2}\oplus \textbf{false od} \tag{8.3}$$

is just $b := \textbf{false}$, since retraction acts on the program's *meaning*, not on its text — and the meaning our probabilistic semantics gives to (8.3) is just that assignment as stated. (In fact it is an example of a program text in which probabilistic choice appears but whose overall effect is standard.)   ...

Definition 8.6.7 STANDARD-PRESERVING TRANSFORMERS
Expectation transformer $t: \mathbb{T}S$ is said to be *standard-preserving* iff it takes standard postconditions to standard preconditions.   □

Lemma 8.6.8 If $t: \mathbb{T}S$ is semi-linear, then it is standard-preserving.
    Proof   The proof is immediate from the fact that any expectation α in $\mathbb{E}S$ is standard iff $2\alpha \ominus \underline{1} \equiv \alpha$.   □

Lem. 8.6.8 allows us to show that t^- is continuous whenever t is continuous and semi-linear.

Lemma 8.6.9 If t in $\mathbb{T}S$ is continuous and semi-linear then t^- is continuous.
    Proof   Let $\mathcal{P}$ be a directed set of subsets of S. Then

$$s \in t^-.(\cup\mathcal{P})$$
iff	$t.[\cup\mathcal{P}].s = 1$	Def. 8.6.6
iff	$(\sqcup P: \mathcal{P} \cdot t.P.s) = 1$	t continuous; $\mathcal{P}$ directed
iff	$(\exists P: \mathcal{P} \cdot t.P.s = 1)$	P standard; t standard-preserving
iff	$(\exists P: \mathcal{P} \cdot s \in t^-.P)$	Def. 8.6.6
iff	$s \in (\cup P: \mathcal{P} \cdot t^-.P)$.	

    □

We have just shown that t^- is defined whenever t is semi-linear and continuous; but we can also show that $(\cdot)^-$ acts as the inverse of $(\cdot)^+$ between $\mathbb{S}S$ and $\mathbb{T}_s S$:

Lemma 8.6.10 For all t in $\mathbb{S}S$ we have

$$t^{+-} \quad = \quad t \ .$$

    Proof   For arbitrary s in S, and P subset of S, we reason

$$s \in t^{+-}.P$$
iff	$t^+.P.s = 1$	Def. 8.6.6

...[25]On the other hand, the program

$$\text{do } b \ \rightarrow \ b := \textbf{true} \sqcap \textbf{false od} \ ,$$

obtained from (8.3) by the suggested syntactic replacements, is different because when b is initially *true* its termination is not guaranteed. (This is why we were careful to refer to a *syntactic* transformation in Footnote 14 on p. 44 concerning nontermination of Monte-Carlo algorithms.)

Mathematically, the effect we are seeing is that $(\cdot)^-$ is not continuous; for, if it were, we would be able to distribute it through iteration — which is precisely what the above example shows us we cannot do.

The issue is particularly relevant when dealing with "abstract probabilistic choice," where we can conclude that programs terminate almost certainly while knowing only that their explicit probabilistic choices $_p\oplus$ all satisfy $0 < p < 1$. This kind of reasoning has been built in to an abstract-probabilistic version of the B Method called qB [MMH03].

iff $t.P.s = 1$ Lem. 8.6.3
iff $s \in t.P$.

□

Lemma 8.6.5 showed that only continuous and semi-linear transformers lie in $\mathbb{T}_s S$; the proof that $\mathbb{T}_s S$ contains *all* continuous and semi-linear elements of $\mathbb{T}S$ can now be completed, and is given in two parts.

Lemma 8.6.11 If expectation transformer $t : \mathbb{T}S$ is continuous and semi-linear then

$$t^{-+} \quad \sqsubseteq \quad t .$$

Proof   Because t is semi-linear, it is feasible and so t^- is defined. Now for arbitrary α in $\mathbb{E}S$ and s in S, we have

$\qquad t^{-+}.\alpha.s$
$=\qquad (\sqcup P : \mathbb{F}S \mid s \in t^-.P \cdot \sqcap_P \alpha)$ Def. 8.6.2
$=\qquad (\sqcup P : \mathbb{F}S \mid t.P.s = 1 \cdot \sqcap_P \alpha)$ Def. 8.6.6
$\le\qquad (\sqcup P : \mathbb{F}S \cdot (\sqcap_P \alpha) * (t.P.s))$
$=\qquad (\sqcup P : \mathbb{F}S \cdot t.((\sqcap_P \alpha) * P).s)$ t semi-linear, hence scaling
$\le\qquad (\sqcup P : \mathbb{F}S \cdot t.\alpha.s)$ t continuous, hence monotonic
$=\qquad t.\alpha.s$.

□

For the second part we address the opposite inequality.

Lemma 8.6.12 If expectation transformer $t : \mathbb{T}S$ is continuous and semi-linear then

$$t^{-+} \quad \sqsupseteq \quad t .$$

Proof   Take arbitrary α in $\mathbb{E}S$ and s in S, and define

$$c \quad := \quad t^{-+}.\alpha.s \quad = \quad (\sqcup P : \mathbb{F}S \mid t.P.s = 1 \cdot \sqcap_P \alpha) ,$$

so that we must show $c \ge t.\alpha.s$.

If we had

$$t.(\alpha \ominus \underline{c}).s \quad = \quad 0 , \tag{8.4}$$

then semi-linearity of t would give us $t.\alpha.s \ominus c = 0$ as required. Thus we establish (8.4), arguing for a contradiction; note that if $\alpha \equiv \underline{0}$ then the argument is trivial. If $\alpha \ne \underline{0}$ we have

$\qquad 0$
$<\qquad t.(\alpha \ominus \underline{c}).s$ assume for contradiction
$=\qquad (\sqcup\alpha) * t.((1/\sqcup\alpha) * (\alpha \ominus \underline{c})).s$ t scaling; $\sqcup\alpha \ne 0$
$\le\qquad (\sqcup\alpha) * t.\{s' : S \mid \alpha.s' > c\}.s$, t monotonic

giving $0 < t.\{s': S \mid \alpha.s' > c\}.s$, and so showing in fact that

$$t.\{s': S \mid \alpha.s' > c\}.s \quad = \quad 1 \, , \tag{8.5}$$

since by Lem. 8.6.8 we know t is standard-preserving.

Since t is continuous as well, we can choose a finite subset P of the set of states $\{s': S \mid \alpha.s' > c\}$ such that $t.P.s = 1$; note however that P cannot be empty (since t is feasible). Then for arbitrary s' in P we have

$$s' \in P$$

implies $\alpha.s' > (\sqcup Q : \mathbb{F}S \mid t.Q.s = 1 \cdot \sqcap_Q \alpha)$ definitions of P and c

implies $\alpha.s' > \sqcap_P \alpha$, $t.P.s = 1$ allows us to take $Q := P$

a contradiction since we could have chosen s' in P at the point where α is smallest.   □

Corollary 8.6.13 If $t: \mathbb{T}S$ is continuous and semi-linear, then $t^{-+} = t$.

   Proof   Lems. 8.6.11 and 8.6.12 suffice.   □

Theorem 8.6.14 SEMI-LINEARITY CHARACTERISES
                STANDARD TRANSFORMERS

   The standard predicate transformers $\mathbb{T}_s S$ are characterised by continuity and semi-linearity: transformer t in $\mathbb{T}S$ is continuous and semi-linear iff it is in $\mathbb{T}_s S$.

   Proof   The result follows from Lem. 8.6.5 and Cor. 8.6.13.   □

Thus we have our characterisation of standard behaviour, within all of $\mathbb{T}S$. We conclude, however, by noting that we can restrict $\mathbb{T}S$ in ways that make the characterisation of standard behaviour particularly simple. For example, among the continuous and semi-sublinear transformers, standard and standard-preserving are equivalent:

Lemma 8.6.15 If $t: \mathbb{T}S$ is continuous and semi-sublinear, then it is standard exactly when it is standard-preserving.

   Proof   Examination of the proof of Lem. 8.6.12 shows that continuity, semi-sublinearity and standard-preservation are sufficient: thus $t^+ = t$ under those conditions, showing that t is standard.

   For the opposite direction, we have immediately from Lem. 8.6.3 that t is standard-preserving if it is standard.   □

Finally we have the following:

Lemma 8.6.16 Any sublinear, continuous and standard predicate transformer is positively conjunctive (distributes $\sqcap$) over all expectations.

   Proof   We showed at (1.24) that if standard-preserving t is sublinear, then for any standard P, Q we have

$$t.(P \sqcap Q) \quad \equiv \quad t.P \sqcap t.Q \, .$$

We now show that the conjunctivity extends to all arguments.

For any state s, either $t.\underline{1}.s$ is zero or it is one. If it is zero then $t.\alpha$ is zero for all α (scaling, α bounded, monotonicity), giving conjunctivity trivially. If it is one, then by continuity there must be a finite and non-empty P with $t.P.s = 1$ and, by conjunctivity, we can choose that P to be the smallest such. Then, because t is standard, we have for any α that

$$t.\alpha.s \;=\; t^{-+}.\alpha.s \;=\; (\sqcup Q \colon \mathbb{F}S \mid t.Q.s = 1 \cdot \sqcap_Q \alpha) \;=\; \sqcap_P \alpha$$

for that (fixed) P, and so t is conjunctive at s over all post-expectations.

□

We note in conclusion however that the converse does not hold in general. That is, a sublinear and conjunctive program is not necessarily standard: consider **skip** $_{\frac{1}{2}}\oplus$ **abort**. The implication does hold, however, for terminating programs.

8.7   Summary

We have explored the structure of the expectation-transformer space, and summarise it in Fig. 8.7.1; and we have extended the earlier theory of Chap. 5, together with the results of this chapter, to infinite state spaces.

Even with infinite state spaces, we need only use discrete probability distributions. That is because under the assumption that any implementation of a probabilistic specification can only use the discrete choice $_p\oplus$ it follows that all implementations will result only in discrete distributions. It is interesting nevertheless to reinstate continuous distributions [Jon90] into our programs and then to investigate their effects on the healthiness conditions [MM01b].

In the standard case, Back and von Wright's decomposition [BvW90] of the transformer space into deterministic, demonic and angelic components explains the complete behaviour of all monotonic predicate transformers; thus in that theory monotonicity remains as the only surviving restriction — and a necessary one — since in the activity of program specification and development monotonicity is paramount.

For expectation transformers the situation is very different — our Thm. 8.5.8 implies that there is a part of the space "left over," namely the monotonic transformers that fail to be semi-sublinear. We call them *exotic*.

An infeasible transformer t is exotic, since the feasibility condition — that for all expectations α we have $t.\alpha \Rrightarrow \sqcup \alpha$ — is trivially preserved by our choice operators, from which the non-exotic subspaces are constructed. As we noted earlier (p. 144) however, allowing infinite-valued transformers could lead to a more general treatment overall, as was found when Dijkstra's

Law of the Excluded Miracle was relaxed for standard programs [Mor87, Mor88b, Nel89].

But there are *two* kinds of feasibility in the probabilistic model, and the Law of the Excluded Miracle deals with only one of them: that programs as relations have non-empty result sets of distributions. This is expressed in arithmetic by our disallowing ∞ from the reals (since it would have to be included otherwise as the infimum of an empty set).

The other feasibility — which does not arise for standard programs — is that distributions sum to no more than one: its arithmetic characterisation is the $\ominus$-subdistributivity introduced by the downwards-facing plane forming the base of the hyper-pyramid of distributions (Sec. 6.9.1). Monotonic transformers like $(\lambda\,\alpha \cdot 2\alpha)$ are exotic also: they do not involve infinite values, and do not seem to correspond to any standard criterion of (in-)feasibility.

One kind of exotic transformer that arises naturally (as opposed to the pathological — but simple — examples above) is the *greatest liberal* pre-expectation of a program: applied to a standard post-condition it returns the probability that either the post-condition is established, or that the program never terminates — though monotonic, it is not semi-sublinear. (The definitions [MM01b] of greatest-liberal pre-expectations are given elsewhere; we used them in Chap. 7.) However the general importance of this part of the space in a programming context is far from clear and it remains a topic for further investigation.

Other examples of exotic transformers arise in the definition of the temporal operators that generalise Kozen's predicate transformer definitions [Koz83], which we examine in Part III. (See for example Footnote 14 on p. 250.)

The practical import of our results is to give laws in the (arithmetic) programming logic that are satisfied by programs in the various levels of the hierarchy. Such laws are the basis for derived program-development rules such as our use of variants and invariants for loops (Chapters 2 and 7) and, more generally, for programming algebras and the probabilistic temporal logics to come in Part III.

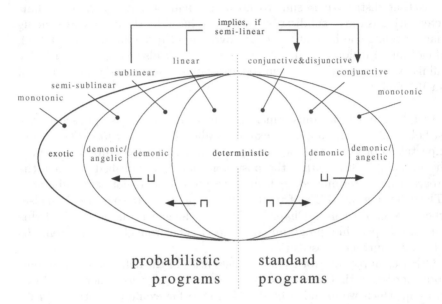

Figure 8.7.1. STRUCTURE OF TRANSFORMER SPACES

The property of *semi-linearity* characterises standard programs; and, on the standard side, the demonic/angelic programs exhaust the space of monotonic transformers [BvW90].

For each standard sub-space there is a corresponding probabilistic subspace, constructed analogously and having a similar — but generalised — algebraic characterisation. However, there are monotonic probabilistic transformers lying outside $\mathbb{T}_{\sqcap} S$: we call them *exotic*.

A summary of the named algebraic characterisations is given in Fig. 8.7.2.

conjunctivity	$t.(\alpha_1 \sqcap \alpha_2)$	$\equiv$	$t.\alpha_1 \sqcap t.\alpha_2$
disjunctivity	$t.(\alpha_1 \sqcup \alpha_2)$	$\equiv$	$t.\alpha_1 \sqcup t.\alpha_2$
semi-linearity	$c(t.\alpha) \ominus \underline{c}'$	$\equiv$	$t.(c\alpha \ominus \underline{c}')$
continuity	$t.(\sqcup\mathcal{A})$	$\equiv$	$(\sqcup\alpha\colon\mathcal{A}\cdot t.\alpha)$

---

semi-sublinearity		$t.(c\alpha)$	$\equiv$	$c(t.\alpha)$	scaling
	and	$t.\alpha \ominus \underline{c}$	$\Rrightarrow$	$t.(\alpha \ominus \underline{c})$	$\ominus$-subdistributivity

---

sublinearity   *semi-sublinearity*

and $t.\alpha_1 + t.\alpha_2 \;\Rrightarrow\; t.(\alpha_1 + \alpha_2)$ sub-additivity

---

linearity   *semi-sublinearity*

and $t.(\alpha_1 + \alpha_2) \;\equiv\; t.\alpha_1 + t.\alpha_2$ additivity

Expectations α are bounded, real c is non-negative and set $\mathcal{A}$ of expectations is bounded and directed (and therefore has a limit).

Figure 8.7.2. SUMMARY OF NAMED ALGEBRAIC PROPERTIES OF TRANSFORMERS

Chapter notes

In standard predicate transformers a careful taxonomy of properties has led to a rich theory of program refinement. Back and von Wright [BvW98] have examined the structure of monotonic predicate transformers, while applications of non-standard constructs have been found to be extremely useful: Gardiner and Morgan [GM93] exploited angelic choice in their demonstration of a complete rule for data refinement, and Cohen has used *miracles* very effectively in his *omega algebra* [Coh00].

Similarly in the probabilistic setting there are many more expectation transformers than the sublinear ones — angelic nondeterminism models the choices available to a maximising player (see Chap. 11), and the more exotic *miracles* turn out to have implications in the specification of discounted games [MM01c].

Part III

Advanced topics:

Quantitative modal logic and game interpretations

244

9

Quantitative temporal logic:
An introduction

9.1   Modal and temporal logics

Modal logics extend classical predicate logic by dealing (roughly speaking) with a number of classical interpretations all at once. Each interpretation is called a "possible world," and specially introduced modal operators of "necessity" and "possibility" are used in the logic to allow formulae in a "current" world to express properties that might hold in other worlds accessible from that one [Tur84, Gol03]. Special cases include logics of knowledge [FHMV95] and — our topic here — temporal logics.

Temporal logics specialise modal logic to the case in which the worlds
are states of a computation and the accessibility relation between worlds
is determined by the state-to-state evolution brought about by execution;
a temporal formula interpreted "here and now," in the current state, can
express properties of (future) states which the computation might or might
not reach later [BAPM83, Sti92, Eme90].[1]

Probability can be added to temporal logic in a variety of ways. At
one extreme the underlying computation is probabilistic, but the temporal
formulae remain standard and do not refer to the probabilities directly.
However the notion of validity is altered from "is true" to "is almost certain" [LS82]. At the other extreme are logics which allow virtually complete
access to explicit probabilities, introducing for example a real-valued term
"$\Pr \phi$" giving for any formula ϕ the probability of its holding over the
distribution of possible computations [FH84, for dynamic logic]. Further
variation is provided in both cases by the extent to which demonic nondeterminism is allowed and, if it is, to what fairness constraints it must
conform.

Our approach, introduced in this chapter, takes a middle course — but
also a more general one. Rather than use temporal logic we take its superset,
Kozen's *modal μ-calculus* [Koz83]; and rather than introduce probability
explicitly into the logic, we replace Booleans "wholesale" by real numbers
just as we did for transformers in Parts I and II.

What we get is a modal logic of quantitative expectations, rather than
of truths.[2]

In summary, we use our expectations — and their transformers — to
generalise the standard modal μ-calculus from Booleans to reals, and we
then take the temporal subset of that.

A satisfactory operational interpretation of the new logic cannot however be based on the notion of "achieving a predicate," as it is in the
standard case, since that is not meaningful for formulae whose value might
lie somewhere strictly between zero and one. Instead, for the quantitative generalisation we take an operational view in which the underlying
computation is a gambling game: the player(s) seek to maximise or minimise their expectations using strategies that determine whether or not to
continue playing.[3]

---

[1] "Past-tense" temporal logics allow formulae to refer as well to states from which the
current one might have come [LPZ85].

[2] We recover explicit probabilities, when we need them, by observing as we have before
(p. 13) that the probability of an event is the expected value of its characteristic function.

[3] This interpretation is related to Back and von Wright's use of games to motivate the
interaction of angelic and demonic nondeterminism expressed by predicate transformers
[BvW96]; Stirling suggests a similar game-based interpretation of validity in the modal
μ-calculus [Sti95], which is the main topic of Chap. 11. In both of those cases only
standard Boolean predicates are used.

Throughout Part III we restrict expectations to the unit interval $[0,1]$ — that is (as in Chap. 7) they are all one-bounded.

9.2 Standard temporal logic: a review

Standard temporal logic [BAPM83] extends classical predicate logic with modal operators based on distinguished "instants of time" called "states"; the instants are determined by an underlying computation that "steps" from one state to the next, according to some given relation between initial and final states.

Conventional models for standard branching-time temporal logic extend models for standard predicate logic so that classical formulae are interpreted within the states separately, and the "step" relation between states is used to interpret the temporal operators.[4],[5]

The conventional approach thus sees the underlying computation as a relation between initial and (possibly several, with nondeterminism) final states. In this section we review an alternative but no less powerful model in which the computation is a predicate transformer rather than a relation.[6]

9.2.1 Syntax

For classical predicate logic, to which the temporal operators will be added, we use the conventional syntax over variables $a, \cdots, z$, with constant, function and relation symbols taken directly from mathematics as needed.

Late Greek letters ϕ, ψ are used as meta-variables standing for (syntactic) formulae; late Roman letters P, Q are used for (both syntactic and semantic) predicates, except where noted explicitly — *i.e.* we use them both for Boolean-valued formulae and for the predicates they denote.

The propositional connectives are false, true, $\neg$ (highest precedence), $\wedge$, $\vee$, $\Rightarrow$ and $\Leftrightarrow$ (lowest precedence); the quantifiers are $\forall, \exists$ with (as usual) their scope always indicated by explicit parentheses, as in $(\forall x \bullet \phi)$ for formula ϕ.

---

[4] In this presentation we use only universal branching operators (*i.e.* not existential as well); see Footnote 11 on p. 249. A major alternative to branching-time temporal logic is *linear-time* temporal logic [Lam80, Lam83, EH86].

[5] By CLASSICAL we mean "non-temporal"; by *standard* we continue to mean "non-probabilistic" or, more generally "quantitative but restricted to the two values zero and one."

[6] The two approaches are of equivalent power, as we saw in Chap. 5, if we use conjunctive transformers (in the standard case) or sublinear ones in general. The possibility of angelic transformers adds more expressive power in principle.

The symbols we use for the temporal operators are as follows:

- $\bigcirc\phi$ (*next-time* ϕ),
- $\Diamond\phi$ (*eventually* ϕ),
- $\Box\phi$ (*always* ϕ) and
- $\phi \rhd \psi$ (ϕ *unless* ψ).

They have highest precedence of all and associate to the right.[7]

9.2.2   Classical interpretation

To interpret classical formulae we fix a universe $\mathcal{V}$ of values and a meaning over $\mathcal{V}$ for any constant, function and relation symbols used. The *state space*

$$S \quad := \quad Var \to \mathcal{V}\ ,$$

is then the set of functions from a fixed set of variables *Var* to values in $\mathcal{V}$.

We write $\|\phi\|_{\mathcal{V}}$ for the meaning of a formula ϕ, a predicate, which is in each case a function from S to the Booleans, or equivalently a subset of S; we then say that a standard formula *is true* (resp. *false*) in a state s — similarly *holds* (resp. *does not hold*) in s, or that s *satisfies* (resp. *does not satisfy*) a formula — exactly when $\|\phi\|_{\mathcal{V}}.s$ gives true (resp. false) when applied to that state.[8]

9.2.3   Informal temporal semantics

For the temporal operators we add to the interpretation above an *underlying computation*, represented by a *predicate transformer* over the state space — the one transformer gives for each state the possible "single steps" that can take that state to each of its successor states. Usually we call the transformer *step*, and so we have

$$step\colon \mathbb{P}S \leftarrow \mathbb{P}S\ ,$$

in which as usual we write the functional arrow backwards to emphasise that the transformer takes final predicates to initial ones.[9] If *step* is deterministic then for each state there is just one successor; where it is nondeterministic there may be more than one.

---

[7]Alternative syntax for the temporal operators includes X for *next-time*, F for *eventually* and G for *always*.

[8]For those who might have skipped straight to this chapter, we recall that "." as in "$f.x$" is FUNCTION APPLICATION and associates to the left.

[9]Strictly speaking therefore we should now take the universe $\mathcal{V}$ and the transformer *step* jointly as the interpretation, and write $\|\phi\|_{(\mathcal{V}, step)}$; but shortly we will suppress all the subscripts for neatness, and will rely on context to tell us what they should be.

When convenient, we assume *step* is the meaning of some (syntactic) program *Step* written in the language of guarded commands over the variables *Var*: thus we have $step = wp.Step$.

We assume also that *step* satisfies the usual *healthiness conditions* for predicate transformers [Dij76], in particular that it is "feasible" and "positively conjunctive," that is

$$step.\text{false} \equiv \text{false} \qquad \textit{feasible}$$
$$step.(\phi \wedge \psi) \equiv step.\phi \wedge step.\psi \qquad \textit{positively conjunctive,} \quad {}^{10}$$

and we can now give an operational intuition for the temporal operators by referring to the execution of various program fragments.

- Next-time *informally*

The formula $\circ\phi$, pronounced "next-time ϕ", holds in the current state if exactly one transition, *i.e.* a single execution of the underlying computation *step*, is guaranteed to establish ϕ. Remembering that ϕ is (interpreted as) a predicate over the state space, and that *step* is just *wp.Step*, we therefore can say that

$$\circ\phi \equiv wp.Step.\phi . \tag{9.1}$$

We note the following:

1. If nondeterminism is present in *Step* it is understood *demonically*, so that $\circ\phi$ holds only if ϕ holds in *all* of the next states that *Step* can reach from the current state.[11]

2. Divergence — or "aborting" behaviour — in *Step* is also interpreted demonically: if *Step* can diverge in the current state, then $\circ\phi$ does not hold for any ϕ. (Even $\circ$true does not hold when *Step* diverges.)

- Eventually *informally*

The formula $\diamond\phi$, pronounced "eventually ϕ", holds if repeated execution of *Step* (including possibly no executions at all) is guaranteed to establish ϕ if sufficiently many steps are taken.

Using programming idioms we have

$$\diamond\phi \equiv wp.(\textbf{do } \neg\phi \rightarrow Step \textbf{ od}).\text{true} , \tag{9.2}$$

so that $\diamond\phi$ holds if execution of the above loop is guaranteed to terminate (necessarily in a state satisfying ϕ, since termination occurs only when the guard is false).[12]

---

[10]Recall that our probabilistic healthiness conditions (Sec. 5.6) generalise these.

[11]This is the "universal" form of the operator, sometimes written ∀X if the logic is expressive enough to include the existential form ∃X as well [BAPM83].

[12]This is why the syntactic form *Step* is convenient — though of course it is not necessary, since the semantics of the loop can be given in terms of *step* directly.

Thus if Program (9.2) can loop forever then $\Diamond\phi$ is false; similarly $\Diamond\phi$ is false if *Step* can diverge before ϕ is established.

- Always *informally*

The formula $\Box\phi$, pronounced "always ϕ", holds if repeated execution of *Step* (including no executions at all) is guaranteed to preserve the truth of ϕ no matter how many steps are taken. We have

$$\Box\phi \quad\equiv\quad wnp.(\textbf{do } \phi \rightarrow Step \textbf{ od}).\text{false} , \qquad (9.3)$$

where *wnp* (weakest *nonterminating* precondition) differs from *wp* only in interpreting looping as success rather than failure, so that for example

$$wnp.(\textbf{do true} \rightarrow \textbf{skip od}).\text{false} \quad\equiv\quad \text{true} . \quad ^{13} \qquad (9.4)$$

Thus $\Box\phi$ holds — for all ϕ — whenever execution (9.3) from the current program state is guaranteed *not* to terminate: that would be the case if *Step* were **skip**, as (9.4) illustrated.[14]

Nondeterminism and divergence continue to be interpreted demonically, making $\Box\phi$ false if any of the execution steps diverges or can reach a state not satisfying ϕ.

- Unless *informally*

The formula $\phi \triangleright \psi$, pronounced "ϕ unless ψ", holds if repeated execution of *Step* is guaranteed to preserve ϕ unless ψ is established.[15] We have

$$\phi \triangleright \psi \quad\equiv\quad wnp.(\textbf{do } (\phi \wedge \neg\psi) \rightarrow Step \textbf{ od}).\psi , \qquad (9.5)$$

so that $\phi \triangleright \psi$ holds if execution of the above loop is guaranteed either to loop forever or to terminate establishing ψ.

---

[13]In Sec. 9.2.4 we see that *wnp* simply takes the greatest- rather than least fixed-point for loops, acting like *wp* otherwise; it differs from the weakest *liberal* precondition (*wlp*) — which also takes the greatest fixed-point — in that *wnp* applies *wp* (rather than *wnp* again "hereditarily") to the loop body. Thus *wnp* treats divergence in the body as failure, whereas *wlp* treats it as success.

[14]We note in passing that *wnp.step* is not healthy in general, even when *step* itself is: it might not be feasible.

[15]This operator is also called *weak until*.

Note that if termination can occur when ψ does not hold then $\phi \rhd \psi$ is false. Comparing the programs (9.3) and (9.5) shows that we have the equivalence

$$\Box\phi \quad \equiv \quad \phi \rhd \mathsf{false} \,,$$

so that *always* is a special case of *unless*.

9.2.4 Formal semantics

Using the informal program-based descriptions of the previous section as a guide, we now give the precise meaning of the temporal operators. Once the underlying computation is given they, like classical formulae, denote predicates over the state space.

In each case the conventional *wp*-semantics gives the definition directly (perhaps after simplification) from Programs (9.1)–(9.5) above, except that as noted we use greatest- (ν) rather than least- (μ) fixed-points for loops under *wnp*.

To avoid clutter, we fix the underlying computation *step* and the values $\mathcal{V}$, and write just $\|\cdot\|$; the ordering with respect to which the least- and greatest-fixed-points are taken is as usual the pointwise extension of $\mathsf{false} \sqsubseteq \mathsf{true}$.

Definition 9.2.1 NEXT-TIME (STANDARD)
For convenience in this and the following definitions, we define $\|\circ\|$ to be the transformer *step* itself. Then for the definition of *next-time* we have

$$\|\circ\phi\| \quad := \quad \|\circ\|.\|\phi\| \,,$$

in which we see the predicate transformer $\|\circ\|$ applied to a predicate $\|\phi\|$ (effectively a postcondition) to give another predicate $\|\circ\phi\|$ (the weakest precondition).[16] $\Box$

Definition 9.2.2 EVENTUALLY (STANDARD)
For *eventually* we appeal to the informal formulation (9.2) above, which was an iteration and so refers us in turn to the semantics at (7.11) on p. 198; we have thus

$$\|\Diamond\phi\| \quad := \quad (\mu P \bullet \|\phi\| \vee \|\circ\|.P) \,.$$

The bound variable P is of type "predicate over S". $\Box$

---

[16] A more conventional "forward" definition of *next-time* would be along the lines of

$$s \models \circ\phi \quad := \quad (\forall s' \mid s \text{ can reach } s' \bullet s' \models \phi) \,,$$

where "s can reach s'" might be "$s' \in rp.\|\circ\|.s$" with "rp" being the standard retraction, operating between STS and SRS in Fig. 5.9.1 on p. 158, that converts a transformer back to a relation.

It is equivalent to our Def. 9.2.1 — which has the characteristic "backwards" flavour of predicate transformers — except that our definition automatically includes the extra possibility of aborting behaviour.

Similarly we can use (9.3) and (9.5) to suggest definitions for the remaining operators:

Definition 9.2.3 ALWAYS (STANDARD)

$$\|\Box\phi\| \quad := \quad (\nu P \cdot \|\phi\| \wedge \|\circ\|.P) \,.$$

□

Definition 9.2.4 UNLESS (STANDARD)

$$\|\phi \triangleright \psi\| \quad := \quad (\nu P \cdot \|\psi\| \vee (\|\phi\| \wedge \|\circ\|.P)) \,.$$

□

All the definitions in this section are based on those of Kozen [Koz83], and are related to similar definitions later given by Morris [Mor90], by Lukkien and van de Snepscheut [LvdS92], and by Hesselink [Hes95]. How to discover and prove general properties of temporal operators, when defined this way, will be examined when we treat their probabilistic versions.

9.3   Quantitative temporal logic

Our generalisation is to use the expectation transformers of Parts I and II, rather than predicate transformers, to give new *quantitative* definitions of the temporal operators in the style of the previous section. We then explore the algebra and interpretation of the resulting quantitative temporal calculus.

To define quantitative versions of the temporal operators we merely adjust Defs. 9.2.1–9.2.4 so that they act over expectations rather than predicates: operators $\wedge, \vee$ are replaced by $\sqcap, \sqcup$ respectively, and we take fixed points of real- rather than Boolean-valued functions.[17,18]

We use early Roman letters A, B for (semantic) expectations.

Definition 9.3.1 NEXT-TIME (QUANTITATIVE)
For the definition of quantitative *next-time* we have

$$\|\circ\phi\| \quad := \quad \|\circ\|.\|\phi\| \,,$$

in which we see the *expectation* transformer $\|\circ\|$ applied to an expectation $\|\phi\|$ to give the greatest pre-expectation $\|\circ\phi\|$. □

---

[17]Our choice of $\sqcap, \sqcup$ rather than say $*, +$ is justified by the operational interpretation we are able to give subsequently, as we shall see.

[18]Recall that our expectations are one-bounded in this (and subsequent) chapters.

Definition 9.3.2 EVENTUALLY (QUANTITATIVE)

For quantitative *eventually* we cannot appeal to the informal "iteration" semantics, because the subject ϕ of the temporal operator $\Diamond$ — which in the standard case was the guard of the motivating iteration — is now possibly a proper expectation.[19] Instead we proceed directly from Def. 9.2.2, whence generalisation gives

$$\|\Diamond\phi\| \;:=\; (\mu A \cdot \|\phi\| \sqcup \|\circ\|.A) \,.$$

Here the bound variable A is of type "one-bounded expectation over S".

□

In the same style we generalise the last two definitions.

Definition 9.3.3 ALWAYS (QUANTITATIVE)

$$\|\Box\phi\| \;:=\; (\nu A \cdot \|\phi\| \sqcap \|\circ\|.A) \,.$$

□

Definition 9.3.4 UNLESS (QUANTITATIVE)

$$\|\phi \triangleright \psi\| \;:=\; (\nu A \cdot \|\psi\| \sqcup (\|\phi\| \sqcap \|\circ\|.A)) \,.$$

□

In subsequent sections we give both an operational interpretation of the quantitative temporal operators and an investigation of the laws they satisfy. Among other things, that will explain why we have chosen the arithmetic $\sqcap$ and $\sqcup$ to generalise the Boolean $\wedge$ and $\vee$.

As a start, however, we exercise our intuition by recalling the programs from Sec. 9.2.3 and allowing the computation *Step* to be probabilistic — but we stipulate for the time being that the formula ϕ remain standard (that is, be zero or one only). That is, we restrict ϕ to be $[P]$ for some predicate P, which allows us to get a feel for the meaning of the quantitative operators in a familiar setting. Program (9.2) for *eventually* becomes

$$wp.(\mathbf{do}\ \neg P\ \to\ Step\ \mathbf{od}).[P]\,, \tag{9.6}$$

where we have used the explicit (and embedded) post-expectation $[P]$, rather than **true**, simply to emphasise that the "aim" of the loop is to establish P.[20]

The expression (9.6) then gives $\Diamond[P]$ in agreement with Def. 9.3.2, because ϕ is in fact $[P]$, and it is easily checked by arithmetic that we can reason

---

[19]The "probabilistic guards" of Def. 7.7.1 do not help, and are not related: use of a proper expectation ϕ as a guard, in that style, would not give the meaning we want here.

[20]It makes no difference here because P must be true on termination anyway, given the guard $\neg P$ — but in the fully quantitative case (to come) we use the post-expectation for the "goal" and the guard for a "strategy."

body of Def. 9.3.2: fixed point

$\equiv$ $\quad\| [P] \| \sqcup \|\circ\|.A$ $\hfill \phi$ is $[P]$

$\equiv$ $\quad$[true] **if** P **else** $\|\circ\|.A$ $\hfill$ arithmetic

$\equiv$ $\quad$body of loop semantics (7.11): fixed point , $\hfill G$ is $\neg P$

where in fact the middle equality holds for *any* (one-bounded) expectation on the right of $\sqcup$ (*i.e.* whether equal to some $\|\circ\|.A$ or not). And since the post-expectation of the iteration is a characteristic function, we can as usual interpret its expected value as given by (9.6) as a probability: in a given state the value of $\diamondsuit [P]$ is indeed just the probability that, starting from that state, repeated execution of *Step* will establish P eventually.

Similar reasoning applies to $\circ [P]$, to $\Box [P]$ and to $[P] \rhd [Q]$, based on Programs (9.1), (9.3) and (9.5) applied to standard expectations, showing that standard temporal formulae over probabilistic computations may be interpreted as probabilities directly, as one would hope: for example, formula $\Box P$ is the probability that P will never become false in future steps of the computation. That interpretation is sufficient for programs containing no demonic nondeterminism.

As noted however in the concluding remarks of Sec. 1.3.3 (p. 15), and explained in detail in Sec. A.1, standard formulae have insufficient resolution when probabilistic and demonic nondeterminism interact: one must use proper expectations. And the above "probability of establishing P" analogies fail — for all four programs — in that more general case, because it does not make sense to speak of "establishing" a proper expectation.

To interpret quantitative temporal formulae in general, therefore, we will rely on a shift of viewpoint described in the next section: we consider gambling games.

9.4   Temporal operators as games

If establishing a postcondition is "winning a game," then our generalisation to expectations simply adds the extra information of how much is won: the standard case embeds as "win *one* if the postcondition is established; win *zero* if it is not." We took this view earlier, in our discussion of card games in Sec. 1.3.2.

Here however we consider *poker machines*.[21] They have a window in which various symbols are visible; pulling a handle changes the symbols probabilistically; and each configuration of visible symbols is associated with a certain reward (possibly — in fact, usually — *zero*). In computational terms, each configuration of symbols shown in the window is a state;

---

[21]They are also called *slot* machines (US) and *fruit* machines (UK).

the underlying computation is the effect of one pull of the handle; and the post-expectation is the "pay function" written on the front of the machine, giving the worth to the player of each symbol configuration should he press *pay* at that point.

Each of the temporal operators corresponds to a different way of gambling with the machine, and we consider them in turn.

9.4.1 The next-time *game*

The simplest game is *next-time* where, directly from Def. 9.3.1, we can see that the game consists of pulling the handle exactly once and then pressing *pay*: the expectation $\circ\phi$ evaluated over the initial configuration (the configuration showing *before* the handle is pulled) gives the win determined jointly by the expectation ϕ and the probabilistic distribution of final configurations (that could show *after* the handle is pulled). The formula ϕ is the pay function written, as we said, on the front of the machine (usually as pictures of possible outcomes and their corresponding payout-values).

Suppose for example the configurations are just natural numbers $\mathbb{N}$, the effect of a handle pull is the computation $n := (n{+}1 \, {}_p\oplus \, 0)$,[22] and the pay function ϕ is $n/(n{+}1)$. Then the expected win from a single handle-pull is

$$\circ(n/(n{+}1))$$
$$\equiv \quad wp.(n := (n{+}1 \, {}_p\oplus \, 0)).(n/(n{+}1)) \qquad\qquad \text{Def. 9.3.1}$$
$$\equiv \quad p * (n{+}1)/(n{+}2) + \bar{p} * 0 \qquad \text{probabilistic choice; assignment}$$
$$\equiv \quad p(n{+}1)/(n{+}2) \ ,$$

from which it is evident that the expected win does depend on the initial configuration: a gambler starting at zero every time would expect an average win per game of $p/2$ in the long run; but the larger the initial configuration, the larger the expected win from a single handle pull would be (approaching p in the limit).

9.4.2 The eventually *game*

For *eventually* the player pulls the handle repeatedly (or not at all), and decides on the basis of the current configuration whether to stop and press *pay*, or to pull again. Thus for $\diamond\phi$ we consider an expectation

$$wp.(\textbf{do } G \rightarrow Step \ \textbf{od}).\phi \ , \qquad\qquad (9.7)$$

---

[22]This underlying computation resembles a quiz show where consecutive correct answers yield steadily higher prizes, but one wrong answer loses it all.

in which the Boolean guard G — a predicate — represents the gambler's *strategy*: if it is true (in a configuration) then he plays again; if it is false, he presses *pay* and stops playing.[23]

For the gambler some strategies G are clearly worse than others. The strategy $G := $ true continues the game forever, which we interpret as losing unconditionally (*i.e.* as winning *zero* no matter what ϕ is) since *pay* is never pressed; the strategy false represents pressing *pay* without pulling the handle at all, in which case the amount won is determined by the current configuration.[24]

For standard pay function $[P]$, *i.e.* a function which pays *one* for configurations satisfying P but *zero* for those that don't, an optimal strategy is given by the predicate $\neg P$ itself that continues play as long as the current configuration pays zero — because when P is false, things can only get better. Whenever P is true, however, the gambler should stop immediately — he can't win more than *one*, anyway, and could well lose it if he carried on.

Indeed (recall Program (9.6)) for the strategy $\neg P$ the expected win is $\Diamond [P]$ and remarkably, as the following lemma shows, even for proper expectations ϕ the expectation $\Diamond\phi$, applied to the initial state, gives the best that can be obtained for *any* strategy. That is,

Lemma 9.4.1 For underlying computation *Step*, expectation ϕ and *any* (memoriless) strategy G we have

$$wp.(\mathbf{do}\ G \to Step\ \mathbf{od}).\phi \quad \Rightarrow \quad \Diamond\phi\ ,$$

showing that $\Diamond\phi$ is an upper bound over all strategies G for the eventually game.

Proof   Compare Def. 9.3.2 with the usual least-fixed-point definition (7.11) on p. 198 of **do** $\cdots$ **od**, and use monotonicity of μ. $\square$

More significant however is the complementary result, that for any computation *Step* and expectation ϕ there is a strategy G whose expected win approaches $\Diamond\phi$ arbitrarily closely — in fact if the state space is finite, there is a G that realises $\Diamond\phi$ exactly. That is,

---

[23]Note we consider only strategies based on the current configuration, so-called MEM-ORILESS strategies: the predicate G cannot express strategies like "Play five times then stop." The use of memory in strategies is discussed at length in Chap. 11.

[24]On a real poker machine, of course, the handle must be pulled at least once after inserting your money: the normal game would thus be $\bigcirc\Diamond\phi$, *i.e.* "pull once, then decide repeatedly whether to pull again." Otherwise, once a winning configuration appeared the players would simply keep putting in more money and pressing the *pay* button.

Note the difference however between $\bigcirc\Diamond\phi$ and the similar-looking $\Diamond\bigcirc\phi$; the latter would be "repeatedly decide now (before pulling) whether to stop after just one more try" — a game whose rules most gamblers find hard to stick to.

Lemma 9.4.2 For deterministic and non-divergent *Step*, and finite state space, we have

$$\Diamond\phi \quad \equiv \quad wp.(\textbf{do } (\phi{<}\Diamond\phi) \to Step \textbf{ od}).\phi \, ,$$

so showing that $\phi < \Diamond\phi$ is the optimal strategy.   □

We sketch a proof of the above, relying on informal operational arguments rather than the *wp*-definitions.[25]

The putatively optimal strategy is the predicate $\phi < \Diamond\phi$ — note that the inequality is a Boolean function of the state — and our first lemmas establish that it guarantees *termination* of the game.

We begin by showing that if the probability of termination ever becomes zero then it remains so (since that probability could not be zero if there were any guaranteed chance of escape).

Lemma 9.4.3 For any (probabilistic) iteration with deterministic and non-divergent body, the (standard) predicate "has probability zero of termination" is invariant.

Proof   Let T be a function from the state space S to the interval $[0,1]$, with $T.s$ for state s being the probability of the iteration's eventual termination from that state; and consider a single execution of the iteration body, supposing that it takes some initial state s to a final state s'. We show the contrapositive of our claim, *i.e.* that $T.s' > 0$ implies $T.s > 0$ also.

Since the body is deterministic and non-divergent, the transition from s to s' has some nonzero probability p — otherwise it could never be taken by an execution of the loop body.[26] That gives $T.s \geq p * T.s' > 0$ immediately, as required, since the probability of termination from s generally cannot be less than the probability of termination from s via s' in particular.   □

Lemma 9.4.4 If a non-empty (standard) predicate I is preserved by the computation *Step*, and the state space is finite, then $\phi = \Diamond\phi$ holds for some state satisfying I.

Proof   Suppose first that I is just true, *i.e.* that it describes all of S. From Def. 9.3.2 we have $\phi \Rightarrow \Diamond\phi$ for any ϕ, and feasibility of *eventually* gives $\Diamond\phi \Rightarrow \sqcup\phi$.[27] Thus because the state space is finite we can choose some state $\widehat{s}$ at which ϕ attains its largest value, *i.e.* so that $\|\phi\|.\widehat{s} = \sqcup\|\phi\|$,

---

[25]The full proof of the more general result, using *wp*-definitions for rigour and allowing both nondeterminism and divergence, has the same structure as the following but depends substantially on technical material from our earlier Chapters 5 and 7 of Part II.

In the finite case the *existence* of such a strategy is a special case of Thm. 11.6.5 to come; but here we are actually saying what that strategy is.

[26]This is where determinism is used. If nondeterminism or divergence were present, a transition *might* be taken even though its minimum *guaranteed* probability was zero.

[27]Feasibility of $\Diamond$ is proved in Lem. B.6.4.

and we have then a "squashing" argument

$$\sqcup \|\phi\| \;=\; \|\phi\|.\widehat{s} \;\leq\; \|\Diamond\phi\|.\widehat{s} \;\leq\; \|\sqcup\phi\|.\widehat{s} \;=\; \sqcup\,\|\phi\|$$

which establishes $\|\phi\|.\widehat{s} = \|\Diamond\phi\|.\widehat{s}$, $i.e.$ that $\|\phi\| = \|\Diamond\phi\|$ at $\widehat{s}$. And since any state satisfies I when I is true, certainly $\widehat{s}$ does.

Now if I is not true, but still is non-empty and invariant, we simply restrict the whole system to I — the above argument repeated, "within I" in that sense, gives our $\widehat{s}$ just as before. □

We now have our termination result for the optimal strategy.

Lemma 9.4.5 For deterministic and non-divergent $Step$, the loop

$$\mathbf{do}\ (\phi{<}\Diamond\phi) \to Step\ \mathbf{od} \tag{9.8}$$

terminates with probability one from every initial state.

Proof   Let Z be the predicate denoting the states at which the probability of termination is zero. From Lem. 9.4.3 we have invariance of Z; and so from Lem. 9.4.4 we have that $\phi = \Diamond\phi$ is attained at some $\widehat{s}$ in Z provided Z is non-empty. But that would be a contradiction, since $\|\phi\|.\widehat{s} = \|\Diamond\phi\|.\widehat{s}$ is the negation of the loop guard — thus termination is with probability one (in fact, is immediate) from $\widehat{s}$, and is not with the probability zero that its membership of Z implies.

Hence Z is empty after all, and so the probability of (9.8)'s termination is everywhere nonzero. Since the state space is finite, we conclude from the Zero-One Law that the probability of termination is in fact everywhere one.
□

We have now shown that the strategy $\phi < \Diamond\phi$, the very reasonable "keep going as long as taking ϕ now is strictly worse than waiting for ϕ later," is terminating at least. We finish off as follows.

Lemma 9.4.6 For deterministic and non-divergent $Step$, and finite state space, we have

$$\Diamond\phi \;\Rightarrow\; wp.(\mathbf{do}\ \phi{<}\Diamond\phi \to Step\ \mathbf{od}).\phi \tag{9.9}$$

Proof   First we show that $\Diamond\phi$ is an invariant of the iteration: we have

$$
\begin{array}{lll}
& [\phi < \Diamond\phi] * \Diamond\phi & \text{guard and invariant}\\
\Rightarrow & [\Diamond\phi = \mathrm{o}\Diamond\phi] * \Diamond\phi & \text{Def. 9.3.2 gives } \Diamond\phi \equiv \phi \sqcup \mathrm{o}\Diamond\phi;\ \text{arithmetic}\\
\Rightarrow & \mathrm{o}\Diamond\phi & \text{arithmetic}\\
\equiv & wp.Step.\Diamond\phi\ . & \text{definition } \|\mathrm{o}\|
\end{array}
$$

But we know from Lem. 9.4.5 that the loop terminates, so we use the probabilistic loop rule (2.4.1) to continue

$$\Diamond \phi$$
$$\Rightarrow \qquad\qquad\qquad\qquad\qquad\qquad\qquad\qquad \text{termination; (2.4.1)}$$
$$wp.(\textbf{do } (\phi \geq \Diamond \phi) \rightarrow Step \textbf{ od}).([\phi \geq \Diamond \phi] * \Diamond \phi)$$

$$\Rightarrow \qquad wp.(\textbf{do } (\phi < \Diamond \phi) \rightarrow Step \textbf{ od}).\phi \;, \qquad\qquad [\phi \geq \Diamond \phi] * \Diamond \phi \;\Rightarrow\; \phi$$

as required.   □

Putting Lemmas 9.4.1 and 9.4.6 together gives our proof of Lem. 9.4.2, that for deterministic and non-divergent *Step*, and finite state space, we have

$$\Diamond \phi \quad \equiv \quad wp.(\textbf{do } (\phi < \Diamond \phi) \rightarrow Step \textbf{ od}).\phi \;.$$

Before moving to infinite state spaces, we give an example showing that finiteness is indeed necessary for Lem. 9.4.6. Let *Step* be $n := n+1$ over the infinite state space $\mathbb{N}$, and define as before $\phi := n/(n+1)$. From Def. 9.3.2 and the $\sqcup$-formulation of least fixed-points we have, for the eventuality, the equality $\Diamond \phi \equiv (\sqcup N : \mathbb{N} \cdot (\Diamond \phi)_N)$, where we define

$$(\Diamond \phi)_0 \quad \equiv \quad 0$$

$$\begin{aligned}(\Diamond \phi)_1 \quad &\equiv \quad n/(n+1) \sqcup wp.(n := n+1).0 \\ &\equiv \quad n/(n+1)\end{aligned}$$

$$\begin{aligned}(\Diamond \phi)_2 \quad &\equiv \quad n/(n+1) \sqcup wp.(n := n+1).(n/(n+1)) \\ &\equiv \quad (n+1)/(n+2)\end{aligned}$$

$$\vdots$$

$$\begin{aligned}(\Diamond \phi)_N \quad &\equiv \quad n/(n+1) \sqcup wp.(n := n+1).(\Diamond \phi)_{N-1} \\ &\equiv \quad (n+N-1)/(n+N) \;,\end{aligned}$$

and hence — taking the limit — we see that $\Diamond \phi \equiv 1$. But then $\phi < \Diamond \phi$ is identically true — because ϕ itself is nowhere one — and thus with that strategy the expectation reduces to

$$wp.(\textbf{do true} \rightarrow n := n+1 \textbf{ od}).(n/(n+1)) \;,$$

which is zero everywhere due to nontermination.

Thus for infinite state spaces we must reconsider Lemmas 9.4.4 and 9.4.5, where the finiteness assumption is used. Let some predicate F over the state space be true for only finitely many states,[28] and define

$$\begin{aligned}Step_F \quad &:= \quad \textbf{if } F \textbf{ then } Step \textbf{ fi} \\ \Diamond_F \phi \quad &:= \quad \text{``}\Diamond \phi \text{ interpreted over computation } Step_F\text{''} \;.\end{aligned}$$

---

[28]This is again the "finitary" technique from Sec. 8.2.

Then it is immediate from Def. 9.3.2 that $\|\Diamond_F\phi\|.s = \|\phi\|.s$ for any state s not satisfying F, because the underlying computation $Step_F$ acts as **skip** outside of F — and that suffices to recover both lemmas. For Lem. 9.4.4 note that if $I \not\subseteq F$ then $\|\Diamond_F\phi\|.s = \|\phi\|.s$ is attained for any $s \in I-F$, and that otherwise I is finite; for Lem. 9.4.5 we have that the probability of termination is bounded away from zero, since it is one except in finitely many states, and nonzero for all of those.

Thus from Lem. 9.4.2 we have

$$\Diamond_F\phi$$
$$\equiv \quad wp.(\textbf{do } (\phi{<}\Diamond_F\phi) \rightarrow Step_F \textbf{ od}).\phi \qquad (9.10)$$
$$\equiv \quad wp.(\textbf{do } (\phi{<}\Diamond_F\phi) \rightarrow Step \textbf{ od}).\phi \ ,$$

where in the last line we can replace $Step_F$ in the body by $Step$ because (recall above that ϕ and $\Diamond_F\phi$ are equal outside F) the body is executed only within F, where $Step$ and $Step_F$ do not differ. That gives our principal theorem for *eventually*:

Theorem 9.4.7 For deterministic, non-divergent and continuous computation $Step$, and expectation ϕ, the expectation $\Diamond\phi$ is the supremum over all standard memoriless strategies G of

$$wp.(\textbf{do } G \rightarrow Step \textbf{ od}).\phi \ , \qquad (9.11)$$

and if the state space is finite the supremum is attained with strategy $G := \phi < \Diamond\phi$.

Proof   From (9.10) we have an explicit set of memoriless strategies $G_F := (\phi{<}\Diamond_F\phi)$ whose limit in (9.11) attains $(\sqcup F \cdot \Diamond_F\phi)$; from Def. 9.3.2 and the assumed continuity of $Step$, however, it is routine to show that

$$(\sqcup F \cdot \Diamond_F\phi) \quad \equiv \quad \Diamond\phi \ .$$

$\square$

It is instructive to specialise Lem. 9.4.2 (or Thm. 9.4.7) to the case of a standard payoff (but still probabilistic computation): we then see that

$$\Diamond[P]$$
$$\equiv \quad wp.(\textbf{do } [P]{<}\Diamond[P] \rightarrow Step \textbf{ od}).[P]$$
$$\equiv \quad wp.(\textbf{do } [\neg P \wedge (\Diamond[P]{\neq}0)] \rightarrow Step \textbf{ od}).[P] \ . \qquad [P] \text{ is standard}$$

Thus for standard $[P]$ the "recommended" strategy for *eventually* is to

continue playing as long as P does not hold but only *provided* it has some chance of holding eventually $(\Diamond[P] \neq 0)$, $\qquad$ (9.12)

since otherwise there is no point in continuing — if $\Diamond[P]$ is zero, one might as well give up now.

Note the earlier operational interpretation of $\Diamond[P]$ in Program (9.6) remains valid, but its strategy is missing the second conjunct. If that gambler

is ignorant of $\Diamond [P] = 0$ he continues to play, but still can never win — so
it makes no difference in the end. Thus the optimal strategy is simply not
unique in this case.

In another sense, however, our current strategy is better: gamblers aware
of Lem. 9.4.2 can get to the pub earlier and spend their remaining money
there.

9.4.3   The always game

It can be shown that $\Box\phi$ is given as an *infimum* of strategies of the gambling
game.[29] The analogue of Lem. 9.4.1 shows that it is in fact the infimum
over *all* strategies G' of

$$wnp.(\textbf{do } G' \to Step \textbf{ od}).\phi \ ,$$

in which the decision to stop or continue is taken by a *demon*, say the casino
manager, with the gambler's worst interests at heart. The casino manager
will not force the gambler to play forever, however, since *wnp* interprets
nontermination as success — and the manager strives for (the gambler's)
failure.

9.4.4   The unless game

Finally, the game for $\phi \triangleright \psi$ combines the eventually and always games: the
gambler tries to maximise the expectation, while the casino manager tries
to minimise it. Let the gambler's strategy be G and the manager's G'; the
game is then

$$
\begin{aligned}
&(\nu A \cdot \\
&\quad \textbf{if } \neg G \textbf{ then resultis } \psi \\
&\quad \textbf{elsif } \neg G' \textbf{ then resultis } \phi \\
&\quad \textbf{else } Step; A) \ .
\end{aligned}
\tag{9.13}
$$

[30]

On each step the gambler decides whether to stop and accept ψ in the
current state; if the gambler did not stop then the manager decides whether
to force him to stop and accept ϕ instead; if neither stopped, one step is
taken and the game continues (with nontermination interpreted as success
for the gambler).

---

[29]Chap. 11 proves a very general result of which this is a special case, showing also
that the restriction to memoriless strategies has no force in the finite case.

[30]We abuse notation here by mixing syntactic elements (like ϕ and *Step*) with semantic
elements (like A), and (here only) we use a quasi-functional programming notation
including *e.g.* **resultis** and a recursion construct that calculates the greatest fixed-point.
That last is of course not computable in general, since it solves the Halting Problem.

For strategies G, G' and expectations ϕ, ψ we denote by $U_{G,G'}.(\phi, \psi)$ the result of playing the unless game (9.13) above. It can be shown that

$$(\sqcap G' \cdot U_{G,G'}.(\phi, \psi)) \quad \Rightarrow \quad \phi \triangleright \psi \quad \Rightarrow \quad (\sqcup G \cdot U_{G,G'}.(\phi, \psi)) \, ,$$

which states (on the left) that for all gambler's strategies G the manager can ensure that he wins no more than $\phi \triangleright \psi$, and (on the right) that for all manager's strategies G' the gambler can win at least $\phi \triangleright \psi$. In the limit we have

$$(\sqcup G \cdot (\sqcap G' \cdot U_{G,G'}.(\phi, \psi))) \quad \equiv \quad \phi \triangleright \psi \quad \equiv \quad (\sqcap G' \cdot (\sqcup G \cdot U_{G,G'}.(\phi, \psi))) \, ,$$

showing that if both players use their best strategy, the result is $\phi \triangleright \psi$. In the language of game theory [vNM47, Kuh03] we are stating a "minimax" result, that this game "has a value." Chap. 11 treats that issue generally, and in more detail.

9.5 Summary

We call the quantitative temporal logic qTL.

We have shown that the semantics of qTL establishes the agreement between the simple fixed-point definitions in Sec. 9.3 and the operational interpretations as games in Sec. 9.4 — that for example $\diamond\phi$ is the least upper bound of all "seek to maximise" game strategies based on ϕ. Such a correspondence is needed to justify our fixed-point definitions, in spite of their simplicity: we must be sure they describe some actual behaviour. Having found that correspondence is why we feel justified in our choice of generalising $\vee, \wedge$ to $\sqcap, \sqcup$ in this context.

Back and von Wright [BvW96] and Stirling [Sti95] suggest similar uses for games, but over predicates (Boolean-valued) rather than the more general expectations (real-valued).

In the next chapter, we use elementary fixed-point properties of the operators *eventually*, *always* and *unless* to establish quantitative analogues for many if not most of the axioms for standard branching-time temporal logic. A crucial part of that will be the use of sublinearity of *next-time* which — as Def. 5.6.1 in temporal livery — is expressed

$$\circ(a * \phi + b * \psi \ominus c) \quad \Leftarrow \quad a * \circ\phi + b * \circ\psi \ominus c \, .$$

As we have seen, that property specialises to conjunctivity when the expectations are standard, the usual property associated with a (standard) modal algebra [Eme90].

Finally, Chap. 11 gives a very general result relating formulae and gambling games, of which all the examples here are special cases.

Chapter notes

As with standard temporal logic there are many formulations of probabilistic temporal logic, each one having its own merits — the probabilistic version is usually interpreted as the proportion of computation paths that satisfy the corresponding standard temporal logic formula. Bérard *et al.* [BFL+99] give a balanced survey of the strengths and weaknesses of the various styles.

The first presentations of probabilistic temporal logic in terms of *threshold probabilities* seem to be those of Hansson and Jonsson [HJ94], and Aziz *et al.* [ASBSV95]; the former introduced it as a logic of *time and probability*.

Our *qTL* does not use thresholds (though they can be expressed within it),[31] and is similar to *pCTL* as it provides quantitative information at each state. The expression of more complex path properties (including fairness conditions) requires the more distinguishing *qMμ* discussed in Chap. 11.[32]

The idea of using games to give an operational understanding of the operators arose quite naturally from the gambling analogy described in Chap. 1. In the present context the game format (including payoffs and player strategies) provides a neat formalisation of the probability distributions over computation paths — these concepts are similar to the *policies* of Markov decision processes [FV96].

---

[31]See (10.13) on p. 290 for a discussion of this.

[32]The logic *qTL* suffers from the usual inability to express the formula "the proportion of all paths that satisfy $\Diamond\Box\,[P]$ as a linear-time formula." Consider for example the transition system

$$
\begin{aligned}
x = 2 &\quad\rightarrow\quad \textbf{skip} \sqcap x := x - 1 \\
x = 1 &\quad\rightarrow\quad x := x - 1 \\
x = 0 &\quad\rightarrow\quad \textbf{skip} ,
\end{aligned}
$$

and choose P to be "x is even." All (infinite) paths satisfy $\Diamond\Box\,[P]$ individually, but that formula's value at $x = 2$ is zero (*i.e.* [false]) because the system *as a game* cannot be forced eventually to move from there to a state in which $\Box\,[P]$ holds — it could (demonically) take the **skip** alternative forever.

Curiously, though, if we restrict to purely deterministic probabilistic systems, as some authors do, this is no longer a problem. In the counter-example above, for instance, the demonic choice would be replaced by ${}_p\oplus$ for some p strictly between zero and one, and the demon's wrecking strategy of remaining forever at $x = 2$ would become almost impossible.

10

The quantitative algebra of qTL

10.1   The role of algebra

Algebra represents a third tier of reasoning, justified by semantics for formulae (the second) which in turn is given in terms of an interpretation
over an operational model (the first). The focus of this chapter will be on
algebraic laws for the quantitative temporal logic qTL that we introduced
in Chap. 9.

This chapter is based on an earlier work *An expectation-based model for probabilistic temporal logic* [MM99a], by permission of Oxford University Press.

$$
\begin{array}{rcl}
\Box(P \Rightarrow Q) & \Rightarrow & \Box P \Rightarrow \Box Q \\
\circ(P \Rightarrow Q) & \Rightarrow & \circ P \Rightarrow \circ Q \\
\Box P & \Rightarrow & \circ P \wedge \circ \Box P \\
P \wedge \Box(P \Rightarrow \circ P) & \Rightarrow & \Box P \\
\Box(P \Rightarrow Q) \wedge \Diamond P & \Rightarrow & \Diamond Q \\
P \vee \circ \Diamond P & \Rightarrow & \Diamond P \\
\Box P & \Rightarrow & \neg \Diamond(\neg P) \\
\Diamond P \wedge \Box(\circ P \Rightarrow P) & \Rightarrow & P
\end{array}
$$

These axioms are based on those of Ben-Ari *et al* [BAPM83]. Predicates over the state space are written P, Q.

Figure 10.1.1. AXIOMS FOR STANDARD BRANCHING-TIME TEMPORAL LOGIC

We have seen such layered reasoning structures already with respect to our sequential programming language $pGCL$, which was the topic of Parts I and II. There, the first tier was the relational model (as illustrated in Fig. 1.3.1 and formalised in Sec. 5.4); the second tier was the wp-semantics (Fig. 1.5.3) and associated loop rules (*e.g.* Sec. 2.4); and the third tier is represented by equalities and inequalities between programs themselves (as in Sec. B.1).[1]

We begin by obtaining an extension of Ben-Ari, Pnueli and Manna's axiomatisation [BAPM83] of branching-time temporal logic, an extract of which appears in Fig. 10.1.1. The structure of their formulae will be largely preserved as we extend them for reasoning over properly quantitative values.[2]

For example, we will see later that our earlier definitions 9.3.1–9.3.3 in the previous chapter have all the properties listed in the figure, when specialised to the standard case, and as an illustration of algebraic reasoning to come we recall the well-known algebraic proof of the following familiar fact about standard *eventually*.

---

[1]Layers or no, we of course take the view that semantics is the essential framework within which all else is built, which fixes not only the "terms of reference" for the studied phenomena — what is noticed, and what is ignored — but also their precise (if idealised) properties.

But reasoning directly at the semantic level is often complex, detailed and even error-prone. Thus it is natural — and for utility, essential — to contain the semantic reasoning to "just once" proofs of theorems in a logic interpreted over it, creating tools for others (including oneself) that are easier to use and operate at a higher level. The resulting collection of reasoning tools is what we are calling the algebra.

[2]Ben-Ari et al. [op. cit.] show those axioms to be complete for standard temporal logic. Here we use them as examples of what is reasonable for a model.

Lemma 10.1.2 STANDARD DOUBLE-EVENTUALLY   For any predicate P
we have the equality

$$\Diamond\Diamond P \quad \equiv \quad \Diamond P \ .$$

Proof   Directly from Def. 9.2.2 (unfolding it) we have $P \Rrightarrow \Diamond P$ and
$\circ\Diamond P \Rrightarrow \Diamond P$. Taking $P := \Diamond P$ in the former gives $\Diamond P \Rrightarrow \Diamond\Diamond P$ trivially;
from the latter we have that $X := \Diamond P$ satisfies

$$X \quad \equiv \quad \Diamond P \vee \circ X \ ,$$

of which however $\Diamond\Diamond P$ is the $\Rrightarrow$-least solution.[3] $\square$

Our aim now is to show how to carry out similar proofs for quantitative
transformers. Fig. 10.3.5 below gives the quantitative equivalents of the
standard axioms which we will use as a basis for that.

The main insight will be that probabilistic conjunction "&" enables
modular reasoning that is applicable *even for probabilistic information*. Although the probabilistic conjunction arose from novel axioms characterising
probabilistic sequential programs (*i.e.* sublinearity and its consequences),
where the distributions were over states, here we see it is equally apt when
the probabilities derive instead from distributions over paths.

In obtaining our extension of the laws we observe that much of
the intuition underlying standard temporal logic is valid for probability as well: because the laws are so similar to the standard ones, many
proofs of probabilistic temporal properties will merely be replayings
of their standard counterparts. Both observations encourage us in the
expectation-transformer approach.

Sec. 10.2 fills in some of the gaps left by the introductory but non-
exhaustive approach of Chap. 9 and uses for illustration a running example
of demonically-nondeterministic coin-flipping. Then in Sec. 10.3 we present
the (beginnings of an) algebra and give short examples of its use.

In Sec. 10.4 we address larger examples, returning to the random walk
(first mentioned in Sec. 2.11.1, and the subject of an extended case study in
Sec. 3.3): it is elementary but nevertheless is completely beyond the reach
of logics that do not quantify probabilities explicitly. Even the "universal"
conclusion "the walker moves left eventually with probability one" requires
properly numeric premises, that the one-step probability is 1/2 in either
direction. But in the more general case where those specific probabilities
are not 1/2, even the conclusion is no longer universal: the probability of
an eventual move is somewhere strictly between zero and one.

Thus we will "flex our algebraic muscles" by reproducing the result that
a probabilistic demonic walker whose probability of moving left or right is

---

[3]We recall Footnote 4 on p. 39 and continue with our practice of using $\Rrightarrow$ and $\equiv$ for
both Boolean- and quantitative reasoning, with the obvious correspondence.

bounded below by $1/3$ (*i.e.* it is not known exactly) will return eventually with probability at least $1 - \sqrt{5}/3$ to his starting point — and we do that without referring explicitly to the underlying probabilistic/demonic computation tree at all. The tree, the operational level, is hidden "beneath" the logic.

In Sec. 10.5 we examine the benefits and limitations of using expectations and discuss how our contributions relate to other approaches, in particular to the logics $pCTL$ and $pCTL^*$ [ASBSV95]. We also discuss model checking briefly.

10.2   Quantitative temporal expectations

Chapter 9 reminded us that the temporal model enables reasoning about "evolving-" rather than just sequential "start-here-finish-there" computations. Its corresponding Boolean-valued logic includes demonic- and angelic nondeterminism and treats standard programs [BAPM83]: we can talk about intermediate states, and we can even deal with computations that are unending.

A significant difference however between qTL and standard temporal logic was that the temporal operators could no longer be operationally motivated in the traditional way, *i.e.* given purely in terms of the existence and/or properties of sequences of states, or even of computation trees; this was primarily because the notion of a path or a tree "satisfying" a proper expectation does not make sense.[4] For the full quantitative logic we need to use *games* as our underlying operational semantics.

Thus one advantage of using the expectation-based approach is that it side-steps explicit mention of path distributions, which can be very complicated in specific cases. Instead it relies "once-and-for-all" on the correspondence summarised in Fig. 5.9.2 (p. 160) between program logic and (fixed-point) transition semantics. And there are further benefits: as we have found before, the general expectations allow us to treat more than probabilities — indeed often an *expected quantity* such as "number of steps to termination" is required, rather than a specific probability. A logic based on expectations allows us to calculate the desired value directly [SPH84, McI02].[5]

We begin by presenting the quantitative generalisation of the modal operators more systematically than we did in the previous chapter, giving

---

[4] For the same reason, earlier at (1.20) on p. 24 we took a different view of sequential programs as well.

[5] Recall how we approached this in Sec. 2.10 and, later, at Fig. 7.7.9 for the expected number of iterations of a loop.

complete tabulations of their expectations over a simple coin-flipping game. Because we are working over a fixed system throughout, we will reduce clutter by omitting the semantic brackets $\| \cdot \|$.

10.2.1 next-time *tabulated*

Consider two fair coins: a *thin* coin that gives *heads* or *tails* with probability $1/2$ each; and a *fat* coin that gives heads, tails or *edge* with probability $1/3$ each.[6] At most one coin is flipped at a time, and the state space is $\{h, t, e\}$, representing the result of the most recent flip. The computation *step* is defined to be [7]

current state	action of *step*	
heads h	flip *thin* or *fat*	(10.1)
tails t	flip *fat* or do nothing	
edge e	do nothing,	

where "or" in an action represents demonic nondeterminism and "do nothing" means the computation remains in that state (remains stuck there, in fact, if "do nothing" is the only option). Fig. 10.2.1 tabulates $\circ P$ for that *step* over all standard P: note how the demon acts to *minimise* the probability of achieving the postcondition, when there is a choice.

For an illustration of proper expectations (*i.e.* rather than standard) consider the formula $\circ(\circ\{t, e\}).h$, that is, applying $\circ$ to the expectation $\circ\{t, e\}$ and evaluating the whole thing at state h.[8] We have

$$\circ\circ\{t, e\}.h$$

$$= \quad (\circ\{t, e\}.h)/2 + (\circ\{t, e\}.t)/2 \qquad \text{Def. 9.3.1}$$
$$\sqcap \quad (\circ\{t, e\}.h)/3 + (\circ\{t, e\}.t)/3 + (\circ\{t, e\}.e)/3$$

---

[6] A reviewer of this volume wrote here "informal testing reveals it takes 8 US quarters to make a reasonably fair 3-sided coin."

An alternative eight-fold approach to making a reasonably fair three-sided coin — say 99% accurate — is to refer to the program of Fig. 7.7.10 (also given in the Preface, on p. v). Set p to $1/3$, and use up to seven flips of a real quarter to decide whether our virtual coin has come up *edge*. Since the probability of exhausting the seven flips with no result is only $1/2^7$, less than 1%, an arbitrary decision at that point — should it occur — will introduce little error.

Either way, if the result is *edge* then we stop there; if not, then we use the final, possibly eighth flip to decide between *heads* and *tails*.

[7] We use this same *step* throughout the running example.

[8] We will omit parentheses for temporal operators by associating them to the right, and giving them higher precedence than *e.g.* ".h".

Strictly speaking only $\circ(\circ\{t, e\})$ is a formula, and the ".h" is a meta-notation meaning "evaluate it at h."

The final predicate P is given explicitly as a subset of $\{h, t, e\}$; the expression $\circ P.s$ gives the value of formula $\circ P$ at initial state s; if ".s" is omitted then the expectation is the same at all three states.

$$\circ \emptyset \quad \equiv \quad \underline{0} \qquad \text{Excluded miracle, since } \sqcup\{\} = 0.$$

	$\circ\{h\}.h$	$=$	$1/3$	Demon flips *fat*; flipping *thin* would give $1/2$.
	$\circ\{h\}.t$	$=$	0	Demon does nothing; flipping would give $1/3$.
	$\circ\{h\}.e$	$=$	0	Cannot leave e.

	$\circ\{t\}.h$	$=$	$1/3$	Demon flips *fat*; flipping *thin* would give $1/2$.
	$\circ\{t\}.t$	$=$	$1/3$	Demon flips *fat*; doing nothing would give 1.
	$\circ\{t\}.e$	$=$	0	Cannot leave e.

$$\circ\{e\} \quad \equiv \quad \{e\} \qquad \text{At } \{h\} \text{ flips } \textit{thin}, \text{ at } \{t\} \text{ does nothing: both avoid } e.$$

	$\circ\{h, t\}.h$	$=$	$2/3$	Demon flips *fat*; flipping *thin* would give 1.
	$\circ\{h, t\}.t$	$=$	$2/3$	Demon flips *fat*; doing nothing would give 1.
	$\circ\{h, t\}.e$	$=$	0	Cannot leave e.

$*$	$\circ\{t, e\}.h$	$=$	$1/2$	Demon flips *thin*; flipping *fat* would give $2/3$.
	$\circ\{t, e\}.t$	$=$	$2/3$	Demon flips *fat*; doing nothing would give 1.
	$\circ\{t, e\}.e$	$=$	1	Cannot leave e.

	$\circ\{e, h\}.h$	$=$	$1/2$	Demon flips *thin*; flipping *fat* would give $2/3$.
	$\circ\{e, h\}.t$	$=$	0	Demon remains at t.
	$\circ\{e, h\}.e$	$=$	1	Cannot leave e.

$$\circ\{h, t, e\} \quad \equiv \quad \underline{1} \qquad \text{Termination guaranteed.}$$

As an example of using the semantics to back up our intuition, at $*$ we calculate

$$\circ\{t, e\}.h$$

$=$	$(\{t, e\}.h)/2 + (\{t, e\}.t)/2$	Def. 9.3.1
	$\sqcap \quad (\{t, e\}.h)/3 + (\{t, e\}.t)/3 + (\{t, e\}.e)/3$	

$=$	$(0/2 + 1/2) \sqcap (0/3 + 1/3 + 1/3)$	$\{t, e\}.h = 0$ *etc.*
$=$	$1/2$.	

Figure 10.2.1. COIN-FLIPPING ILLUSTRATION OF $\circ$

$$= \qquad (1/2)/2 + (2/3)/2$$
$$\sqcap \quad (1/2)/3 + (2/3)/3 + (1)/3 \qquad \text{Fig. 10.2.1}$$

$$= \qquad 7/12 \ ,$$

which is the largest probability that can be guaranteed for reaching state t or e in exactly two steps from state h.

In fact this is a replay of our sequential example (1.20) where we pointed out that a standard conclusion (here $\{t, e\}$) can generate non-standard intermediate formulae (here $\circ\{t, e\}$) yet still give us probabilistic conclusions overall (here, that with probability $7/12$ we achieve $\{t, e\}$ in exactly two steps). Again the proper expectation is acting as "glue."

10.2.2 eventually *tabulated*

In Def. 9.3.2 we generalised Boolean $\vee$ to arithmetic $\sqcup$ (and later in Def. 9.3.3 we took $\wedge$ to $\sqcap$). The interaction of angelic and demonic non-determinism in $\Diamond$ is significant: the choice of whether to stop or to step is made to *maximise* the expectation; but the resolution of nondeterminism during a step is made to *minimise* it. Recalling the poker machines of Sec. 9.4, we would say that the player is seeking to maximise the payout (he decides whether to go on or to stop), while the machine itself (viewed pessimistically) is trying to minimise it.[9]

Fig. 10.2.2 gives a complete tabulation of $\Diamond P.s$ for standard $P \subseteq S$ and $s \in S$, again in the coin example.

For an example of proper expectations consider $\Diamond\Diamond\{t\}$. At state t we have

$$\Diamond\Diamond\{t\}.t \quad = \quad \Diamond\{t\}.t \ \sqcup \ \circ\Diamond\Diamond\{t\}.t \quad = \quad 1 \ ,$$

since $\Diamond\{t\}.t = 1$. At state e we have

$$\Diamond\Diamond\{t\}.e \quad = \quad \Diamond\{t\}.e \ \sqcup \ \circ\Diamond\Diamond\{t\}.e \quad = \quad \Diamond\Diamond\{t\}.e \ ,$$

of which the least solution is $\Diamond\Diamond\{t\}.e := 0$. Finally, we calculate

$$\Diamond\Diamond\{t\}.h$$
$$= \qquad \Diamond\{t\}.h \ \sqcup \ \circ\Diamond\Diamond\{t\}.h \qquad \qquad \text{Def. 9.3.2}$$

$$= \qquad 1/2 \qquad\qquad\qquad\qquad\qquad \text{Fig. 10.2.2; definition } step$$
$$\sqcup \qquad (\Diamond\Diamond\{t\}.h)/2 + (\Diamond\Diamond\{t\}.t)/2$$
$$\sqcap \qquad (\Diamond\Diamond\{t\}.h)/3 + (\Diamond\Diamond\{t\}.t)/3 + (\Diamond\Diamond\{t\}.e)/3$$

---

[9]Note however that the poker machine cannot "seek to minimise" purely probabilistic outcomes: it "has a choice" only if its mechanism contains (presumably legal) demonic nondeterminism. See also Footnote 12 on p. 275 to come.

$$\Diamond\emptyset \;\equiv\; \underline{0} \qquad \text{Excluded miracle.}$$
$$\Diamond\{h\} \;\equiv\; \{h\} \qquad \text{Do nothing at } t \text{ and } e.$$

* $\quad \Diamond\{t\}.h \;=\; 1/2 \qquad$ Flipping either coin repeatedly is guaranteed to leave h; repeating *thin* would guarantee reaching t. Repeating *fat* splits the eventual departure fairly between arriving at t and at e.

$$\Diamond\{t\}.t \;=\; 1 \qquad \text{Already at } t.$$
$$\Diamond\{t\}.e \;=\; 0 \qquad \text{Cannot leave } e.$$

$$\Diamond\{e\} \;\equiv\; \{e\} \qquad \text{At } h \text{ flip } \textit{thin}, \text{ at } t \text{ do nothing; both avoid } e.$$
$$\Diamond\{h,t\} \;\equiv\; \{h,t\} \qquad \text{If at } e, \text{ cannot leave it.}$$
$$\Diamond\{t,e\} \;\equiv\; \underline{1} \qquad \text{Flipping either coin repeatedly is guaranteed to leave } h \text{ eventually.}$$
$$\Diamond\{e,h\} \;\equiv\; \{e,h\} \qquad \text{Forever do nothing at } t.$$
$$\Diamond\{h,t,e\} \;\equiv\; \underline{1} \qquad \text{At } h,t,e \text{ already: termination not required.}$$

At * we have for example

$$\Diamond\{t\}.h$$
$$= \qquad \{t\}.h \;\sqcup\; \circ\Diamond\{t\}.h \qquad\qquad\qquad\qquad \text{Def. 9.3.2}$$

$$= \qquad \begin{aligned} &(\Diamond\{t\}.h)/2 + (\Diamond\{t\}.t)/2 \qquad\qquad \{t\}.h = 0; \text{ definition } \textit{step}\\ \sqcap\;\; &(\Diamond\{t\}.h)/3 + (\Diamond\{t\}.t)/3 + (\Diamond\{t\}.e)/3 \end{aligned}$$

$$= \qquad \begin{aligned} &(\Diamond\{t\}.h)/2 + 1/2 \qquad\qquad\qquad \Diamond\{t\}.t = 1; \; \Diamond\{t\}.e = 0\\ \sqcap\;\; &(\Diamond\{t\}.h)/3 + 1/3 \end{aligned}$$

$$= \qquad (\Diamond\{t\}.h)/3 + 1/3 \;,$$

whence by arithmetic we have $\Diamond\{t\}.h = 1/2$.

Figure 10.2.2. COIN-FLIPPING ILLUSTRATION OF $\Diamond$

$$
=\qquad\begin{array}{ll} 1/2 & \qquad\qquad \Diamond\Diamond\{t\}.t = 1;\ \Diamond\Diamond\{t\}.e = 0 \\ \sqcup \quad (\Diamond\Diamond\{t\}.h)/2 + 1/2 & \\ \sqcap \quad (\Diamond\Diamond\{t\}.h)/3 + 1/3 & \end{array}
$$

$$
=\qquad 1/2 \ \sqcup \ (\Diamond\Diamond\{t\}.h + 1)/3 \ ,
$$

whose only solution is $\Diamond\Diamond\{t\}.h := 1/2$. Thus we have

$$
\Diamond\Diamond\{t\} \quad\equiv\quad \Diamond\{t\}\ , \qquad ^{10}
$$

which agrees with our double-eventually law Lem. 10.1.2 but generalises it
to the quantitative case (proved as Lem. B.6.1 in Sec. B.6, and illustrated
some time ago by Footnote 18 on p. 94).

10.2.3   always *tabulated*

Fig. 10.2.3 tabulates $\Box P.s$ in the same style as for the other operators.

For an example of proper expectations define $A := \{e\} \sqcup \Diamond\{t\}$, and consider $\Box A$. Expectation A is the probability that the last flip was edge or
— if it was not — that tails will be flipped eventually. To calculate $\Box A$ we
first consider state e, where we have

$$
\Box A.e \quad=\quad 1 \sqcap o\Box A.e \quad=\quad \Box A.e\ ,
$$

whose greatest solution is $\Box A.e := 1$. At state h we have

$$
\Box A.h
$$

$$
=\qquad\begin{array}{ll} 1/2 & \qquad\text{Def. 9.3.3; Fig. 10.2.2; definition } step \\ \sqcap \quad (\Box A.h)/2 + (\Box A.t)/2 & \\ \sqcap \quad (\Box A.h)/3 + (\Box A.t)/3 + (\Box A.e)/3 & \end{array}
$$

$$
=\qquad 1/2 \ \sqcap \ (\Box A.h + \Box A.t)/2 \ , \qquad \Box A.e = 1;\ \Box A.h, \Box A.t \le 1
$$

whence we have trivially $\Box A.h \le 1/2$. And at state t we have

$$
\Box A.t
$$

$$
=\qquad\begin{array}{ll} 1 & \qquad\text{Def. 9.3.3; Fig. 10.2.2; definition } step \\ \sqcap \quad (\Box A.h)/3 + (\Box A.t)/3 + (\Box A.e)/3 & \\ \sqcap \quad \Box A.t & \end{array}
$$

$$
\le\qquad (1/2 + (\Box A.t)/3) \ \sqcap \ \Box A.t \ , \qquad \Box A.e = 1;\ \Box A.h \le 1/2
$$

---

[10]Here and below we will be calculating extremal solutions pointwise (in this case at
state e); and we are aware that such pointwise solutions are not guaranteed to be an
overall solution when collected together. But if the collection *is* a solution, then in fact
it must be an extremal one as required: in our examples all pointwise calculations do
give overall solutions.

$$
\begin{array}{rcll}
\Box\emptyset & \equiv & \underline{0} & \text{Excluded miracle.} \\
\Box\{h\} & \equiv & \underline{0} & \text{Repeated flips of either coin eventually leave } h. \\
\Box\{t\} & \equiv & \underline{0} & \text{Repeated flips eventually leave } t. \\
\Box\{e\} & \equiv & \{e\} & \text{Cannot leave } e. \\
\Box\{h,t\} & \equiv & \underline{0} & \text{Repeated flips leave } h,t. \\
\end{array}
$$

$$
\begin{array}{rcll}
\Box\{t,e\}.h & = & 0 & \text{Already not in } t,e. \\
\ast \quad \Box\{t,e\}.t & = & 1/2 & \text{Repeated flips are guaranteed to leave } t \text{ eventually,} \\
& & & \text{then reaching } e \text{ with probability } 1/2. \\
\Box\{t,e\}.e & = & 1 & \text{Cannot leave } e. \\
\end{array}
$$

$$
\begin{array}{rcll}
\Box\{e,h\} & \equiv & \{e\} & \text{Repeated flips of } \textit{thin} \text{ at } h \text{ are guaranteed to reach } t. \\
\end{array}
$$

$$
\begin{array}{rcll}
\Box\{h,t,e\} & \equiv & \underline{1} & \text{Termination guaranteed.} \\
\end{array}
$$

At $\ast$ we have for example

$$
\begin{aligned}
& \Box\{t,e\}.t \\
= \quad & \{t,e\}.t \,\sqcap\, \mathrm{o}\Box\{t,e\}.t \hspace{4cm} \text{Def. 9.3.3} \\[2mm]
= \quad & \hspace{5cm} \{t,e\}.t = 1; \text{ definition } \textit{step} \\
& (\Box\{t,e\}.h)/3 + (\Box\{t,e\}.t)/3 + (\Box\{t,e\}.e)/3 \\
& \sqcap \quad \Box\{t,e\}.t \\[2mm]
= \quad & (1/3 + (\Box\{t,e\}.t)/3) \,\sqcap\, \Box\{t,e\}.t \,, \hspace{1cm} \Box\{t,e\}.h = 0;\ \Box\{t,e\}.e = 1
\end{aligned}
$$

whose greatest solution is $\Box\{t,e\}.t = 1/2$.

Figure 10.2.3. COIN-FLIPPING ILLUSTRATION OF $\Box$

---

whence by arithmetic we have $\Box A.t \leq 3/4$. It is easily checked from the definition that the pointwise maxima above

$$
\Box A.h, \Box A.t, \Box A.e \quad := \quad 1/2,\ 3/4,\ 1 \tag{10.2}
$$

are a solution collectively, and thus give the maximal one.

10.2.4 Combined operators

The operational meaning of (10.2) above, in particular e.g. of the fact that

$$
\Box(\{e\} \sqcup \Diamond\{t\}).t \quad = \quad 3/4 \,, \tag{10.3}
$$

is not obvious at first. What it does *not* mean however is that

$$
\textit{(i.e. not this)} \quad \left\{ \begin{array}{l} \text{3/4 of all paths in the probabilistic compu-} \\ \text{tation tree rooted at } t \text{ satisfy "either the} \\ \text{current flip-result is } e \text{ or there is a flip-result} \\ t \text{ further on,"} \end{array} \right. \tag{10.4}
$$

and that might be somewhat of a surprise.

One of the complicating factors is of course that there is not a single *"the"* tree rooted at t — because of the non-determinism in *step*, there will be potentially a large (and generally infinite) number of them. And just which tree of probabilistic choices, from all the possibilities, is actually carried out depends for example on the resolution of the non-determinism as the coin-flipping proceeds. So the question is really "3/4 of what?"

An answer to that is given by recognising *strategies* as essential and explicit components of any model combining demons/angels and probability. The players decide *in advance* what their choices will be subsequently in the game, taking every possibility of previous coin-flip outcomes into account; but, at the same time, each player's knowledge of the other's strategy must be limited to those choices already made. Once those (fairly complicated) details are settled, we can say that *for each pair of maximising/minimising strategies, there is just one probabilistic tree that represents the game to be played*; and then (10.4) at least makes sense as an hypothesis.[11]

The upshot of these complications is that our game interpretation is more important than ever: a correct informal treatment of the above formula must be in terms of games, of strategies and payoffs for the opposing players. The point of our logic is therefore that it allows precise and reliable reasoning about such games.

In the case (10.3), we reason informally as follows. The outer operator $\Box$ represents an agent deciding repeatedly whether to take another step, or to stop and evaluate its inner argument in the current state: its "aim" (demonic) is to minimise the overall outcome.[12]

---

[11]But it turns out even then that interpretation (10.4), while meaningful once the strategies are filtered out, is not always appropriate over the single trees separately, even though in many simple cases — like $\Diamond P$ for standard P — it is correct. Footnote 22 on p. 281 gives another example in which that interpretation is correct; Footnote 32 on p. 263 is a case where it is *not* correct; and we return to the issue in more detail in Chap. 11 at Footnote 20 on p. 309.

[12]Following standard game-theory terminology [Kuh03] we will use the word AGENT in these anthropomorphic informal discussions to represent "something that chooses," whether demonically or angelically.

Probabilistic choice however is *not* considered to be implemented by an agent since — paradoxically — there is actually no choice, no "free will" involved: the die must be rolled or the coin flipped; its result distribution is fixed in advance; and there is no way that any postulated agent could "cheat" by attempting to influence the result.

However, we have seen imprecise or "variable" distributions like the $\geq 1/2 \oplus$ discussed in the Miller-Rabin primality test (p. 50), and they do seem to allow some influence to be applied to the distribution. Recall that in our approach (p. 50) they are in fact choices made by a demonic agent between fixed probability distributions, *i.e.* effectively two choices successively: first demonic, then probabilistic. But only the demonic choice could be called a "decision" made by an "agent."

Its inner argument contains the formula $\Diamond\{t\}$, representing a second agent — an angel — seeking to maximise *its* argument $\{t\}$; and, as we have seen (p. 256), an optimal strategy for that is "keep stepping until t".[13] In state e that strategy yields zero (since from (10.1) we see that *step* never leaves e); in state t the optimal strategy is obviously "stop now," and yields one. In state h however, where the angel decides to take a step because remaining at h can only yield zero, the demon "in *step*" (a third agent) decides whether to flip the fat or the thin coin; flipping *fat* is worse for the angel (since e is one of the outcomes, not possible with *thin*); and repeatedly flipping *fat* gives probability $1/2$ of reaching t when almost certainly the outcome is eventually not h.

Having followed the angel's thoughts,[14] the demon can see that its own job will be to try to minimise the composite expectation

$$\{e\} \quad \sqcup \quad (e \mapsto 0, t \mapsto 1, h \mapsto 1/2)$$

formed from the angel's best-possible strategy above and the explicit choice "$\{e\}\sqcup$", *viz.* $(e \mapsto 1, t \mapsto 1, h \mapsto 1/2)$. It starts in state t (the ".t" in the expression we are evaluating), and sees clearly there is no point in staying there, with payoff one: it aims instead for the lower $1/2$ that would result from reaching h — and so its strategy at t is "keep stepping." The demon in *step* — an ally — obligingly chooses "flip *fat*" rather than "do nothing" from (10.1), and with probability $1/2$ for each the demon ends up immediately in either h or e, where the payoff (handing over to the angel) is, as we calculated above, either $1/2$ or 1 respectively.

Thus the overall value of $\Box(\{e\} \sqcup \Diamond\{t\}).t$ is

$$1/2*1/2 + 1/2*1 \quad = \quad 3/4 \,,$$

as we saw above directly from the semantics.

10.2.5 Alternation-free formulae

In the previous section we discussed the operational interpretation of a formula containing "nested" temporal operators, in that case $\Diamond$ within $\Box$, and mentioned that the $\Box$-demon "was aware" of the $\Diamond$-angel's plans (recall Footnote 14 above). This is possible — informally — because the inner, $\Diamond$ formula is wholly contained within the outer, $\Box$ formula and does not in any way "refer back" to it. Thus in the conventional mathematical way (called

---

[13]Recall however (p. 261) that it is not the only optimal strategy in this case; it is just better for our purposes here, since the proof-generated optimal strategy $\{t\} < \Diamond\{t\}$ given by (9.12) requires knowing $\Diamond\{t\}$ — which, inconveniently, is just what we are trying to find out. But this simple $\neg P$ strategy for $\Diamond P$ of course works only for standard P: the more general $\neg A$ for expectation A would be ill-defined.

[14]See the next section for a discussion of why the demon can read the angel's thoughts in this example.

referential transparency) we can evaluate the inner formula on its own: the angel's decisions, and the formula's value, do not depend on whether or how it is embedded in another.

In the more general modal μ-calculus where we allow ourselves to write fixed-point formulae directly (*i.e.* not just the ones encapsulated for us by definitions 9.3.2–9.3.4), it is possible to construct formulae that do not have this inside-to-outside reasoning property in such a convenient way: a very simple example is

$$(\mu X \cdot A \ \sqcup \ (\nu Y \cdot B \sqcap (X \sqcup Y))) \ .$$

Here, the inner formula $(\nu Y \cdot B \sqcap (X \sqcup Y))$ "loops back", via its use of X, to the outer fixed point, and the inner and outer agents alternate their turns.

Because formulae written using the temporal operators we have defined cannot loop back that way, they are called *alternation-free*. In alternation-free formulae the strategies can be handled much more simply than otherwise.

The treatment of fully alternating fixed-point formulae — *i.e.* not alternation-free — is given in the next chapter.

10.3   Quantitative temporal algebra

Given the structural correspondence between the standard and probabilistic models, one would expect them to share many temporal laws, and indeed they do. Although we do not attempt a complete axiomatisation, in this section we give a systematic presentation of the basic temporal laws, and show how with suitable choices of probabilistic "propositional" operators ($\sqcup, \sqcap, \&$, and the $\Rrightarrow, \multimap$ added below) the standard axioms (Fig. 10.1.1) can be generalised (Fig. 10.3.5) and — most important of all — they can be easily proved.

The basic properties of the temporal operators are summarised in Fig. 10.3.1, whose structure is as follows. The healthiness conditions generate the properties for *next-time*, ensuring (Chap. 5) the existence of a corresponding relational probabilistic/demonic computation to whose single step *next-time* refers.[15] The pairs of properties for *eventually* and *always* are the usual for extremal fixed-points, determining them uniquely.

Since the probabilistic definitions of the temporal operators replace the standard $\lor, \land$ with $\sqcup, \sqcap$ respectively, whose properties are so similar, many

---

[15]The list includes *subdistributivity of* &, called "probabilistic conjunction," for convenience; it is implied by *scaling, weighted sum* and *truncated subtraction*. It does not allow us to remove the last of those, unfortunately, even in the context of the other healthiness conditions. (See Footnote 34 on p. 148.)

healthiness conditions

monotonicity

$$A \quad \Rrightarrow \quad B$$
$$\text{implies} \quad \circ A \quad \Rrightarrow \quad \circ B$$

excluded miracle	$\circ A$	$\Rrightarrow$	$\sqcup A$
scaling	$\circ(pA)$	$\equiv$	$p(\circ A)$
weighted sum	$\circ(A \,_p\!\oplus B)$	$\Lleftarrow$	$\circ A \,_p\!\oplus \circ B$
probabilistic conjunction	$\circ(A \,\&\, B)$	$\Lleftarrow$	$\circ A \,\&\, \circ B$
truncated subtraction	$\circ(A \ominus \underline{p})$	$\Lleftarrow$	$\circ A \ominus \underline{p}$

least fixed-point

fixed point

$$\Diamond A \quad \equiv \quad A \sqcup \circ \Diamond A$$

least

$$A \sqcup \circ B \Rrightarrow B \quad \text{implies} \quad \Diamond A \Rrightarrow B$$

greatest fixed-point

fixed point

$$\square A \quad \equiv \quad A \sqcap \circ \square A$$

greatest

$$A \sqcap \circ B \Lleftarrow B \quad \text{implies} \quad \square A \Lleftarrow B$$

The properties labelled *least*- and *greatest fixed-point* state first that the formula *is* a fixed point, and second that it is least- or greatest. Those second properties we have used before, in the case *e.g.* of least calling it the "general $f.x \leq x$ property" that if $f.x \leq x$ then $\mu.f \leq x$.

Figure 10.3.1. BASIC PROPERTIES OF PROBABILISTIC TEMPORAL OPERATORS

---

standard results — and their proofs — are trivially retained: Lem. 10.1.2 "double eventually" generalises easily to Lem. B.6.1 for example.

For probabilistic implication there are several choices, however, only one of which is the "obvious" embedding of standard implication.[16]

Definition 10.3.2 STANDARD-EMBEDDED IMPLICATION
For scalars a, b define

$$a \Rrightarrow b \quad := \quad [a \leq b] \ .$$

When lifted to expectations the definition implies the correspondence $[P \Rightarrow Q] \equiv ([P] \Rrightarrow [Q])$ for standard P, Q, which is why we call it "standard-embedded" implication.   □

---

[16]This is unfortunate and deserves to be improved.

A more interesting form of implication is found as the $\Rightarrow$-"adjoint" of conjunction &; it is suggested by the fact that standard conjunction and standard implication have this adjoint property for any predicates P, Q, X:

$$(P \wedge X) \Rightarrow Q \quad \text{iff} \quad P \Rightarrow (X \Rightarrow Q) . \quad ^{17} \tag{10.5}$$

The advantage of defining a probabilistic implication $\dashrightarrow$ by analogy is that it acquires a $\circ$-superdistributivity property (see Fig. 10.3.5), which the naively embedded $\Rrightarrow$ does not have.

Definition 10.3.3 &-ADJOINT IMPLICATION   For scalars a, b define

$$a \dashrightarrow b \; := \; 1 - (a \ominus b) \,,$$

so that for expectations A, B, X we have the adjoint property

$$(A \,\&\, X) \;\Rightarrow\; B \quad \text{iff} \quad A \;\Rightarrow\; (X \dashrightarrow B) \,.$$

$\square$

We will see the value of this definition very shortly.

With those two definitions we can now write down a number of probabilistic properties that are analogues of the standard axioms; they are listed in Fig. 10.3.5. All can be proved from the basic properties Fig. 10.3.1 and the definitions of the two implications;[18] possibly the longest proof — but still not very long — is of **. It is a good example of the use of our adjoint property.

Lemma 10.3.4 PROBABILISTIC ALWAYS-EVENTUALLY LAW
    For all expectations A, B we have (informally) that if A always "implies" B, "and" in fact eventually A, "then" so too eventually B.[19] Put formally, that is

$$\square(A \dashrightarrow B) \;\&\; \Diamond A \;\Rightarrow\; \Diamond B \,.$$

---

[17] "Adjointness" is a general property relating partial orders: given two partial orders $(U, \sqsubseteq_U)$ and $(V, \sqsubseteq_V)$, functions $f\colon U \to V$ and $g\colon V \to U$ are said to be ADJOINT, or to form a GALOIS CONNECTION [Mor94a], if for all $u \in U$ and $v \in V$ we have

$$f.u \sqsubseteq_V v \quad \text{if and only if} \quad u \sqsubseteq_U g.v \,.$$

In (10.5) the two orders U, V are the Booleans with false $\sqsubseteq$ true, and the functions f, g are $(\wedge X)$, i.e. "conjoin with X", and $(X \Rightarrow)$, i.e. "imply by X".

[18] This presentation is backwards: in fact we started from the standard proofs of the axioms in Fig. 10.1.1, and from them synthesised definitions for the probabilistic operators that would allow similar proofs over expectations.

[19] We write these informal words to bring out the connection with the standard version of this law.

Proof

$$\Box(A \rightarrowtriangle B) \ \& \ \Diamond A \ \Rightarrow \ \Diamond B$$

iff $\quad \Diamond A \ \Rightarrow \ \Box(A \rightarrowtriangle B) \rightarrowtriangle \Diamond B$ $\hfill$ adjoint

if $\quad A \sqcup \circ(\Box(A \rightarrowtriangle B) \rightarrowtriangle \Diamond B)$ $\hfill$ *least* property of $\Diamond A$
$\quad \Rightarrow \ \Box(A \rightarrowtriangle B) \rightarrowtriangle \Diamond B$

iff $\quad A \sqcup \circ(\Box(A \rightarrowtriangle B) \rightarrowtriangle \Diamond B)$ $\hfill$ adjoint
$\qquad \& \ \Box(A \rightarrowtriangle B)$
$\quad \Rightarrow \ \Diamond B$

iff $\quad A \ \& \ \Box(A \rightarrowtriangle B)$ $\hfill$ distribute $\sqcup$ through $\&$
$\qquad \sqcup \ \circ(\Box(A \rightarrowtriangle B) \rightarrowtriangle \Diamond B) \ \& \ \Box(A \rightarrowtriangle B)$
$\quad \Rightarrow \ \Diamond B$

iff $\quad A \ \& \ \Box(A \rightarrowtriangle B)$ $\hfill$ *fixed-point* property of $\Diamond B$
$\qquad \sqcup \ \circ(\Box(A \rightarrowtriangle B) \rightarrowtriangle \Diamond B) \ \& \ \Box(A \rightarrowtriangle B)$
$\quad \Rightarrow \ B \sqcup \circ\Diamond B$

if $\qquad\qquad\qquad\qquad A \& \Box(A \rightarrowtriangle B) \Rightarrow A \& (A \rightarrowtriangle B) \Rightarrow B$
$\quad \circ(\Box(A \rightarrowtriangle B) \rightarrowtriangle \Diamond B) \ \& \ \Box(A \rightarrowtriangle B)$
$\quad \Rightarrow \ \circ\Diamond B$

if $\qquad \circ(\Box(A \rightarrowtriangle B) \rightarrowtriangle \Diamond B) \qquad \Box(A \rightarrowtriangle B) \Rightarrow \circ\Box(A \rightarrowtriangle B)$
$\qquad \& \ \circ\Box(A \rightarrowtriangle B)$
$\quad \Rightarrow \ \circ\Diamond B$

if $\hfill$ $\circ$-subdistributivity of $\&$
$\quad \circ((\Box(A \rightarrowtriangle B) \rightarrowtriangle \Diamond B) \ \& \ \Box(A \rightarrowtriangle B))$
$\quad \Rightarrow \ \circ\Diamond B$

if $\quad \circ\Diamond B \Rightarrow \circ\Diamond B$. $\hfill$ monotonicity; adjoint

$\hfill \Box$

Our generalisation of Ben-Ari's laws thus establishes that the operational intuitions underlying standard branching-time temporal logic are also useful for the quantitative logic.

The real surprise however lies in our choice of operators; specifically our need for two kinds of "probabilistic implication" reveals the operational principles communicated by the laws in Fig. 10.1.1, and we conclude this section with a brief examination of what those principles are.

The axioms in Fig. 10.3.5 (and in Fig. 10.1.1) essentially fall into two classes: those which combine quantities "conjunctively" (indicated by the $\lhd$ symbol in Fig. 10.3.5) and those which are designed to be more specific

$$\Box(A \to B) \quad \Rightarrow \quad \Box A \to \Box B \qquad \lhd^{20}$$
$$* \qquad \circ(A \to B) \quad \Rightarrow \quad \circ A \to \circ B \qquad \lhd$$
$$\Box A \quad \Rightarrow \quad \circ A \sqcap \circ\Box A \qquad \lhd$$
$$A \,\&\, \Box(A \Rrightarrow \circ A) \quad \Rightarrow \quad \Box A$$
$$** \qquad \Box(A \to B) \,\&\, \Diamond A \quad \Rightarrow \quad \Diamond B \qquad \lhd$$
$$A \sqcup \circ\Diamond A \quad \Rightarrow \quad \Diamond A \qquad \lhd$$
$$\Box A \quad \Rightarrow \quad \underline{1} - \Diamond(\underline{1}-A) \qquad \lhd$$
$$\Diamond A \,\&\, \Box(\circ A \Rrightarrow A) \quad \Rightarrow \quad A$$

For $*$ — that is, $\circ$-superdistributivity of $\to$ — we reason

$$\circ(A \to B) \quad \Rightarrow \quad \circ A \to \circ B$$
iff $\quad \circ(A \to B) \,\&\, \circ A \Rightarrow \circ B \qquad\qquad$ adjoint
if $\quad \circ((A \to B) \,\&\, A) \Rightarrow \circ B \qquad$ $\circ$-subdistributivity of &
if $\quad \circ B \Rightarrow \circ B$. $\qquad\qquad$ adjoint: $(A \to B) \,\&\, A \Rightarrow B$

The standard axioms were given in Fig. 10.1.1 on p. 266.

Figure 10.3.5. PROBABILISTIC GENERALISATION
OF STANDARD TEMPORAL AXIOMS

---

about the operational meaning of the temporal operators in relation to "next-time."

Considering the former kind we find exclusive use of conjunction & and its adjoint implication $\to$. Recall from Fig. 10.3.1 that & generalises ordinary (Boolean) conjunction to the probabilistic context; thus its appearance here is explained once we notice that the purpose of the $\lhd$-indicated laws is to set out appropriate ways of combining probabilities.

The remaining two laws (without $\lhd$ in the figure) require the introduction of $\Rrightarrow$, our alternative generalisation of implication; and by taking a closer look at the laws' operational motivation we can see why. Considering the fourth law from Fig. 10.3.5 we notice that the expression "$A \Rrightarrow \circ A$" defines a standard expectation (and thus corresponds to a predicate). Used as a guard in the loop [21]

$$\textbf{do } (A \leq \circ A) \to \textit{Step } \textbf{od} \;,$$

we obtain the operational interpretation of $\Box(A \Rrightarrow \circ A)$ if we use a *greatest* fixed-point semantics for the loop. With A standard, the effect is to select the proportion of paths in which $A \Rrightarrow \circ A$ ("the probability that if A holds now then next time it certainly holds") is always true.[22] The law then states

---

[20]See p. 280 for the meaning of this symbol.

[21]Converting it to Boolean we use $\leq$ rather than $\Rrightarrow$.

[22]Recall Footnote 11 on p. 275: *always* of a standard formula is one of the cases in which the "proportion of paths" interpretation is correct.

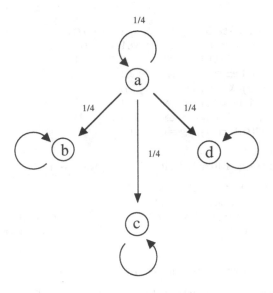

Figure 10.3.6. TRANSITION SYSTEM JUSTIFYING CHOICE OF OPERATORS

that if, in addition, A holds initially ("A &" in the formula) then A itself must *always* hold along the selected paths, and as such correctly encodes the relationship between "next-time" unfoldings and the desired temporal interpretation for $\Box A$. Similar operational interpretations were explored in Chap. 5. Attempting to find any operational relationship between the alternative $A \,\&\, \Box(A \rightarrow \circ A)$ and $\Box A$, however, has so far proved to be fruitless.

In spite of the above explanations, one might nevertheless be tempted to experiment by replacing & by $\sqcap$ and $\rightarrow$ by $\Rrightarrow$ in Fig. 10.3.5; the hope would be to obtain a set of stronger laws using fewer operators, since the inequalities

$$A \,\&\, B \;\Rrightarrow\; A \sqcap B \quad \text{and} \quad A \Rrightarrow B \;\Rrightarrow\; A \rightarrow B \qquad (10.6)$$

(holding for all expectations A, B) would supply tighter relationships between probabilistic expressions than those set out in Fig. 10.3.5. We present a trivial example to illustrate the failure of some of those experiments.

Consider the transition diagram set out in Fig. 10.3.6 which describes a single computation step; we interpret $\circ$ as that step, in the temporal formulae. We take as our example an attempted strengthening of ** in Fig. 10.3.5 by showing that

$$\Box(\{b,c\} \rightarrow \{c\}) \;\sqcap\; \Diamond\{b,c\} \;\Rrightarrow\; \Diamond\{c\}$$

does *not* hold for the system of Fig. 10.3.6. We compare the left-hand side, a lower bound on the probability that both $\{b,c\}$ "implies" $\{c\}$ always holds "and" $\{b, c\}$ eventually holds, with the right-hand side, the probability that $\{c\}$ eventually holds. We now reason

$$
\begin{array}{ll}
& \square(\{b,c\} \rightarrowtail \{c\}) \sqcap \diamondsuit\{b,c\} \\
\equiv & \square(\underline{1} - \{b\}) \sqcap \diamondsuit\{b,c\} & \text{Def. 10.3.3} \\[4pt]
\equiv & & \text{Fig. 10.3.6; Defs. 9.3.2, 9.3.3} \\
& (2\{a\}/3 + \{c,d\}) \sqcap (2\{a\}/3 + \{b,c\}) \\[4pt]
\equiv & 2\{a\}/3 + \{c\} \\
\not\Rrightarrow & \{a\}/3 + \{c\} \\
\equiv & \diamondsuit\{c\} . & \text{Fig. 10.3.6; Def. 9.3.2}
\end{array}
$$

On the other hand combining the probabilities with & on the left, as Lem. 10.3.4 requires, is correctly comparable to $\diamondsuit\{c\}$:

$$
\begin{array}{ll}
& (2\{a\}/3 + \{c,d\}) \ \& \ (2\{a\}/3 + \{b,c\}) \\
\equiv & \{a\}/3 + \{c\} & \text{definition \&} \\
\equiv & \diamondsuit\{c\} .
\end{array}
$$

Similarly, replacing $\Rrightarrow$ by $\rightarrowtail$ in the fourth axiom in Fig. 10.3.5 is incorrect: the implication

$$
\{a,b\} \ \& \ \square(\{a,b\} \rightarrowtail \circ\{a,b\}) \quad \Rrightarrow \quad \square\{a,b\}
$$

does *not* hold for the system of Fig. 10.3.6. We have

$$
\begin{array}{ll}
& \{a,b\} \ \& \ \square(\{a,b\} \rightarrowtail \circ\{a,b\}) \\
\equiv & \{a,b\} \ \& \ \square(\{a,b\} \rightarrowtail \{a\} + \{b\}/2) & \text{Fig. 10.3.6; Def. 9.3.1} \\
\equiv & \{a,b\} \ \& \ \square(\underline{1} - \{b\}/2) & \text{Def. 10.3.3} \\
\equiv & \{a,b\} \ \& \ (\underline{1} - \{b\}/2) & \text{Fig. 10.3.6; Def. 9.3.3} \\
\equiv & \{a\} + \{b\}/2 & \text{definition \&} \\
\not\Rrightarrow & \{a\} + \{b\}/3 \\
\equiv & \square\{a,b\} . & \text{Fig. 10.3.6; Def. 9.3.3}
\end{array}
$$

10.4   Examples: demonic random walkers and stumblers

We finish this chapter with two, more substantial examples based on the random walk: they include both demonic nondeterminism and fully quantitative reasoning.

The *unbounded random walk* [GW86] was the subject of Sec. 3.3: it concerns a (Markov) process that moves up or down in discrete steps with certain probabilities.[23] In our earlier treatment we "encoded" the process as a sequential program whose termination depended on its having the property we sought (*i.e.* its transition to a certain part of the number line); proof of the property then boiled down to proof of termination.

In temporal logic we do not need such a coding trick; rather we can treat the walker as a continuing process directly.

We consider a "demonic" walker whose transition probabilities are uncertain:[24] to illustrate the role of specific numbers our premise here will be that the up- and the down transition each are taken with probability at least 1/3; our conclusion will be that the walker moves one position up (or down) *eventually* with probability u at least 0.382 (approximately), and that the *recurrence* probability r of the walker's eventually returning to his starting position is at least 0.255.

In summary we use the elementary argument that u is at least 1/3, for an immediate move up, plus $1/3 * u^2$ for an immediate move down followed by two eventual moves up: then the least solution of $u \geq (u^2 + 1)/3$ is our quoted result above; and finally r is calculated from u. Our treatment below with the temporal operators thus will formalise

1. that the demonic nondeterminism need not be "factored out," [25]

2. that the probabilities may be added in the way suggested above,

3. that u is independent of the starting position,

4. that the eventualities may be composed and

5. that r is determined by u.

We note before beginning that Point (1) is especially important in that it justifies Point (3). Many, indeed most of the demonic walkers satisfying the constraints above will be non-homogeneous, *i.e.* they will have *actual* transition probabilities that vary from place to place, although all will satisfy the "no less than 1/3" criterion; thus u will not be constant for each

---

[23]A MARKOV PROCESS [GW86, GS92] is one whose probabilistic behaviour depends only on its current state and is not influenced by the history of previous states it might have passed through.

[24]Strictly speaking the demonic choice makes the process potentially non-Markovian, because that choice is completely unconstrained and could in principle depend on the history of the process. But our analysis here is unaffected by that.

One of the results of the next chapter — the sufficiency of memoriless strategies — shows that "memory" makes no difference in the game interpretation when the state space is finite; here however it is infinite.

[25]This should by now be a familiar point, as it is a major feature of the approach we promote in this volume overall. With abstraction and refinement built into the model we are not limited to single deterministic (probabilistic) programs or processes; rather we treat whole groups of them at once as "first-class citizens." The demonic choice between programs is itself a program, and it satisfies the same general laws.

of those separately. But the demonically "worst" walker, representing all possibilities at the same time, is in fact homogeneous — as we will see.

10.4.1   The demonic walker

We begin with the premise, a specification of the walker. Take the state space to be the integers $\mathbb{Z}$, so that expectations are of type $\mathbb{Z} \to [0, 1]$. The specification of the walker will be that it is any refinement of the program

$$
step \quad := \quad \left|
\begin{array}{llll}
s := & s - 1 & @ & 1/3 \\
s := & s + 1 & @ & 1/3
\end{array}
\right. \tag{10.7}
$$

where $s: \mathbb{Z}$ is the current position.

Recall that the deficit $1 - (1/3 + 1/3) = 1/3$ in the probabilities above represents the probability of aborting, and that **abort** is refined by any behaviour at all. Thus a walker, refining *step*, is free to use the deficit $1/3$ to add additional behaviours that go up or down in any combination, and for any distance.

In order to use our temporal tools to reason about eventuality, we "capture" the above specification (10.7) as a qTL formula in which ○ stands for *step* in the usual way — we thus have for any expectation A and state s that

$$
○A.s \quad = \quad A.(s-1)/3 + A.(s+1)/3 . \tag{10.8}
$$

We now calculate a guaranteed lower bound for the probability that the specific (so to speak "worst") walker *step* eventually moves one step up; note that we are reasoning about *step* itself, not about any particular refinement of it. Let $\Diamond\{s+1\}.s$ be u (for *up*). We have

	u	
=	$\Diamond\{s+1\}.s$	definition u
=	$\{s+1\}.s \ \sqcup \ ○\Diamond\{s+1\}.s$	Def. 9.3.2
=	$\Diamond\{s+1\}.(s-1)/3 + \Diamond\{s+1\}.(s+1)/3$	$\{s+1\}.s = 0$; from (10.8)
=	$\Diamond\{s+1\}.(s-1)/3 + 1/3$	$\Diamond\{s+1\}.(s+1) = 1$
=	$\Diamond\Diamond\{s+1\}.(s-1)/3 + 1/3$	Lem. B.6.1

$$
\begin{array}{lll}
\geq & \Diamond u\{s\}.(s{-}1)/3 + 1/3 & \text{see ($\ddagger$) below; Lem. B.6.3: } \Diamond \textit{ monotonicity} \\
= & u(\Diamond\{s\}.(s{-}1))/3 + 1/3 & \text{Lem. B.6.5: } \Diamond \textit{ scaling} \\
= & u^2/3 + 1/3 \ , & \Diamond\{s\}.(s{-}1) \text{ is } u \text{ also }^{26}
\end{array}
$$

giving $u \geq (u^2 + 1)/3$, so that as promised we have

$$
u \ \geq \ \frac{3 - \sqrt{5}}{2} \ \simeq \ 0.382 \ . \quad ^{27}
$$

$\ddagger$ For the deferred justification we simply note that

$$
\Diamond\{s{+}1\} \ \Leftarrow \ \Diamond\{s{+}1\}.s * \{s\} \ \equiv \ u * \{s\} \ .
$$

A similar proof establishes the same inequality for $\Diamond\{s{-}1\}.s$; and we can then bound the probability r of eventual return by reasoning

$$
\begin{array}{lll}
& r & \\
= & \circ\Diamond\{s\}.s & \text{write "eventual return" in temporal logic} \\
= & \Diamond\{s\}.(s{-}1)/3 + \Diamond\{s\}.(s{+}1)/3 & \text{(10.8) with } A{:=} \Diamond\{s\} \\
\geq & 2/3 * (3 - \sqrt{5})/2 & \text{above} \\
= & 1 - \sqrt{5}/3 & \\
\simeq & 0.255 \ . \quad ^{28} &
\end{array}
$$

We conclude by noting that the formula $\circ\Diamond\{s\}$ is monotonic as a function of the underlying *step* — that is, if we consider any refinement *step'* of it, as allowed by our original specification, then $\circ\Diamond\{s\}$ over that refined system cannot be less than it was over the original.[29] Thus our conclusion holds for all the walkers allowed by the specification, homogeneous or not, and not only for the "worst" one — even though we used homogeneity in the argument.

---

[26]That inequality is a form of data refinement, and follows from Lem. B.6.2 by considering monotonic "shift by k" transformers $(\cdot\diagup k)$ defined so that for any expectation A and integer k we have

$$
(A\diagup k).s \ := \ A.(s{-}k) \ .
$$

Using $t := (\cdot\diagup s)$ in the lemma shows that $\Diamond\{s{+}1\}.s = \Diamond\{1\}.0$ for all s.

[27]As a rough check of this answer we note that u should slightly exceed the infinite summation $1/3 + 1/3^3 + \cdots = 0.375$, the sum of the probabilities of the disjoint events "up, down-up-up...".

[28]Here our rough check suggests r should exceed $2(1/3^2 + 1/3^4 \cdots) = 0.25$.

[29]For extensive arguments of this kind we would parametrise $\circ$ by the transition it denotes, as we do in Chap. 11. Here we could write $\circ'\Diamond\{s\}$.

10.4.2   The demonic stumbler

Now we consider a generalisation: suppose that instead of (10.7) we have a random *stumbler* who

> on each step may remain where he is — he is not obliged
> to move with any probability at all — but *eventually* he
> must move and then, as before, he reaches the one-step-up
> or one-step-down position with probability at least 1/3 each.

(10.9)

It is tempting to start from

$$\Diamond\{s{+}1\}.s \ \geq\ 1/3 \quad \text{the stumbler } eventually \text{ moves up,}$$
with probability at least 1/3 and

(10.10)

$$\Diamond\{s{-}1\}.s \ \geq\ 1/3 \quad \text{the stumbler } eventually \text{ moves}$$
down, with probability at least 1/3,

but there are some pitfalls here, as we now illustrate.

For example, recall that our earlier calculations showed the specification "move one step up *eventually* with probability at least 0.382" to be satisfied by a random walker who moves up/down *immediately* with probability at least 1/3 in each case, with the 1/3 slightly lower than 0.382 as one would expect.

Similar calculations therefore show that the specification "move one step up *eventually* with probability at least 1/3" is satisfied by a random walker who moves up/down *immediately* with probability at least 0.3 in each case, with the 0.3 slightly lower than 1/3 again as one would expect. *But does a 0.3-walker satisfy our original specification (10.9)?* It is not clear.

†   The problem with (10.10) is that it would be satisfied by a walker who from s with probability at least 1/3 eventually visits $s{-}1$ — and then goes on to $s{+}1$, "reusing" his 1/3 for the latter. More likely what we have in mind is "eventually the stumbler will stop delaying, and take a step: with probability at least 1/3 it will be up, and with probability at least 1/3 it will be down."

At some point, however, arguing about a specification in natural language becomes ineffective.[30]

---

[30]Specifications need to be debugged, just as programs do. A formal, or at least semi-formal notation is important for that — one writes a "guess," from the natural language or other information, of a formal specification that captures the essential requirements, and then one "runs" the specification by proving (or failing to prove) consequences of it. The proofs are carried out in mathematics (possibly a specialised form, as here); and whether the proof is by logic, or model checking or even animation is irrelevant — the point is that the specification under test is written in a language with precise semantics.

For programs, serious debugging similarly can not take place until the program is actually coded, written in a language with a precise meaning (as *e.g.* provided by the implementation platform).   ...

Accordingly we will bite the bullet and follow the example of (10.8), specifying

$$\Diamond A.s \;\geq\; A.(s{-}1)/3 \;+\; A.(s{+}1)/3 \;, \qquad (10.11)$$

in which the original $\circ A$ is simply replaced by $\Diamond A$. Note that this specification is stronger than $(10.10)^{31}$ — but it is still satisfied by the stumbler we probably had in mind, one who waits for some time and then takes a single step either way with probability 1/3 each. It is *not* satisfied by the "reusing" walker at (†) above, however.

Our analysis of the stumbler is now almost as for the walker (and is shorter than the discussion above...) It begins however with

$$
\begin{aligned}
&\quad \Diamond\{s{+}1\}.s \\
=\;&\quad \Diamond\Diamond\{s{+}1\}.s && \text{Lem. B.6.1} \\
\geq\;&\quad \Diamond\{s{+}1\}.(s{-}1)/3 + \Diamond\{s{+}1\}.(s{+}1)/3 \;. && \text{from (10.11)}
\end{aligned}
$$

We then reason as before to the same conclusion "at no extra cost."[32]

...[30] One of the aims of formality therefore is to move the debugging as far as possible from the program back to the specification, as it is cheaper the earlier it is done. At the same time, however, one must retain the precision of the notation.

[31] It is stronger because (10.10) is satisfied by the 0.3-walker, who could with probability $1 - 2*0.3 = 0.4$ abort on his very first step. Taking $A{:}=$ [true] in (10.11) however gives $\Diamond$ [true] $.s \geq 2/3 > 1 - 0.4$, which constraint is therefore not satisfied by the 0.3-walker.

[32] We are again assuming homogeneity of the worst solution, but in this case our reference to it is implicit — and it requires somewhat more work, beyond the logic, to establish its existence. We sketch the argument below.

Basically the worst solution to (10.11) is constructed as the demonic choice of all solutions; but there are some details to settle. The random walker from the previous section is a solution, so there is at least one. The *worst* one we construct formally as a transformer by taking the arithmetic minimum of all solutions; that it too is a solution is shown by distributing **min** through $\Diamond$, a straightforward fixed-point argument based on the definition of eventually.

We say "formally" however because an infinitary infimum of transformers over an infinite state space is not necessarily continuous (refer Footnote 13 on p. 225). In spite of that, an arithmetic argument shows that the formal minimum retains sublinearity (as on p. 225), and thus continues to satisfy the algebraic laws of Fig. 10.3.1. Those laws do not rely on continuity; in particular their characterisation of least- and greatest fixed-points does not use the $\mathbb{N}$-limit formulation.

Finally, because the specification (10.11) is translation invariant, the worst transformer — constructed as above by pointwise minimum over the initial states — will be translation invariant as well, and we then have translation invariance of $\Diamond\{s{+}1\}$ etc. by the same argument used at Footnote 26 above.

Thus the worst sublinear transformer satisfying (10.11) satisfies $r \geq 1 + \sqrt{5}/3$ as at the end of Sec. 10.4.1; and so all sublinear *and continuous*, *i.e.* regular transformers that refine it — those "real" solutions lying above it but properly in our computational model — do so as well. We have in effect used an "ideal" element outside the model (the possibly non-continuous worst solution) to prove properties of elements within it.[33]

10.5 Summary

The contribution of the last two chapters has been the reinterpretation of temporal logic over expectation- rather than predicate transformers, and a determination of how the standard operators should be embedded in the expanded framework: we established (Fig. 10.3.5) the reasoning principles analogous to those for standard temporal logic, when the operators are restricted to the usual modalities ○, ◇ and □. The postulated axioms for ○ — characterising probabilistic transitions — facilitate proofs of the laws, whilst our use of probabilistic conjunction rather than some other generalisation of ordinary conjunction retains modular reasoning, even for quantitative information.

Sec. 10.4 demonstrated those principles in practice, especially where the result depends on explicit probabilities. A logic built over that interpretation thus has the advantages of specialising smoothly to the standard case, giving quantitative results where desired yet via the *Zero-One Law* (Chapters 2 and 7) allowing explicit probabilities to be discarded [MM03] if irrelevant,[34] having reasoning principles that generalise familiar ones and finally offering the possibility of new applications, beyond strict temporal logic, to probabilistic games, together with a framework for the direct calculation of expected quantities such as space or time complexity [McI02].

Finally the greater distinguishing power of $[0,1]$ over $\{0,1\}$ also has compelling implications for specification — the use of (5) for example illustrates how quantitative temporal properties may be expressed in a manner both pleasingly succinct and well suited for reasoning.

---

[33]It is tempting in the argument of Footnote 32 to opt for safety by staying within the model, "closing up" after the infinitary minimum to retain continuity: by taking the Cauchy closure of the infimum, we could construct the greatest regular transformer below all solutions to the specification (10.11) and would have no need for the formal — but possibly non-continuous — ideal element.

But that does not work in general, and indeed we have no assurance that it does in this case. The specification

$$\bigcirc 1 \ \Lleftarrow\ 1 \qquad\qquad (10.12)$$

is satisfied by all terminating programs; the formal infimum of those is the unboundedly demonic but terminating programs that can reach every final state — sometimes called **chaos** [BvW98, p. 195] — and it too satisfies (10.12). But **chaos** is not continuous (refer Footnote 13 on p. 335), and its Cauchy closure — the most refined *regular* program below it, that therefore is refined by every solution of (10.12) — is in fact **abort**. Of course **abort** does *not* satisfy (10.12).

Continuity is a convenient — and conventional — restriction to impose on our computations; but at times it can thus be an advantage to drop it [Boo82].

[34]We use quantitative *unless* ▷, and the *qTL* version of the *Zero-One Law* becomes

If P is standard and $p * (A \triangleright P) \Rrightarrow \Diamond P$ for some nonzero p, then we have $A \triangleright P \Rrightarrow \Diamond P$.

The approach of $pCTL$ [BdA95] is close to ours in the following sense. Omit $pCTL$'s operators **A** and **E** (expressing absolute judgements rather than extremal probabilities), and consider only the $\mathbb{P}_{\geq p}$-form of threshold-probability judgements.[35] Then a $pCTL$ state-formula is true in just those states in which the "equivalent" expectation formula takes the value one, where the expectation formula is obtained by

- replacing the propositional operators $\vee, \wedge, \neg$ by the arithmetic $\sqcup, \sqcap, (1-)$ respectively,

- leaving the modal operators as they are and

- replacing $\mathbb{P}_{\geq p}$ by $(\underline{p} \Rrightarrow)$, as in Def. 10.3.2,

at all levels of the $pCTL$ formula. Thus for example the $pCTL$ judgement $\mathbb{P}_{\geq p}\Diamond A$ — expressing that (standard) A will be established eventually with probability at least p — becomes $\underline{p} \Rrightarrow \Diamond A$ (where we interpret A as a characteristic function).

Huth and Kwiatkowska establish similar correspondences, but for different sets of formulae: in their Theorem 1 [HK97] they show that the "optimistic" view (in our terms the ceiling $\lceil \cdot \rceil$) distributes through any positive existential formula, and that the "pessimistic" view ($\lfloor \cdot \rfloor$)) distributes through universal formulae.

Moving to $pCTL^*$ however we find that there the logics diverge. For example we cannot translate

$$\mathbb{P}_{\geq p} \left(\Diamond P \wedge \Diamond Q \right) \tag{10.13}$$

directly into expectations, because there is no equivalent operator acting between expectations (just numbers) in the way that $\wedge$ acts between the path formulae that contain so much more information. To write as above "the probability is at least p that both P will be established eventually and Q will be established eventually" we would proceed from first principles, thinking of the probability of winning a suitable game:

- if P and Q are both false then take a step;

- if one is true but not the other, then take a step but subsequently seek only the other; and

- if both are true then stop (and win).

The expectation formula is thus

$$\underline{p} \quad \Rrightarrow \quad \Diamond((P \sqcap \Diamond Q) \sqcup (Q \sqcap \Diamond P)) \,.$$

---

[35]The fact that we use *eventually* and *always* rather than *until* is not important.

There is a more general translation process, however, based on explicit access to the game interpretations of the modal operators (Chap. 9); and it does seem that simple propositional-free formulae carry over even if they are not in $pCTL$. For example

$$\mathbb{P}_{\geq p}\Box\Diamond P \quad \text{and} \quad \underline{p} \rightrightarrows \Box\Diamond P$$

are equivalent for standard P, both expressing that with probability at least p the predicate P is established infinitely often. De Alfaro and Majumdar [dAM01] gave a general procedure for such translations, based on Büchi automata.

Finally, our treatment of nondeterminism does not explicitly mention a scheduler [SL95, BdA95] because it does not have to: the (demonic) schedule is built-in to the semantics [MMS96] of expectation transformers. (Recall the infimum used in Def. 5.5.2 on p. 144.) A *probabilistic* scheduler [BdA95], able to interpolate between discrete choices, corresponds to the probabilistic (convex) closure condition of He [HSM97].

To treat the existential duals $\exists X$, $\exists F$ and $\exists G$, we would allow both demonic and angelic nondeterminism in *step*: as in the standard case, the existential modalities are obtained via a double complementation. The weakening effect of that on the healthiness conditions in that case was discussed in Chap. 8.

Chapter notes

The proof system for qTL is a generalisation of the system of Ben-Ari *et al.* for branching-time temporal logic. Hart and Sharir [HS86] also generalised their logic to probabilistic deterministic systems, though they restricted to probability-one properties. Similarly Rao's $UNITY$-based probabilistic temporal framework is for probability-one properties.

Other probabilistic properties are amenable to algebraic treatments. McIver [McI01] uses an algebraic approach to generalise the usual approach to *stationary-distribution* properties of Markov processes: using the demonic/probabilistic framework allows an "extended stationarity" to be defined which is achieved even for periodic processes while agreeing with the conventional definition otherwise.

Standard treatments of the random walk use generating functions of probability theory [GW86], rather than algebraic properties of the appropriate fixed-point definitions relating to reachability; Hurd *et al.*'s formalisation of $pGCL$ in HOL [HMM04] also takes a fixed-point approach.

11

The quantitative modal μ-calculus, [1] and gambling games

11.1 Introduction to the μ-calculus

The modal μ-calculus, introduced by Kozen [Koz83], generalises the temporal logic we introduced in Chap. 9 and developed in Chap. 10. For our purposes here, we can see it as elevating definitions such as Defs. 9.2.2–9.2.4 so that their right-hand sides become part of the syntax of logical formulae — that is, the modal operators $\diamond$, $\square$ and $\triangleright$ become abbreviations, only syntactic sugar for certain fixed-point expressions that could have been written out in full. The single remaining modal primitive is then the *next-time* operator $\circ$.

---

[1] Although this chapter is based on a conference publication [MM02], more detail may be found in the significantly revised and expanded journal version [MM04b].

The advantage of that — beyond unifying the formerly diverse operators — is that we have a richer language if we step *outside* the abbreviations, one in which it is possible to express behaviours that the temporal operators cannot. We will see examples of that shortly.[2]

In Chap. 9 we gave game-based interpretations of the quantitative temporal formulae expressible in qTL. In this chapter we show for a finite-state system that the game-based interpretation is appropriate for *all* μ-calculus formulae, and we show also that *memoriless strategies suffice* for achieving the *minimax* value of the quantitative games that result.

Taken together with Chap. 10, this equivalence allows us to move smoothly between the operational and algebraic worlds for quantitative μ-expressions in general. The former is important for establishing the suitability of our specifications and for interpreting the formulae that will result from calculations and derivations based on them. The latter is of course essential for doing the calculations themselves.[3]

Even in the standard case, general μ-calculus expressions can be difficult to use — in all but the simplest cases they are not easy on the intuition, especially for example with the nesting of alternating fixed points. (Recall Sec. 10.2.5.) And even the more specialised temporal properties (particularly "branching-time properties") are notoriously challenging for specification [Var01]. Stirling's "two-player-game" interpretation, however, provides an alternative and operational view [Sti95].

Beyond the standard case, things can only get worse. As we have seen, the quantitative modal μ-calculus acts over *probabilistic* transition systems, extending the above; and so it would benefit even more from having two complementary interpretations. That is what we provide in this chapter, and we show that over a finite state-space they are equivalent: our earlier quantitative interpretation (Chapters 9 and 10) generalises Kozen's standard semantics, as we have seen; the other interpretation (introduced below) generalises Stirling's standard games.

In our operational interpretation, a significant complication is that we must distinguish nondeterministic choice from probabilistic choice: the former is represented by the two players' "strategies," as before; but the latter is represented by gambling. In Sec. 11.4.2 we set out the details.

In Sec. 11.5 we give an example of the use of the quantitative calculus for general expectations (*i.e.* not just for probabilities).

---

[2]Here we are drawing together comments made in Footnote 32 on p. 263 and in Sec. 10.2.5.

[3]Footnote 10 on p. 319 discusses further merits of this "double view."

11.2 *Quantitative* μ-calculus for probability

The standard equivalence established by Stirling showed that a formula's Boolean value (in the Kozen interpretation) corresponds to the existence of a winning strategy (in his game interpretation). In the quantitative formulation, strategies in the game must become "optimal" rather than "winning"; and the correspondence is now between a formula's value (since it denotes a real number, in the first interpretation) and the expected winnings from the (zero-sum) gambling game (of the second interpretation). Our main aim is to establish the well-definedness of the latter (which, as a "minimax," cannot be taken for granted), and to show it equals the former — so generalising the standard result. The details are set out in Sec. 11.6, which is where we also show that memoriless strategies suffice.

The benefit of this "dual" approach is that a specifier can build his intuitions into a game and can then use the features of the logic to prove properties about the problem. For example, the *sublinearity* of $qM\mu$ — the quantitative generalisation of the *conjunctivity* of standard modal algebras — was used in Chap. 10 to prove a number of algebraic laws that generalise those of standard branching-time temporal logic.

Preliminary experiments have shown that the proof system is effective for unravelling the intricacies of distributed protocols [Rab82, MM99b]. Moreover it provides an attractive proof framework for Markov decision processes [MM01c, FV96] and indeed many of the problems there have a succinct specification as μ-calculus formulae, as the example of Sec. 11.5 illustrates below. In "reachability-style problems" [dA99], proof-theoretic methods based on $qM\mu$ have produced very direct arguments related to the abstraction of probabilities [MM03], and even more telling is that the logic is applicable even in infinite state spaces [dA99]. All of which is to suggest that further exploration of $qM\mu$ will continue to be fruitful.

In the following we shall assume generally that the state space S is countable (though for the principal result we must restrict to finiteness, in Sec. 11.6).

11.3 Logical formulae and transition systems

In this section we fix our syntax for the logical language, presented to some extent informally in the earlier chapters of Part III, and we discuss some of the details of the probabilistic transition systems over which the formulae are to be interpreted.

Formulae in the language of $qM\mu$, in positive form,[4] are constructed as follows, where A, K and G all represent constant terms in the uninterpreted formulae:

$$\phi \quad := \quad X \mid A \mid \langle K \rangle \phi \mid [K]\phi \mid \phi_1 \sqcap \phi_2 \mid \phi_1 \sqcup \phi_2$$
$$\mid \quad \phi_1 \lhd G \rhd \phi_2 \mid (\mu X \bullet \phi) \mid (\nu X \bullet \phi) \,.$$

Informally, we note here that the A-terms denote (quantitative) rewards, the K-terms denote sets of execution steps resolved either angelically $\langle \cdot \rangle$ or demonically $[\cdot]$, and the G-terms denote Booleans used in conditionals $\cdot \lhd \cdot \rhd \cdot$ representing if-then-else choice between subformulae; the variables X are used only for binding fixed-points. The precise meanings of A, K and G are given in Sec. 11.4 below.

It is well known from the standard μ-calculus that these formulae can be used to express complex path-properties of computational sequences; here however we shall interpret the formulae over sequences based on generalised probabilistic transitions that correspond to the "game rounds" of Everett [Eve57]. They are modelled by the space $\mathcal{R}S$ comprising the "relational style" functions in $S \to \overline{S^{\$}}$ in which $S^{\$}$ is just S with a special "payoff" state \$ adjoined. The transitions in $\mathcal{R}S$ give the probability of passage from initial s to final (proper) s' as $r.s.s'$; any deficit $1 - \sum_{s':S} r.s.s'$ is interpreted as the probability of an immediate halt with payoff $r.s.\$/(1 - \sum_{s':S} r.s.s')$.[5]

The above "quotient" formulation of the payoff has three desirable properties. The first is that the probabilistically *expected* halt-and-payoff is just $r.s.\$$ (and we simply define it to be so in the case $\sum_{s':S} r.s.s' = 1$, in which case the expected immediate payoff $r.s.\$$ is necessarily zero). The second property is that we can consider the probabilities of outcomes from s to sum to one exactly (rather than no more than one), since any deficit is "soaked up" in the probability of transit to payoff. Thus we deal with full- rather than sub-distributions.

The third property (feasibility) is that the transitions preserve boundedness of expectations — in $qM\mu$, generally the set of expectations we consider is one-bounded. Thus if A in $\mathbb{E}S$ gives a reward $A.s'$ that is expected to be

---

[4]The restriction to the positive fragment is for the usual reason: that the interpretation of any expression $(\lambda X \bullet \phi)$, constructed according to the given rules, should yield a monotone function of X.

[5]Thus these transitions differ from our deterministic probabilistic model Def. 5.1.3 only in having the extra "payoff" state. Footnote 18 on p. 309 below locates the explanation of why we need that extra state.

realised at state s' after transition r, then the pre-expectation at s before transition r is

$$r.s.\$ + \int_{r.s} A \, ,$$

where the domain of the function $r.s$ under $\int$ is restricted to (final) states in S proper.[6]

It is the expected value realised by making transition r from s to s' or possibly \$, and taking $A.s'$ in the former case. That this pre-expectation is also one-bounded allows us to confine our work to the real interval $[0,1]$ throughout this chapter.

Hence computation trees can be constructed by "pasting together" applications of transitions $r_0, r_1, \ldots$ drawn from $\mathcal{R}S$, with branches to \$ being tips. The probabilities attached to the individual steps then generate a distribution over computational paths, defined for the σ-algebra of extensions of finite sequences within the tree.[7,8]

In our interpretations we will use *valuations* in the usual way. Given a formula ϕ, a valuation $\mathcal{V}$ will do four things here:

1. it will map each A in ϕ to a fixed expectation in $\mathbb{E}S$;

2. it will map each occurrence of K to a fixed finite set of probabilistic transitions in $\mathcal{R}S$;

3. it will map each occurrence of G to a predicate over S; and

4. it will keep track of the current instances of "unfoldings" of fixed-points, by including mappings for bound variables X.

---

[6]To avoid clutter we will assume this restriction where necessary in the sequel.

[7]A σ-ALGEBRA over a set is a collection of its subsets that is closed under countable unions and complements (and thus countable intersections also), which operations correspond to the basic techniques of reasoning about probability: the probabilities of events' unions, intersections and complements. Thus the algebra contains "just enough" sets to include all those events whose probability one could ask for: taking "too many" (*e.g.* all) subsets is avoided, because it can lead to difficulties.

For example, if *every* subset of $[0,1]$ were to be assigned a probability, then still only a countable number — *i.e.* almost none — of the singleton sets could be assigned more than zero — otherwise summing them up would assign an infinite probability to $[0,1]$ as a whole.

For the unit interval, the usual approach is to use a *continuous* distribution by assigning a probability of $b-a$ to every open interval (a,b) with $0 \le a \le b \le 1$, and then to form the smallest σ-algebra containing those intervals [GS92]. (Singleton sets are still assigned probability zero by this procedure, but cannot be countably combined to form any interval (a,b) with $a < b$, so there is no contradiction caused by the fact that the interval's probability $b-a$ is not the sum of the zeroes contributed by its uncountably many individual points.) We do not use continuous distributions here, since without a continuous-distribution primitive our programs can generate only countably many outcomes; in other work however we have made that extension [MM01b].

When the σ-algebra is generated by a topology (whose basis in this case is the open intervals) it is called a BOREL algebra. Our path-distributions here are an example of that technique, applied to the "Smyth topology" on sequences.[8]

(For notational economy, in (4) we are allowing $\mathcal{V}$ to take on as well the role usually given to a separate "environment" parameter.)

We make one simplification to our language, without compromising expressivity. Because the valuation $\mathcal{V}$ assigns finite sets to all occurrences of K, we can replace each modality $\langle K \rangle \phi$ ($[K]\phi$) by an explicit maxjunct $\bigsqcup_{k:K} \{k\} \phi$ (minjunct $\bigsqcap_{k:K} \{k\} \phi$) of (symbols k denoting) transitions k in the set (denoted by) K. Then we have only one kind of modality $\{\cdot\}$, instead of the two $\langle \cdot \rangle, [\cdot]$.

In the rest of this chapter we shall therefore use the *reduced language* given by

$$\phi \quad := \quad X \mid A \mid \{k\}\phi \mid \phi_1 \sqcap \phi_2 \mid \phi_1 \sqcup \phi_2 \mid \phi_1 \lhd G \rhd \phi_2 \mid (\mu X \cdot \phi) \mid (\nu X \cdot \phi) \,.$$

We replace (2) above in respect of $\mathcal{V}$ by:

> 2! it maps each occurrence of $\{k\}$ to a fixed probabilistic transition in $\mathcal{R}S$.

11.4   Two interpretations of $qM\mu$

11.4.1   The logical, denotational interpretation: a generalisation of Kozen's logic

In this section we recall the interpretation introduced less formally in Chapters 9 and 10, where our quantitative logic for demonic/probabilistic programs suggested a generalisation of Kozen's logical interpretation of μ-calculus for probabilistic transition systems.

Let ϕ be a formula and $\mathcal{V}$ a valuation. We write $\|\phi\|_{\mathcal{V}}$ for its meaning, an expectation in $\mathbb{E}S$ determined by the following rules:

1. $\|X\|_{\mathcal{V}} \quad := \quad \mathcal{V}.X$.

2. $\|A\|_{\mathcal{V}} \quad := \quad \mathcal{V}.A$.

---

[8]Take the *prefix* order over sequences, and generate the Smyth topology from that.[9] A basis for the topology is then the set of all extensions of any given finite sequence, which we close under countable unions and intersections to form the Borel algebra on which our probability measure is defined.

[9]In general, the SMYTH TOPOLOGY [Smy89] is formed from a partial order $\sqsubseteq$ by taking all subsets which are "up-closed" and "inaccessible"; it has the striking property that $\sqcup$- and topological continuity agree.

For our computation sequences, the up-closed subsets are those closed under extension. A subset is INACCESSIBLE just when no chain can "reach" the subset unless in fact the chain lies partially within the subset already: that is, no supremum of any chain is in the subset unless some element of the chain is. Here that means our extension-closed sets of sequences must be generated from finite prefixes.

3. $\|\{k\}\phi\|_{\mathcal{V}}.s \quad := \quad \mathcal{V}.k.s.\$ + \int_{\mathcal{V}.k.s} \|\phi\|_{\mathcal{V}}$.

4. $\|\phi' \sqcap \phi''\|_{\mathcal{V}}.s \quad := \quad \|\phi'\|_{\mathcal{V}}.s \ \sqcap \ \|\phi''\|_{\mathcal{V}}.s$; and
   $\|\phi' \sqcup \phi''\|_{\mathcal{V}}.s \quad := \quad \|\phi'\|_{\mathcal{V}}.s \ \sqcup \ \|\phi''\|_{\mathcal{V}}.s$.

5. $\|\phi' \lhd G \rhd \phi''\|_{\mathcal{V}}.s \quad := \quad \|\phi'\|_{\mathcal{V}}.s \ \text{if} \ (\mathcal{V}.G.s) \ \text{else} \ \|\phi''\|_{\mathcal{V}}.s$.

6. $\|(\mu X \cdot \phi')\|_{\mathcal{V}} \quad := \quad (\mu x \cdot \|\phi'\|_{\mathcal{V}[X \mapsto x]})$ where, in the semantics on the *rhs*, by $(\mu x \cdot exp)$ we
   mean the least fixed-point of the function $(\lambda x \cdot exp)$.

7. $\|(\nu X \cdot \phi')\|_{\mathcal{V}} \quad := \quad (\nu x \cdot \|\phi'\|_{\mathcal{V}[X \mapsto x]})$.

Note that in the valuation $\mathcal{V}[X \mapsto x]$, the variable X is mapped to the expectation x in $\mathbb{E}S$; other mappings in $\mathcal{V}$ are unaffected.

Lemma 11.4.1 The quantitative logic $qM\mu$ is well defined: for any ϕ in the language, and valuation $\mathcal{V}$, $\|\phi\|_{\mathcal{V}}$ is a well-defined expectation in $\mathbb{E}S$.

   Proof   As in Chapters 9 and 10: structural induction, arithmetic and that $(\mathbb{E}S, \Rrightarrow)$ is a complete partial order when one-bounded.   □

11.4.2   The operational, gambling-game interpretation: a generalisation of Stirling's game

In this section we give our novel alternative account of formulae ϕ (of the reduced language), extending Stirling's turn-based game for the standard case [Sti95]. It is played between two players, whom we refer to respectively as *Max* and *Min*. As we did in Sec. 11.4.1, we assume a probabilistic transition system $\mathcal{R}S$ and a valuation $\mathcal{V}$. The game progresses through a sequence of *game positions*, each of which is either a pair (ϕ, s) where ϕ is a formula and s is a state in S, or a single (p) for some non-negative real number p representing a payoff. Following Stirling, we will use the idea of "colours" to represent placeholders for possible return to a fixed point.

A sequence of game positions is called a *game path* and is of the form (ϕ_0, s_0), (ϕ_1, s_1), ... with (if finite) a payoff position (p_n) at the end. The initial formula ϕ_0 is the given ϕ, and s_0 is an *initial* state in S. A move from position (ϕ_i, s_i) to (ϕ_{i+1}, s_{i+1}) or (p_{i+1}) is specified by the following rules.

1. If ϕ_i is $\phi' \sqcap \phi''$ (resp. $\phi' \sqcup \phi''$) then *Min* (*Max*) chooses one of the minjuncts (maxjuncts): the next game position is $(\widehat{\phi}, s_i)$, where $\widehat{\phi}$ is the chosen 'junct ϕ' or ϕ''. (That is, the new formula ϕ_{i+1} is $\widehat{\phi}$, and the new state s_{i+1} is the same as the current one s_i.)

2. If ϕ_i is $\phi' \lhd G \rhd \phi''$, the next game position is (ϕ', s_i) if $\mathcal{V}.G.s_i$ holds, and otherwise it is (ϕ'', s_i).

3. if ϕ_i is $\{k\}\phi'$ then the distribution $\mathcal{V}.k.s_i$ is used to choose either a next state s' in S or the payoff state $\$$. If a state s' is chosen, then the

next game position is (ϕ', s'); if \$ is chosen, then the game terminates in position (p) where p is the payoff $\mathcal{V}.\mathsf{k}.s.\$/(1 - \sum_{s':S} \mathcal{V}.\mathsf{k}.s.s').$[10]

4. If ϕ_i is $\nu X \cdot \phi'$ $(\mu X \cdot \phi')$ then a fresh "colour" C is chosen from some infinite supply, and is bound to the formula $\phi'[X \mapsto \mathsf{C}]$ for later use; the next game position is (C, s_i).[11]

5. If ϕ_i is C the next game position is (ϕ', s_i), where ϕ' is the formula previously bound to C. (Note from (4) that the formula bound to C probably has another C within it, put there at the time of binding: that is what causes the fixed-point to "loop.")

6. If ϕ_i is A then the game terminates in position (p) where $p = \mathcal{V}.\mathsf{A}.s_i$.

A game path is said to be *valid* if it can occur as a sequence according to the above rules. Note that along any game path at most one colour can appear infinitely often:

Lemma 11.4.2 All valid game paths are either finite, terminating at some payoff (p), or infinite; if infinite, then exactly one colour appears infinitely often.

Proof   Stirling [Sti95].   □

To complete the description of the game, one would normally give the winning/losing conditions. Here however we are operating over real- rather than Boolean values, and we speak of the ($minimax$) "value" of the game. In the choices $\phi' \sqcup \phi''$ ($\phi' \sqcap \phi''$) player Max (Min) tries to maximise (minimise) a real-valued "payoff" associated with the game,[12] defined as follows. There are a number of cases:

- The play is finite, terminating in a game state (p); in this case the payoff is p.

- The play is infinite and there is a colour C appearing infinitely often that was generated by a greatest fixed-point ν; in this case the payoff is one.

- The play is infinite and there is a colour C appearing infinitely often that was generated by a least fixed-point μ; in this case the payoff is zero.

---

[10]Note that if the denominator is zero then the probability of that branch is zero as well, and it cannot be selected. In that case we just leave it out.

[11]This use of *colours* is taken from Stirling's original paper [Sti95]. The device allows easy determination, later on, of which recursion operator actually "caused" an infinite path, and is why μ and ν need not be distinguished at this point; they generate the same tree.

[12]In fact attributing the wins/losses to the two players makes it into a ZERO-SUM game.

11.5   Example: strategic software development

We now give an example of a quantitative game and its associated formula.

Typical properties of probabilistic systems are usually cost-based. The following problem is an illustration of a property that concerns general expected values, and so lies strictly outside the scope of "plain" probabilistic temporal logic.

> A company X (for maX) has discovered a market opportunity for new software: and it could publish a low-quality version early, to capture that market and lock in its users; or it could publish later and at higher quality, but then running the risk that an alternative supplier N(for miN) will beat it to market. Company X's marketing strategists have to determine the timing of their product launch.
>
> The marketeers' pressure for early publication comes from two sources — they know that the internal cost of the product (a bounded integer-valued variable c) will rise the longer it's left with the developers, and there is also the increasing risk that N will publish first. They estimate that N will publish (recorded by Boolean variable n) with probability no more than some constant p per unit time.[13]
>
> The complementary argument, for delaying publication, is that the more time spent on development the better the quality (variable q) of the final product will be. That will save on maintenance costs later and will enhance the company's reputation.
>
> Finally, because X is a much bigger company than N is, they can be sure that if they publish first then N won't publish at all. When X does publish, its profit is given by a function $profit.c.q.n$ into $[0,1]$ of the cost (c), the quality (q) and whether N has published already (n).

The situation is summed up by the transition system set out in Fig. 11.5.1: at each time step X can choose either to publish immediately (by selecting the left-hand branch at $\sqcup_X$, and terminating), or to postpone publication and continue development for another time step (by selecting the right-hand branch at $\sqcup_X$). If X chooses the latter option then it risks N's publishing first: in the worst case (from X's point of view, the left-hand branch at $\sqcap_N$) N does so with probability p (the left-hand probabilistic branch at $_p\oplus$). These steps are repeated until X publishes or — the nonterminating case — forever, if X never publishes at all. In that latter case, the payoff is 0.

---

[13]Note the nondeterminism introduced by *no more than*.

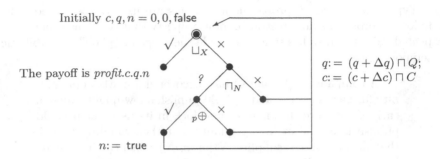

Initially $c, q, n = 0, 0, \text{false}$

The payoff is *profit.c.q.n*

$\sqcup_X$

$\sqcap_N$

$_p\oplus$

$n := \text{true}$

$q := (q + \Delta q) \sqcap Q;$
$c := (c + \Delta c) \sqcap C$

If repeated forever, then the payoff is zero.

$\sqcup_X$ — Angelic choice ($\exists$ modality) of whether X publishes ($\checkmark$) or not ($\times$). (Chosen by *Max* in the game.)

$\sqcap_N$ — Demonic choice ($\forall$ modality) of whether N even considers (?) publishing. (Chosen by *Min* in the game.)

$_p\oplus$ — Probabilistic choice of whether N publishes. ("Chosen" by *chance*.)

Function *profit.c.q.n* determines the payoff realised by X from publication of a product with quality q and development costs c, and depends on whether N has already published or not (Boolean n).

In general, given a definition of *profit.c.q.n*, Player X's optimal payoff can be computed to reveal the best timing for the launch of the product.

Figure 11.5.1. STRATEGIC SOFTWARE DEVELOPMENT EXAMPLE

The utility of our game interpretation in Sec. 11.4.2 is that we can easily use the intuition it provides to write a formula describing the above system over our given state space (c, q, n): the formula is $(\mu X \bullet \text{profit} \sqcup [K]X)$, where K denotes the set $\{k_0, k_1\}$ of transitions

$$k_0 \quad := \quad q := (q + \Delta q) \sqcap Q;$$
$$c := (c + \Delta c) \sqcap C$$

$$\text{and} \quad k_1 \quad := \quad n := \text{true} \,_p\oplus \text{false};$$
$$k_0 \, .$$

Thus $k_0 \sqcap k_1$ executes k_0 either way, but might assign to n first — the choice is demonic. (We assume a valuation mapping profit to *profit* etc.)

Recall that the choice implicit in [K] is the way we express probability ranges, if the problem demands it: here, it is that the probability of N's publishing is "no more than p" is coded up as a choice between exactly p (if k_1 is chosen at every play) and zero (if k_0 is always chosen).[14]

Since the choice $\sqcap_N$ can be resolved by player N to any probability in $[0,1]$, the overall probability range for N's publishing is $[0,p]$.

Note that our transitions k_0, k_1 take an initial state to a full- (rather than sub-) distribution over final states, and that there is no "immediate payoff" component (*i.e.* it is zero). This is of course a special case of the transitions we allow in $\mathcal{RS}$: in the notation of the introduction we just have $r.s.\$ = 0$ and $\sum_{s':S} r.s.s' = 1$. The role of (and need for) the extra generality is discussed in the conclusion (located by Footnote 18 on p. 309).

We have used the least- (rather than greatest) fixed-point because "never publishing" pays zero.

In the *reduced language* (at end of Sec. 11.3) we would write our formula as

$$Game \quad := \quad (\mu X \bullet \mathsf{profit} \sqcup (\{k_0\}X \sqcap \{k_1\}X)) \,,$$

and then the Kozen interpretation $\|Game\|_{\mathcal{V}}.s_0$, given appropriate values $profit, k_0, k_1$ supplied by $\mathcal{V}$, is a well-defined expectation as set out in Sec. 11.4.1. For example when p is $1/3$, and $profit.c.q.n$ is defined (0 **if** n **else** $q-c$)[15] — a simple definition assuming all payoff is lost if N reaches the market first — then a short calculation gives the value $(8/9)(\Delta q - \Delta c)$, assuming $\Delta q \geq \Delta c$, initial state $q, c, n = 0, 0, \mathsf{false}$, and that the bounds Q, C are not too low. As we argue for the general case in Sec. 11.6 to come, this turns out to be the same as X's *optimal* expected payoff in the game, whatever N's strategy might be.

The alternative interpretation is then a game in which we generate a tree by "unfolding" the transition system in Fig. 11.5.1. At each unfolding, X and N need to select a branch; and their selections could be different each time they revisit their respective decision points. Thus let σ_X and σ_N be sequences (possibly infinite) of the choices to be made by X and N. When they follow those sequences, the resulting game tree generates a well-defined probability distribution over valid game paths [GW86]. Anticipating the next section, let $[\![\phi]\!]_{\mathcal{V}}^{\sigma_X,\sigma_N}$ denote that path distribution: we can now describe X's actual payoff as a function P of the strategies chosen, *viz.*

$$P.\sigma_X.\sigma_N \quad := \quad \int_{[\![\phi]\!]_{\mathcal{V}}^{\sigma_X,\sigma_N}} \textit{"profit applied to the final state"} \,,$$

---

[14]The definition of $\geq_p \oplus$ on p. 21 makes this explicit; see (2.9) on p. 51 for an example.

[15]The profit function should be one-bounded, but to avoid clutter we have not scaled it down here (*e.g.* by dividing by $Q \sqcup C$).

with the understanding that the random variable over paths, in the integral's body, yields zero if in fact there is no final state (due to an infinite path).

As is usual in game theory, when the actual strategies of the two players are unknown (as they are in this case), we must define the *value* of the game to be the *minimax* over all strategy sequences of the expected payoff — but that minimax is well defined only when it is equal to the corresponding "maximin," *i.e.* only when

$$\sqcup_{\sigma_X} \sqcap_{\sigma_N} P.\sigma_X.\sigma_N \quad = \quad \sqcap_{\sigma_N} \sqcup_{\sigma_X} P.\sigma_X.\sigma_N \ .$$

In some cases, the value of a game can be realised by a *memoriless strategy* — roughly speaking, a memoriless strategy is independent of the number of unfoldings of the game tree. Memoriless strategies are particularly important for the efficient computation of expected payoffs [FV96], and in Sec. 11.6 we show they suffice for $qM\mu$ when the state space is finite.

To summarise, in the next section we show that the techniques used in this example are valid in general — *i.e.* that the values of the games described in Sec. 11.4.2 are all well defined, that they can be realised by memoriless strategies if the state space is finite, and that the value corresponds exactly to the denotational interpretation of Sec. 11.4.1.

For the current example, those results justify our using the Kozen interpretation to calculate X's optimal profit in the game, which in this simple case led to a direct calculation. For more complex formulae, the optimal payoff is determined directly using model-checking methods derived from Markov Decision Processes [FV96].[16]

11.6   Proof of equivalence of interpretations

In this section we give our main result, the equivalence of the two interpretations of a $qM\mu$ formula: the operational, "Stirling-game" interpretation of Sec. 11.4.2, and the denotational "Kozen-logic" interpretation of Sec. 11.4.1. In both cases we must address explicitly the question of *strategies*, and whether they can or cannot have "memory" of where the game or transition system has gone so far.

---

[16] A different and more extended example based on the "futures market," with numeric calculations, is given in the journal version of this chapter [MM02]. The results were compared with answers calculated by Gethin Norman using Kwiatkowska's probabilistic model checker *PRISM* [PRI].

11.6.1 Equivalence for fixed strategies

We begin with the Stirling interpretation, and our first step will be to explain how the games can be formalised provided the players' strategies are decided beforehand.

The current position of a game — as we saw in Sec. 11.4.2 — is a formula/state pair. We introduce two *strategy functions* called $\underline{\sigma}$ and $\overline{\sigma}$, to formalise the players' decisions as they go along: the functions are of type "finite game path" to Boolean, and the player *Min* (resp. *Max*), instead of deciding "on the fly" how to interpret a decision point ⊓ (resp. ⊔), takes the strategy function $\underline{\sigma}$ (resp. $\overline{\sigma}$) and applies that to the sequence of game positions traversed so far with, say, result "true" meaning "take the left subformula."

These strategies model full memory, because each is given as an argument the complete history of the game up to its point of use. (That history includes the current state s.)

The formalisation of the Stirling game is then in two stages. In the first stage we construct a (possibly infinite) probabilistically-branching game tree $[\![\phi]\!]_{\mathcal{V}}^{\underline{\sigma},\overline{\sigma}}.s$, using the given formula ϕ, the initial state s and the pre-packaged strategy functions $\underline{\sigma},\overline{\sigma}$.

For the second stage we use a function *Val*, from valid game paths to the non-negative reals, which gives exactly the "payoff" described at the end of Sec. 11.4.2. Then we have

Definition 11.6.1 VALUE OF FIXED-STRATEGY STIRLING GAME   The value of a game played from formula ϕ and initial state s, with strategies $\underline{\sigma},\overline{\sigma}$, is given by the expected value

$$\int_{[\![\phi]\!]_{\mathcal{V}}^{\underline{\sigma},\overline{\sigma}}.s} Val$$

of *Val* over the game tree $[\![\phi]\!]_{\mathcal{V}}^{\underline{\sigma},\overline{\sigma}}.s$.

The argument that this is well defined is the usual one, based on showing that *Val* is a measurable function over the distribution on the σ-algebra defined by the tree.   □

Our second step is to show that the above game corresponds to a Kozen-style interpretation over the same data: that is, we augment the semantics of Sec. 11.4.1 with the same strategy functions. For clarity we use slightly different brackets $\|\phi\|_{\mathcal{V}}^{\underline{\sigma},\overline{\sigma}}$ for the extended semantics.

The necessary alterations to the rules in Sec. 11.4.1 are straightforward, the principal one being that in Case 4, instead of taking a minimum or maximum, we use the argument $\underline{\sigma}$ or $\overline{\sigma}$ as appropriate to determine whether to carry on with ϕ' or with ϕ''. (A technical complication is then that all the definitions have to be changed so that the "game sequence so far" is available to $\underline{\sigma}$ and $\overline{\sigma}$ when required. That can be arranged for example by introducing an extra "path so far" argument and passing it, suitably extended, on every right-hand side.)

We then have our first equivalence:

Lemma 11.6.2 EQUIVALENCE OF GIVEN-STRATEGY GAMES AND LOGIC
For all $qM\mu$ formulae ϕ, valuations $\mathcal{V}$, states s and strategies $\underline{\sigma}, \overline{\sigma}$, we have

$$\int_{[\![\phi]\!]_{\mathcal{V}}^{\underline{\sigma},\overline{\sigma}}.s} Val = \|\phi\|_{\mathcal{V}}^{\underline{\sigma},\overline{\sigma}}.s .$$

Proof   (sketch) The proof is by structural induction over ϕ, straight-forward except when least- or greatest fixed-points generate infinite trees. In those cases we consider longer and longer finite sub-trees of the infinite tree: as well as the valid paths already described, they may contain extra finite paths, ending in μ- or ν-generated colours.

Modify the path-valuation function Val so that it assigns zero (resp. one) to a finite path ending in a μ- (resp. ν-) colour. It can be shown that the expression $\int_{\Delta} Val$ is a continuous function of distributions Δ generated from a   subtree-ordered sequence of partial trees, provided there is a single colour such that every extra finite path in any tree in the sequence terminates in that colour. In that case the game's overall value is the limit of the non-decreasing (resp. non-increasing) sequence of values assigned by the extended Val to the finite trees.

Since, in a single μ- (resp. ν-) structural induction step, the infinite paths assigned zero (resp. one) by Val are exactly those containing infinitely many occurrences of the associated fixed colour, a limit of trees as above can be constructed. And each of any μ- (resp. ν-) generated infinite path's prefixes is assigned zero (resp. one) appropriately by the above extension of Val to finite colour-terminated paths.

A full proof may be found elsewhere [MM, MM02].   □

Lem. 11.6.2 is the key to completing the argument that the value of the Stirling game is the *minimax* over all strategies of the expected payoff: recalling the issues raised at the end of Sec. 11.5, we must do that to show it to be well defined. That is, in the notation of this section we must establish

$$\sqcap_{\underline{\sigma}} \sqcup_{\overline{\sigma}} \int_{[\![\phi]\!]_{\mathcal{V}}^{\underline{\sigma},\overline{\sigma}}.s} Val = \sqcup_{\overline{\sigma}} \sqcap_{\underline{\sigma}} \int_{[\![\phi]\!]_{\mathcal{V}}^{\underline{\sigma},\overline{\sigma}}.s} Val . \qquad (11.1)$$

The importance of Lem. 11.6.2 is now evident, that it allows us to carry out the argument in a denotational rather than operational context — that is, rather than deal with the trees that $\int_{[\![\phi]\!]_{\mathcal{V}}^{\underline{\sigma},\overline{\sigma}}.s}$ implies, we simply show the equality

$$\sqcap_{\underline{\sigma}} \sqcup_{\overline{\sigma}} \|\phi\|_{\mathcal{V}}^{\underline{\sigma},\overline{\sigma}} = \sqcup_{\overline{\sigma}} \sqcap_{\underline{\sigma}} \|\phi\|_{\mathcal{V}}^{\underline{\sigma},\overline{\sigma}} ,$$

and avoid the trees altogether.

In fact we show that value to be the one given by the original — and much simpler — Kozen interpretation $\|\phi\|_\mathcal{V}$ with its $\sqcap$ and $\sqcup$ operators still in place, which therefore is the value of the Stirling game.

11.6.2 Full equivalence, and memoriless strategies

Let formula $\phi_{\underline{G}}$ be derived from ϕ by replacing each operator $\sqcap$ in ϕ by a specific state predicate drawn from a tuple $\underline{G}$ of our choice, possibly a different predicate for each syntactic occurrence of $\sqcap$. Similarly we write $\phi_{\overline{G}}$ for the derived formula in which all instances of $\sqcup$ are replaced one-by-one by successive state predicates in a tuple $\overline{G}$. With those conventions, we appeal to the following lemma concerning the existence of memoriless strategies:

Lemma 11.6.3 MEMORILESS STRATEGIES SUFFICE   For any formula ϕ, possibly containing strategy operators $\sqcap/\sqcup$, and valuation $\mathcal{V}$, there are state-predicate tuples $\underline{G}/\overline{G}$ — possibly depending on $\mathcal{V}$ — such that

$$\|\phi_{\underline{G}}\|_\mathcal{V} \;=\; \|\phi\|_\mathcal{V} \;=\; \|\phi_{\overline{G}}\|_\mathcal{V}\,.$$

Proof   The proof is given elsewhere [MM02].   □

For example, Lem. 11.6.3 tells us that if the formula ϕ is

$$(\mu X \cdot A_1 \sqcup (\nu Y \cdot A_2 \sqcap \{k\}(A_3 \sqcup (X \triangleleft G \triangleright Y))))\,,$$

then we can find tuples $\underline{G} := (\underline{G}_1)$ and $\overline{G} := (\overline{G}_1, \overline{G}_2)$ so that the formulae

$$\phi_{\underline{G}} := (\mu X \cdot A_1 \sqcup (\nu Y \cdot A_2 \triangleleft \underline{G}_1 \triangleright \{k\}(A_3 \sqcup (X \triangleleft G \triangleright Y)))) \quad \text{and}$$
$$\phi_{\overline{G}} := (\mu X \cdot A_1 \triangleleft \overline{G}_1 \triangleright (\nu Y \cdot A_2 \sqcap \{k\}(A_3 \triangleleft \overline{G}_2 \triangleright (X \triangleleft G \triangleright Y))))$$

are both equivalent to ϕ under $\|\cdot\|_\mathcal{V}$.[17]

We now move to our main proof for this chapter.

Lemma 11.6.4 MINIMAX DEFINED FOR KOZEN INTERPRETATION
For all $qM\mu$ formulae ϕ, valuations $\mathcal{V}$ and strategies $\underline{\sigma}, \overline{\sigma}$, we have

$$\sqcap_{\underline{\sigma}} \sqcup_{\overline{\sigma}} \|\phi\|_\mathcal{V}^{\underline{\sigma},\overline{\sigma}} \;=\; \sqcup_{\overline{\sigma}} \sqcap_{\underline{\sigma}} \|\phi\|_\mathcal{V}^{\underline{\sigma},\overline{\sigma}}\,. \tag{11.2}$$

Proof   From monotonicity, we need only prove *lhs* $\leq$ *rhs*. Note that from Lem. 11.6.3 we have predicates $\overline{G}$ and $\underline{G}$ satisfying

$$\|\phi_{\underline{G}}\|_\mathcal{V} \;=\; \|\phi\|_\mathcal{V} \;=\; \|\phi_{\overline{G}}\|_\mathcal{V}\,, \tag{11.3}$$

a fact which we use further below.

To begin with, using the predicates $\underline{G}$, we start from the *lhs* and observe that

$$\sqcap_{\underline{\sigma}} \sqcup_{\overline{\sigma}} \|\phi\|_\mathcal{V}^{\underline{\sigma},\overline{\sigma}} \;\leq\; \sqcup_{\overline{\sigma}} \|\phi_{\underline{G}}\|_\mathcal{V}^{\overline{\sigma}}\,, \tag{11.4}$$

---

[17]In fact it is easy to show that all three formulae are then equivalent to $\phi_{\underline{G},\overline{G}}$, but we do not use that.

(in which on the right we omit the now-ignored $\underline{\sigma}$ argument), because the $\sqcap_{\underline{\sigma}}$ could have selected exactly those predicates $\underline{G}$ by an appropriate choice of $\underline{\sigma}$. We then eliminate the explicit strategies altogether by observing that

$$\sqcup_{\overline{\sigma}} \|\phi_{\underline{G}}\|_{\mathcal{V}}^{\overline{\sigma}} \quad \leq \quad \|\phi_{\underline{G}}\|_{\mathcal{V}} \,, \tag{11.5}$$

because the simpler $\| \ \|$-style semantics on the right interprets $\sqcup$ as maximum, which cannot be less than the result of appealing to some strategy function $\overline{\sigma}$.

We can now continue on our way towards the *rhs* of (11.2) as follows:

$$
\begin{aligned}
& \|\phi_{\underline{G}}\|_{\mathcal{V}} & \\
\dagger \quad = \quad & \|\phi\|_{\mathcal{V}} & \text{first equality at (11.3)} \\
= \quad & \|\phi_{\overline{G}}\|_{\mathcal{V}} & \text{second equality at (11.3)} \\
\leq \quad & \sqcap_{\underline{\sigma}} \|\phi_{\overline{G}}\|_{\mathcal{V}}^{\sigma} & \text{as for (11.5) above} \\
\leq \quad & \sqcup_{\overline{\sigma}} \sqcap_{\underline{\sigma}} \|\phi\|_{\mathcal{V}}^{\sigma, \overline{\sigma}} \,, & \text{as for (11.4) above}
\end{aligned}
$$

and we are done.   □

The proof above establishes the equivalence we seek between the two interpretations.

Theorem 11.6.5 EQUIVALENCE OF qMu AND GAMES
The value of a Stirling game is well defined, and equals $\|\phi\|_{\mathcal{V}}$.
Proof   Lem. 11.6.2 and Lem. 11.6.4 establish the equality (11.1), for well-definedness; the stated equality with $\|\phi\|_{\mathcal{V}}$ occurs at ($\dagger$) during the proof of the latter.   □

Finally, we have an even tighter result about the players' strategies:

Lemma 11.6.6 EXISTENCE OF MEMORILESS STRATEGIES   There exists a memoriless strategy $\overline{G}$ which, if followed by player *Max*, achieves the value of the Stirling game against all strategies of player *Min*.
(A similar result holds for player *Min*.)
Proof   Directly from Lem. 11.6.3 and Thm. 11.6.5.   □

11.7 Summary

Our main achievement in this chapter has been to establish an equivalence between two interpretations for the quantitative μ-calculus $qM\mu$, interpretations that extend the corresponding standard interpretations to allow probabilistic transition systems. The framework is general enough to specify cost-based properties of probabilistic/concurrent systems — and many such properties lie outside standard temporal logic. The game interpretation is close to automata-style approaches, whilst the Kozen-style logic provides an attractive proof system.

Many presentations of probabilistic transitions (including our earlier work) do not include the extra "payoff" state \$, giving instead simply functions from S to $\overline{S}$; that in effect takes the primitive elements, over which the formulae are built, to be probabilistic programs. In contrast, here our primitive elements are small probabilistic *games* (in the sense *e.g.* of Everett), of which the more usual programs are the special case of payoff zero. The full proof [MM] of Lem. 11.6.3 shows that to be necessary, since we treat the G/ν case simply by appealing via a duality over our formulae to the $\overline{\mathsf{G}}/\mu$; but it is a duality under which probabilistic programs are not closed, whereas the slightly more general probabilistic games are.[18] As a consequence of that we have had to prove a slightly more general result.

De Alfaro and Majumdar [dAM01] use $qM\mu$ to address an issue similar to, but not the same as ours: in the more general context of concurrent games, they show that for every *LTL* formula Ψ one can construct a $qM\mu$ formula ϕ such that $\|\phi\|_{\mathcal{V}}$ is the greatest assured probability that Player 1 can force the game path to satisfy Ψ.[19] (See the remarks at end of Chap. 10.)

The difference can be seen by considering the least fixed-point formula

$$\Psi \quad := \quad (\mu X \bullet \{k\}\mathsf{atB} \sqcup \{k\}X)$$

over the transition system

$$\mathcal{V}.k \quad := \quad (s := A \,_{1/2}\!\oplus s := B) \text{ if } (s = A) \text{ else } (s := A)$$

operating on the two-element state space $\{A, B\}$.[20]

Player 1 can force satisfaction of Ψ with probability one in this game, since the only path for which it fails (all A's) occurs with probability zero.

Yet $\|\Psi\|_{\mathcal{V}} = 1/2$, which is the value of the *Stirling* game played in this system. It is "at each step, seek to maximise ($\sqcup$) the payoff, depending on whether after the following step ($\{k\}$) you will accept atB and terminate, or go around again (X)." Note that the decision "whether to repeat after the next step" is made *before* that step is taken. (Deciding *after* the step would be described by the formula $(\mu X \bullet \{k\}(\mathsf{atB} \sqcup X))$.) The optimal strategy for *Max* is of course given by

$$\Psi_{\overline{\mathsf{atA}}} \quad := \quad (\mu X \bullet \{k\}\mathsf{atB} \text{ if } \mathsf{atA} \text{ else } \{k\}X) \,.$$

---

[18] Recall Footnote 5 on p. 296.

[19] By LTL we mean *linear-time temporal logic*.

[20] In fact formula Ψ expresses the notorious $\Diamond\circ\mathsf{atB}$, or $\mathsf{AF\,AX\,atB}$ [Var01] in the temporal subset qTL of $qM\mu$, where $\mathcal{V}.\mathsf{atB}.s := 1$ if $(s = B)$ else 0.

Chapter notes

The original presentation of standard μ-calculus is due to Kozen [Koz83], and probabilistic generalisations of it [HK97, MM97] started to emerge around 1997. One of its primary uses has been to formalise fragments of temporal logic — for example Narasimha *et al.* [NCI99] formalise probabilistic temporal state properties, and de Alfaro and Majumdar [dAM01] use it to formalise reachability and safety properties of probabilistic automata. As mentioned above, their emphasis is different from ours, and one result of their work is to confirm that the $qM\mu$ logic is as expressive as *LTL*.

The game interpretation for the quantitative μ-calculus was inspired by Stirling's similar result for standard μ-calculus [Sti95], and the proof of equivalence draws freely on earlier results of Everett [Eve57] which in turn extended Shapley games [Sha53] to include infinite repetition without applying a discount penalty.

Game frameworks (in the sense used here) appear elsewhere in the computing literature: Dolev *et al.* formulate *scheduler-luck* games [DIM83] to analyse expected termination times of randomised protocols, and de Alfaro and Henzinger [dAH00] use more generalised games to model concurrency.

The framework of partial orders is extremely effective for proving the *existence* of stationary strategies and — once shown to exist — they can then be found using algorithms based on linear-programming techniques. Other proof techniques use various kinds of metrics, for example Shapley [*op. cit.*] and Van Roy [Van98] which are effective for finding good approximations to the optimal strategy.

Part IV

Appendices, bibliography and indexes

Appendix A
Alternative approaches

A.1   Probabilistic Hoare-triples

This section explores the first of two alternative approaches to our use of expectations as the basis for a probabilistic program logic. The alternative turns out to be non-compositional when both probabilistic- and demonic choice are present.

Our point of departure is to generalise Hoare-triples *as a whole* from absolute to probabilistic judgements. That is, instead of changing the "raw material" of our logical statements, *i.e.* changing what they are about (about expectations rather than predicates), we change the nature of the statements themselves.

The standard view is that a precondition *guarantees* some program will establish a postcondition; we generalise that as follows. Continuing with standard predicates, we introduce probability via probabilistic judgements of the form

$$p \vdash \{pre\}\ prog\ \{post\}\ , \qquad (A.1)$$

that mean "from any initial state in *pre* the program *prog* will with probability at least p reach a final state in *post*." In general, probability p can be an expression over the initial state.

A typical Hoare-triple rule in the resulting system would be this one, for sequential composition: when probabilities p, q are *constant*, we have

$$\frac{p \vdash \{pre\} \; prog_0 \; \{mid\} \qquad q \vdash \{mid\} \; prog_1 \; \{post\}}{p * q \;\vdash\; \{pre\} \; prog_0; prog_1 \; \{post\}}$$

for any programs $prog_0, prog_1$ and standard predicates *pre, mid, post*.[1]

It relies on the probabilistic choices in $prog_0$ and $prog_1$ being independent, and on the monotonicity of multiplication. Indeed by defining

$$p \;\vdash\; \{pre\} \; prog \; \{post\} \qquad := \qquad p * [pre] \Rrightarrow wp.prog.[post]$$

such statements become special cases within our current system, and the above sequential composition rule is easily proved from sublinearity.[2] That means that we can use rules like the above safely, if we find them more intuitive than the full expectation-based logic; it also means that the proposal adds no expressive power.

In fact, the problem is that probabilistic Hoare-triples are not expressive enough, and thus we cannot adopt this approach as the *sole* basis for our program logic: not only are the judgements (A.1) too weak, they are not compositional in general. Consider for example the two programs

$$prog_0 \qquad := \qquad n:= 4 \sqcap (n:= 5 \;{}_{\frac{1}{2}}\oplus\; n:= 6)$$
$$prog_1 \qquad := \qquad (n:= 4 \sqcap n:= 5) \;{}_{\frac{1}{2}}\oplus\; (n:= 4 \sqcap n:= 6) \;.$$

(They correspond to executing the game of Fig. 1.3.1 from initial squares 0 and 1 respectively.) In Fig. A.1.1 we set out all eight possible judgements of the form (A.1), showing that in this simpler system $prog_0$ and $prog_1$ would be identified. Are they therefore the same?

No they are not: define a further program

$$prog \quad := \quad (n:= 5 \;{}_{\frac{1}{2}}\oplus\; n:= 6) \textbf{ if } n = 4 \textbf{ else skip} \;,$$

and consider the sequential compositions $(prog_0; prog)$ and $(prog_1; prog)$ with respect to the postcondition $n = 5$: we have

$$1/2 \;\vdash\; \{\textsf{true}\} \; prog_0; prog \; \{n = 5\}$$
$$\text{but} \quad 1/2 \;\nvdash\; \{\textsf{true}\} \; prog_1; prog \; \{n = 5\} \;,$$

---

[1] If q in particular were not constant, we would have to take account of its being evaluated over the *final* state of $prog_0$ (*i.e.* the initial state of $prog_1$) rather than the *initial* state of $prog_0$ as is the case for p. The resulting composite probability would then be

$$p \;*\; (\sqcap v \mid mid \bullet q) \;,$$

where v is the vector of variables that $prog_0$ can assign to, since — taking the demonic view — we would have to assume that any choice inherent in postcondition *mid* for $prog_0$ would be exploited to make q as low as possible.

[2] Use its consequences *scaling* and *monotonicity*.

possible postcondition	$prog_0$ probability	$prog_1$ probability
false	0	0
$n = 4$	0	0
$n = 5$	0	0
$n = 6$	0	0
$n \neq 4$	0	0
$n \neq 5$	1/2	1/2
$n \neq 6$	1/2	1/2
true	1	1

Programs $prog_0$ and $prog_1$ cannot be distinguished with standard postconditions.

Figure A.1.1. COUNTER-EXAMPLE TO COMPOSITIONALITY

postE	$[n = 4] + 2[n = 5]$	$[n = 4] + 2[n = 6]$
$wp.prog_0.postE$	1	1
$wp.prog_1.postE$	1/2	1/2

Programs $prog_0, prog_1$ are distinguished by either of the two post-expectations $postE$.

Figure A.1.2. COMPOSITIONALITY REQUIRES FULL USE OF EXPECTATIONS

and in fact the strongest judgement we can make about $prog_1$ is

$$1/4 \vdash \{\text{true}\}\ prog_1; prog\ \{n = 5\}\ .\quad [3]$$

That is why standard postconditions are not expressive enough — if the programs $(prog_0; prog)$ and $(prog_1; prog)$ are different, then $prog_0$ and $prog_1$ cannot be the same.[4] This lack of compositionality is why we do not use probabilistic Hoare-triples.

Fig. A.1.2 shows that $prog_0$ and $prog_1$ are indeed distinguished by the properly probabilistic post-expectations that we introduced in Chap. 1.

---

[3]For $(prog_0; prog)$ note that it doesn't matter how the initial nondeterministic choice is resolved, since the result is 1/2 either way. For $(prog_1; prog)$ however the probability of establishing $n = 5$ is $1/2 \sqcap 1 = 1/2$ for the left branch of the initial choice $\frac{1}{2} \oplus$, but $1/2 \sqcap 0 = 0$ for the right branch; thus overall it is only $(1/2 + 0)/2 = 1/4$.

[4]A further (but only informal) argument that $prog_0$ and $prog_1$ should be distinguished is the observation that $prog_0$ should terminate in states $5, 6$ "with equal frequency," however low or high that might be — but $prog_1$ does not have that property.

Demonic nondeterminism is to blame for the above effects. In Chap. 8 we saw from Thm. 8.3.5 that deterministic programs are linear. It is clear that Boolean postconditions are enough for those: over a finite state space at least, linearity determines general pre-expectations from the weakest pre-expectations with respect to the standard "point" postconditions that correspond to single states.

Thus it seems that any semantics for the probabilistic language of guarded commands — with its demonic nondeterminism — must be at least as powerful as the system we have proposed. He *et al.* give a more extensive discussion of alternative models [HSM97].

A.2   A programming logic of distributions

A second alternative to our approach is to "lift" the whole semantics, from states to *distributions* over states. We imagine that probabilistic programs move from distributions to distributions (rather than from states to states, as standard programs do), and we reconstruct the whole of the usual weakest-precondition apparatus above that, considering distributions now as "higher-order" states in their own right but with an internal, probabilistic structure.

Thus our pre- and postconditions will be formulae about *distributions*, containing (sub-)formulae like

$$\text{Pr.}(c = \mathsf{heads}) \geq 1/2 \qquad \text{the probability that } c \text{ is} \\ \text{heads is at least } 1/2$$

$$\text{and} \quad \text{Exp.}(n) \leq 3 \qquad \text{the expected value of } n \; [5] \\ \text{is no more than 3.}$$

These hold, or do not hold, over *distributions* of states containing variables like c and n. We would for example have the judgement

$$\{\mathsf{true}\} \quad c\colon= \; \mathsf{heads} \; {}_{1/2}\!\oplus \mathsf{tails} \quad \{\, \text{Exp.}[c = \mathsf{heads}] \; \geq \; 1/2 \,\} \,,$$

about the behaviour of a fair coin c; in the style of Chap. 1 (but as a Hoare triple) we would instead have written that as

$$\{1/2\} \quad c\colon= \; \mathsf{heads} \; {}_{1/2}\!\oplus \mathsf{tails} \quad \{c = \mathsf{heads}\} \,, \quad [6]$$

and in the notation of the previous alternative we would have written

$$1/2 \vdash \; \{\mathsf{true}\} \; c\colon= \; \mathsf{heads} \; {}_{1/2}\!\oplus \mathsf{tails} \; \{c = \mathsf{heads}\} \,.$$

---

[5] In fact the use of Exp is the more general since, as we have seen, we can express probabilities via characteristic functions: the first formula above is equivalently Exp.$[c = \mathsf{heads}] \geq 1/2$.

[6] Recall that to reduce clutter we omit embedding brackets $[\cdots]$ immediately enclosed by assertion brackets $\{\cdots\}$.

But again (as in Sec. A.1) we are in difficulty with demonic nonde-terminism. Consider this example: if *Fair* stands for the predicate over distributions

$$\text{Pr.}(c = \mathsf{heads}) \quad = \quad \text{Pr.}(c = \mathsf{tails}) \,,$$

then we have these two judgements about programs operating over a variable c representing a coin as above: both

$$\{\textit{Fair}\} \quad \mathbf{skip} \quad \{\textit{Fair}\}$$

and $\{\textit{Fair}\} \quad c := \bar{c} \quad \{\textit{Fair}\} \,,$

hold, where $\overline{\mathsf{heads}} = \mathsf{tails}$ *etc.* The first program leaves the state un-changed, and the second permutes it in a way that does not change the given (uniform) distribution.

But we also have the general principle that if two programs satisfy the *same* specification then so does the demonic choice between them,[7] and so from the above we would expect

$$\{\textit{Fair}\} \quad \mathbf{skip} \sqcap c := \bar{c} \quad \{\textit{Fair}\}$$

to hold as well — yet it does not. That demonic choice $\mathbf{skip} \sqcap c := \bar{c}$ is refined for example by the deterministic $c := \mathsf{tails}$ which never establishes postcondition *Fair* at all, whether precondition *Fair* held initially or not.

Because there are several phenomena involved here — and all our pre-conceptions as well — we cannot point to any one of them and say "that causes the contradiction." But one way of describing the situation is as follows.

Our treatment of demonic nondeterminism is the traditional one in which the imagined demon can resolve the choice, at runtime, with full knowledge of the state at the time the choice is to be made. That is inherent in our postulated refinement

$$\mathbf{skip} \sqcap c := \bar{c} \quad \sqsubseteq \quad c := \mathsf{tails}$$

from above, in which we imagine the demon chooses the left-hand **skip** when c is tails, and the right-hand $c := \bar{c}$ otherwise. In effect we are using the law

$$(\cdots \sqcap \cdots) \quad \sqsubseteq \quad (\cdots \mathbf{if}\ G\ \mathbf{else}\ \cdots) \,, \tag{A.2}$$

which holds for any Boolean G and for the test "$c = \mathsf{tails}$" in particular.[8]

---

[7]This is a general property of any approach that relates refinement $\sqsubseteq$ and demonic choice $\sqcap$ in the elementary way we prefer, that is as given by the simple rules for a partial order. See for example Law 6 in Sec. B.1.

[8]See Law 7 in Sec. B.1.

When we lift the whole semantic structure up to distributions, from states, the demonic choice "loses" the ability to see *individual* states: it can only see distributions. Equivalently, the choice ⊓ must be resolved "blind" although still arbitrarily, *i.e.* unpredictably but without looking at the state.

One way of doing that is to insist that all demonic choices are made in advance, as if the demon were required to write its future decisions down on a piece of paper before the program is begun. Once the program is running, the decisions "left now, or right" are carried out exactly, in sequence, and cannot be changed.[9]

There are circumstances in which such *oblivious* nondeterminism, as we call it, is the behaviour we are trying to capture — for example when we are dealing with concurrency or modularity in which separation of processes, or information hiding, can "protect" parts of the state from being read freely by other parts of the system.

For sequential programs, however, the use of laws like (A.2) on p. 317 is so pervasive that we consider it to be the deciding factor in this case.

---

[9]Making the choices in advance is in fact the usual semantic technique for dealing with nondeterminism when it is the principal object of study [Seg95]; we do just that in Sec. 11.6.1 when dealing with demonic, angelic and probabilistic choice all at once. In that case nondeterminism is controlled by whether the decisions made in advance are a sequence of simple Booleans, interpreted "go left" or "go right" (as suggested above: a very weak form of nondeterminism), or are a sequence of predicates over states (so called "memoriless" strategies that can see the current state but have no access to previous states: a stronger form), or are a sequence of predicates over "state histories," which can resolve a nondeterministic choice using knowledge not only of the state the system is in now, but also of the states it has passed through to get there (a stronger form still, and the one used in this text). But this extra semantic power has a cost.[10]

---

[10]An advantage of including strategies explicitly in the mathematical model is that it is then possible to make fine adjustments, as above, to their power; and it is easier to discuss issues related to the strategies themselves. A great disadvantage for practical reasoning, however, is that such models often fail to be "fully abstract," where *full abstraction* means that program fragments are identified in the model exactly when they are operationally interchangeable [Sto88].

Here one loses full abstraction because the strategy sequence contains "too much information," in this case the order in which the strategy elements are used. The two programs

$$c: = \text{ heads} \sqcap \text{tails} \, ; \;\; d: = \text{ heads} \sqcap \text{tails}$$

$$\text{and} \quad\quad d: = \text{ heads} \sqcap \text{tails} \, ; \;\; c: = \text{ heads} \sqcap \text{tails}$$

are equal in their observable behaviour; yet in their semantics — as functions of strategy sequences — they differ. That is, the first program's assignment to c is controlled by the first element of the strategy sequence; but in the second program, the first element of the strategy sequence controls d.

When unwanted distinctions like that occur, it is necessary to use more elaborate techniques to prove algebraic equalities. Kozen rejected a similar sequence-of-choices model for (deterministic) probabilistic programs on just those grounds (among other reasons) [Koz81]; and it is for similar reasons (again, among others) that we use the model we have chosen, an extension of Kozen's [HSM97].

In Chap. 11 we have it both ways, however: we prove the equivalence of two models, one with explicit strategies and one without. The explicit-strategy model is used to establish *e.g.* that memoriless and full-memory strategies are equivalent over finite state spaces (*i.e.* that the "stronger" and "stronger-still" options above are the same); and the implicit-strategy model — our main subject — can be used to formulate algebraic and logical laws (as we did in Chap. 10).

Appendix B
Supplementary material

B.1   Some algebraic laws of probabilistic programs

Many algebraic laws for programs have quite easy proofs from the weakest pre-expectation definitions of the operators involved, and subsequently allow reasoning about programs directly without appealing again to the logic. In effect, they provide a "third layer" of intellectual tools, above the logic which in turn lies above the semantics — and, in practice, we use the algebra if we can. If that fails we appeal to the logic; and in rare cases we must push all the way down to the model, the principal reference from which all else is derived.

Although many of the laws we give apply to all demonic/probabilistic programs, in some cases we make restrictions as indicated by these naming conventions: [1]

- General demonic/probabilistic programs   ... *prog*
as in... $s := 0 _{\geq p}\oplus 1$ for $[0,1]$-valued expression p [2]

- Deterministic (but possibly probabilistic) programs   ... *det*
    $s := 0 _p\oplus 1$

- Standard (but possibly demonic) programs   ... *std*
    $s := 0 \sqcap 1$

- Standard deterministic programs   ... *stdet*
    $s := (0 \text{ if } G \text{ else } 1)$ for Boolean expression G

B.1.1   *A list of algebraic laws...*

We list a number of the laws, below; they are collected into groups based on the healthiness conditions (or other facts) that justify them. In some cases we include "conventional" laws (*i.e.* well-known from standard programming) also.[3]

- Laws following from basic arithmetic of the operators separately:

1. Demonic choice is commutative, associative and idempotent.[4]

2. Probabilistic choice is idempotent and *quasi-commutative*:
$$\begin{aligned} prog _p\oplus prog &= prog \\ prog_1 {}_p\oplus prog_2 &= prog_2 {}_{\overline{p}}\oplus prog_1 \end{aligned}$$

We will usually assume the above two laws without comment.

3. Probabilistic choice is *quasi-associative*: for $0 \leq p, q < 1$ and $p+q \leq 1$ we have
$$prog_1 {}_p\oplus (prog_2 \oplus_{q/\overline{p}} prog_3) \quad = \quad (prog_1 {}_{p/\overline{q}} \oplus prog_2) \oplus_q prog_3$$

---

[1] We haven't included laws here for angelic programs.

[2] Note that the probability p may be an *expression*, *i.e.* it may depend on the current state.

[3] A similar set of probabilistic laws was given by Jifeng He *et al.* in earlier work [HSM97]; they were shown in fact to be complete, but for a slightly different model.

[4] Recall that an operator $\odot$ is IDEMPOTENT whenever $x \odot x = x$ for all x.

4. Probabilistic choice is *quasi-distributive*: for $0 \leq p, q, r \leq 1$ we have

$$(prog_1 \; {}_p\oplus prog_2) \;\; {}_q\oplus (prog_3 \; {}_r\oplus prog_4) \;\; = \;\; \left| \begin{array}{ll} prog_1 & @ \; pq \\ prog_2 & @ \; \bar{p}q \\ prog_3 & @ \; \bar{q}r \\ prog_4 & @ \; \bar{q}\,\bar{r} \end{array} \right.$$

5. (From Law 4; compare Laws 23 and 24 below.)

$$(prog_1 \; {}_p\oplus prog_2) \;\; {}_q\oplus \; (prog_1 \; {}_p\oplus prog_3) \;\; = \;\; prog_1 \; {}_p\oplus \; (prog_2 \; {}_q\oplus prog_3)$$
$$\text{and} \quad (prog_1 \; {}_p\oplus prog_2) \;\; {}_{\geq q}\oplus \; (prog_1 \; {}_p\oplus prog_3) \;\; = \;\; prog_1 \; {}_p\oplus \; (prog_2 \; {}_{\geq q}\oplus prog_3)$$

- Law relating demonic choice and refinement:

6. $\quad prog_1 \; \sqsubseteq \; prog_2 \sqcap prog_3 \qquad$ iff $\qquad \begin{array}{l} prog_1 \sqsubseteq prog_2 \\ \text{and} \quad prog_1 \sqsubseteq prog_3 \end{array}$

- Laws depending on the arithmetic of ${}_p\oplus$ and $\sqcap$ together:

7. For any probabilities $0 \leq p \leq q \leq 1$ and Boolean expression G, both possibly functions of the state, we have

$$prog_1 \sqcap prog_2 \;\; \sqsubseteq \;\; \left\{ \begin{array}{l} prog_1 \; {}_{\geq p}\oplus prog_2 \quad \sqsubseteq \\ \\ prog_1 \; \textbf{if } G \textbf{ else } prog_2 \end{array} \right. \left\{ \begin{array}{l} prog_1 \\ prog_1 \; {}_p\oplus \;\; prog_2 \\ prog_1 \; {}_{\geq q}\oplus prog_2 \end{array} \right.$$

8. $(prog_1 \sqcap prog_2) \;\; {}_p\oplus \;\; prog_3 \;\;\;\; = \;\;\;\; (prog_1 \; {}_p\oplus prog_3) \;\; \sqcap \;\; (prog_2 \; {}_p\oplus prog_3)$

9. $(prog_1 \sqcap prog_2) \;\; {}_{\geq p}\oplus \;\; prog_3 \;\;\;\; = \;\;\;\; (prog_1 \; {}_{\geq p}\oplus prog_3) \;\; \sqcap \;\; (prog_2 \; {}_{\geq p}\oplus prog_3)$

10. $(prog_1 \sqcap prog_2) \;\; {}_{\leq p}\oplus \;\; prog_3 \;\;\;\; = \;\;\;\; (prog_1 \; {}_{\leq p}\oplus prog_3) \;\; \sqcap \;\; (prog_2 \; {}_{\leq p}\oplus prog_3)$

11. $(prog_1 \; {}_p\oplus prog_2) \;\; \sqcap \;\; prog_3 \;\;\;\; \sqsupseteq \;\;\;\; (prog_1 \sqcap prog_3) \;\; {}_p\oplus \;\; (prog_2 \sqcap prog_3)$

- Laws involving sequential composition that are consequences of the way *wp* is applied from right to left:

12. $(prog_1 \; \textbf{if } G \textbf{ else } prog_2); prog_3 \;\;\; = \;\;\; prog_1; prog_3 \; \textbf{if } G \textbf{ else } \; prog_2; prog_3$

13. $(prog_1 \; {}_p\oplus prog_2); prog_3 \;\;\;\;\;\;\;\;\; = \;\;\; prog_1; prog_3 \; {}_p\oplus \;\; prog_2; prog_3$

14. $(prog_1 \sqcap prog_2); prog_3 \;\;\;\;\;\;\;\;\;\;\;\; = \;\;\; prog_1; prog_3 \; \sqcap \;\; prog_2; prog_3$

15. (From Laws 13, 14 and the definition of ${}_{\geq p}\oplus$.)

$$(prog_1 \; {}_{\geq p}\oplus prog_2); prog_3 \;\;\;\; = \;\;\; prog_1; prog_3 \; {}_{\geq p}\oplus \;\; prog_2; prog_3$$

- Laws following from sublinearity (and its consequence, monotonicity):

16. (Compare Law 18 below.)

$$prog_1; (prog_2 \ {}_p\oplus prog_3) \quad \sqsupseteq \quad prog_1; prog_2 \ {}_p\oplus \ prog_1; prog_3$$

17. (Compare Law 19 below.)

$$prog_1; (prog_2 \sqcap prog_3) \quad \sqsubseteq \quad prog_1; prog_2 \sqcap prog_1; prog_3$$

- Laws for sequential composition when the left-hand side is restricted:

18. (Compare Law 16.) $det; (prog_1 \ {}_p\oplus prog_2) \quad = \quad det; prog_1 \ {}_p\oplus det; prog_2$

19. (Compare Law 17.) $std; (prog_1 \sqcap prog_2) \quad = \quad std; prog_1 \sqcap std; prog_2$

20. (From Laws 18, 19 and the definition of $\geq_p\oplus$.)

$$stdet; (prog_1 \ {}_{\geq p}\oplus prog_2) \quad = \quad stdet; prog_1 \ {}_{\geq p}\oplus \ stdet; prog_2$$

21. (From Laws 13, 18 and 4.)

$$(det_1 \ {}_p\oplus det_2); (prog_3 \ {}_q\oplus prog_4) \quad = \quad \left| \begin{array}{ll} det_1; prog_3 & @ \ pq \\ det_1; prog_4 & @ \ p\overline{q} \\ det_2; prog_3 & @ \ \overline{p}q \\ det_2; prog_4 & @ \ \overline{p}\,\overline{q} \end{array} \right.$$

22. (From Laws 8, 13 and 19.)

$$(std_1 \ {}_p\oplus std_2); (prog_3 \sqcap prog_4)$$

$$\begin{array}{ll} = & std_1; prog_3 \ {}_p\oplus \ std_2; prog_3 \\ \sqcap & std_1; prog_3 \ {}_p\oplus \ std_2; prog_4 \\ \sqcap & std_1; prog_4 \ {}_p\oplus \ std_2; prog_3 \\ \sqcap & std_1; prog_4 \ {}_p\oplus \ std_2; prog_4 \end{array}$$

- Quasi-associative/distributive laws with inequalities:

23. (From Laws 7 and 8 and the definition of $\geq_p\oplus$; compare Law 5.)

$$(prog_1 \ {}_p\oplus prog_2) \ {}_{\geq q}\oplus \ (prog_1 \ {}_{\geq p}\oplus prog_3)$$

$$\sqsupseteq \quad prog_1 \ {}_{\geq p}\oplus \ (prog_2 \ {}_{\geq q}\oplus prog_3)$$

$$\sqsupseteq \quad (prog_1 \ {}_{\geq p}\oplus prog_2) \ {}_{\geq q}\oplus \ (prog_1 \ {}_p\oplus prog_3)$$

$$= \quad (prog_1 \ {}_{\geq p}\oplus prog_2) \ {}_{\geq q}\oplus \ (prog_1 \ {}_{\geq p}\oplus prog_3)$$

24. (From Law 8 and the definition of $\geq_p \oplus$; compare Law 5.)

$$(prog_1 \geq_p \oplus prog_2) \geq_q \oplus prog_3 \quad \sqsupseteq \quad \begin{matrix} (prog_1 \geq_q \oplus prog_3) \\ \geq_p \oplus \quad (prog_2 \geq_q \oplus prog_3) \end{matrix}$$

B.1.2 ... and an example of their use

As an example we show the following elementary equivalence, that a 99% reliability for global Boolean a to hold is achieved via two, lower 90% reliabilities for local Booleans b and c.[5] Our first steps (set out in full detail) will be to remove the local variables: we have

$\quad b := \text{true} \geq_{.9} \oplus \text{false};$
$\quad c := \text{true} \geq_{.9} \oplus \text{false};$
$\quad a := b \vee c$

$=$ $b := \text{true} \geq_{.9} \oplus \text{false};$ Law 15
$\quad\quad (c := \text{true}; a := b \vee c) \geq_{.9} \oplus (c := \text{false}; a := b \vee c)$

$=$ $b := \text{true} \geq_{.9} \oplus \text{false};$ standard program algebra; c local
$\quad\quad a := \text{true} \geq_{.9} \oplus b$

$=$ $b := \text{true}; \quad a := \text{true} \geq_{.9} \oplus b$ Law 15
$\quad \geq_{.9} \oplus \quad b := \text{false}; \quad a := \text{true} \geq_{.9} \oplus b$

$=$ $(b := \text{true}; a := \text{true}) \geq_{.9} \oplus (b := \text{true}; a := b)$ Law 20
$\quad \geq_{.9} \oplus \quad (b := \text{false}; a := \text{true}) \geq_{.9} \oplus (b := \text{false}; a := b)$

$=$ $a := \text{true} \geq_{.9} \oplus \text{true}$ standard program algebra; b local
$\quad \geq_{.9} \oplus \quad a := \text{true} \geq_{.9} \oplus \text{false}$

$=$ $a := \text{true} \geq_{.9} \oplus (\text{true} \geq_{.9} \oplus \text{false})$.

Now the local variables are gone and the structure of the program is clear. We finish off by continuing

$\quad a := \text{true} \geq_{.9} \oplus (\text{true} \geq_{.9} \oplus \text{false})$

$=$ $a := \text{true}$ definition $\geq_p \oplus$; Law 8
$\quad \sqcap \quad a := \text{true} _{.9} \oplus \text{true}$
$\quad \sqcap \quad a := \text{true} _{.9} \oplus (\text{true} _{.9} \oplus \text{false})$

---

[5]The example comes from a case study being carried out by Steve Schneider *et al.* on control-system reliability [SHRT04], using a probabilistic version *pAMN* of *Event B* [Abr96b].

$$
\begin{aligned}
=\quad & a\!:=\ \textsf{true}\ \sqcap\ a\!:=\ (\textsf{true}_{\ .9/.99}\oplus \textsf{true})_{\ .99}\oplus \textsf{false} && \text{Law 3}\\
=\quad & a\!:=\ \textsf{true}\ \sqcap\ a\!:=\ \textsf{true}_{\ .99}\oplus \textsf{false}\\
=\quad & a\!:=\ \textsf{true}_{\ \geq.99}\oplus \textsf{false}\ , && \text{definition } {}_{\geq p}\oplus
\end{aligned}
$$

which finally is our 99% reliability. The calculation we have just done suggests these further laws (thus from Laws 8, 3 and the definition of ${}_{\geq p}\oplus$):

25. $prog_1 \oplus_{\leq pq} prog_2 \quad = \quad prog_1 \oplus_{\leq p} (prog_1 \oplus_{\leq q} prog_2)$

26. $prog_1 \oplus_{\geq pq} prog_2 \quad = \quad prog_1 \oplus_{\geq p} (prog_1 \oplus_{\geq q} prog_2)$

They might be useful if the two instances of $prog_1$ on the right were subsequently to be manipulated (*e.g.* refined) in different ways, such as in these laws (thus from Laws 25, 26 and monotonicity):

27. $(prog_1 \sqcap prog_2) \oplus_{\leq pq} prog_3 \quad \sqsubseteq \quad prog_1 \oplus_{\leq p} (prog_2 \oplus_{\leq q} prog_3)$

28. $(prog_1 \sqcap prog_2) \oplus_{\geq pq} prog_3 \quad \sqsubseteq \quad prog_1 \oplus_{\geq p} (prog_2 \oplus_{\geq q} prog_3)$

We finish with some remarks concerning the demonic choice "hidden" within ${}_{\geq p}\oplus$, and its interaction with probabilistic choice. (Similar issues were explored with respect to (2.8) on p. 50.)

The statement $c\!:=\ \textsf{true}_{\geq.9}\oplus\textsf{false}$ above contains demonic choice resolved *after* the probabilistic choice carried out in $b\!:=\ \textsf{true}_{\geq.9}\oplus\textsf{false}$ immediately before it. But since the following statement is $a\!:=\ b \vee c$, the "knowledge" that the c-demon has of the b-outcome is of no use: the best c-strategy for making a true is $c\!:=\ \textsf{true}$; and the best c-strategy for making a false is $c\!:=\ \textsf{true}_{.9}\oplus\textsf{false}$. Both are independent of the value of b, however determined, and in fact c's strategy could be chosen in advance — *i.e.* before the program is run — without in any way reducing c's power to influence the result in a.

But consider the related program

$$
\begin{aligned}
b\!:=&\ \textsf{true}_{\geq 1/2}\oplus\textsf{false};\\
c\!:=&\ \textsf{true}_{\geq 1/2}\oplus\textsf{false};\\
a\!:=&\ (b \Leftrightarrow c)
\end{aligned}
\tag{B.1}
$$

in which we have replaced the disjunction "$\vee$" with equivalence "$\Leftrightarrow$", an operator which (unlike disjunction) is not monotonic in c (or in b).[6] Expressed

---

[6] At the same time we have used $1/2$ instead of $.9$ for the probabilities, because it brings the issues closer to everyday experience, *e.g.* coin flipping.

in English, it suggests the question

if two coins are flipped, each with probability *at least* 1/2
of giving heads, what is the probability that they will come   (B.2)
up showing the same face?

Note that our coins are not fair — rather they are "heads biased."

An informal analysis would suggest that the worst case occurs when the
first coin always gives heads, but the second gives heads only half the time
(in other words, at the extremes of probability which separate the potential
outcomes as much as possible). In that case the answer is that they will be
the same with probability [7]

$$
\begin{aligned}
&1 * 1/2 && \leftarrow \text{for heads/heads} \\
+\ &0 * 1/2 && \leftarrow \text{for\ \ tails/tails} \\
\\
=\ &1/2\ .
\end{aligned}
$$

However for the program (B.1) the informal description above, and the
subsequent analysis, is *quite wrong*.[8] In fact we have a situation in which

one coin is flipped, with probability at least 1/2 of giving heads,
and then a second coin is flipped again with probability at least
1/2 of giving heads *but which probability can be affected by the
outcome of the first coin.*

What really is the probability that they will show the same face?

Suppose b's strategy is to use probability 1/2; subsequently c uses 1/2 if
b chose true, but uses 1 if b chose false. The probability that a will be true
is now $1/2 * 1/2 + 1/2 * 0 = 1/4$, and that is borne out by calculation:

$$
\begin{aligned}
&b := \ \text{true} \ _{\geq 1/2}\oplus \text{false}; \\
&c := \ \text{true} \ _{\geq 1/2}\oplus \text{false}; \\
&a := \ (b \Leftrightarrow c)
\end{aligned}
$$

$=$ $b := \ \text{true} \ _{\geq 1/2}\oplus \text{false};$ Law 15; c local
           $a := \ b \ _{\geq 1/2}\oplus \overline{b}$

$=$ Laws 15, 20; b local
           $a := \ (\text{true} \ _{\geq 1/2}\oplus \text{false}) \ _{\geq 1/2}\oplus (\text{false} \ _{\geq 1/2}\oplus \text{true})$

$= \ldots$

---

[7] More rigorously, if we pick probabilities $p, q \geq 1/2$ for b, c, then the probability the
two coins show the same face is $pq + \overline{p}\,\overline{q}$, an expression whose least value is 1/2 in the
range given for p, q.

[8] Lynch *et al.* discuss this same issue [LSS94, Example 4.1].

$\ldots =$ definition $\geq_p \oplus$; Law 8

$$
\begin{array}{ll}
\quad\ a := \mathsf{true} & \leftarrow\ \textit{true most likely} \\
\sqcap\ \ a := \mathsf{true}\ {}_{1/2}\oplus \mathsf{false} & \\
\sqcap\ \ a := \mathsf{true}\ {}_{1/2}\oplus \mathsf{false} & \\
\sqcap\ \ a := \mathsf{true}\ {}_{1/2}\oplus (\mathsf{false}\ {}_{1/2}\oplus \mathsf{true}) & \\
\sqcap\ \ a := (\mathsf{true}\ {}_{1/2}\oplus \mathsf{false})\ {}_{1/2}\oplus \mathsf{false} & \leftarrow\ \textit{true least likely} \\
\sqcap\ \ a := (\mathsf{true}\ {}_{1/2}\oplus \mathsf{false})\ {}_{1/2}\oplus (\mathsf{false}\ {}_{1/2}\oplus \mathsf{true}) &
\end{array}
$$

$=$ Law 7, *i.e.* $(\sqcap) \sqsubseteq ({}_p\oplus)$

$$
\begin{array}{ll}
\quad\ a := \mathsf{true} & \leftarrow\ \textit{true most likely} \\
\ \vdots\quad \text{(others subsumed)} & \quad\vdots \\
\sqcap\quad a := (\mathsf{true}\ {}_{1/2}\oplus \mathsf{false})\ {}_{1/2}\oplus \mathsf{false} & \leftarrow\ \textit{true least likely}
\end{array}
$$

$=$ $a := \mathsf{true}\ {}_{\geq 1/4}\oplus \mathsf{false}$.   Law 3, definition $\geq_p \oplus$

Thus we confirm that the c-demon's knowledge of b's outcome is important in this case.

A program corresponding to the description (B.2) above would in fact be

$$
\begin{aligned}
&b, c := (\mathsf{true}\ {}_{\geq 1/2}\oplus \mathsf{false}), (\mathsf{true}\ {}_{\geq 1/2}\oplus \mathsf{false}); \\
&a := (b \Leftrightarrow c)\ ,
\end{aligned}
$$

in which the first statement generalises the syntactic sugar for $\geq_p \oplus$, *i.e.* is an abbreviation for

$$
\begin{array}{ll}
\quad\ b, c := \mathsf{true}, \mathsf{true} & \\
\sqcap\ \ b, c := \mathsf{true}, (\mathsf{true}\ {}_{1/2}\oplus \mathsf{false}) & \\
\sqcap\ \ b, c := (\mathsf{true}\ {}_{1/2}\oplus \mathsf{false}), \mathsf{true} & \text{(B.3)} \\
\sqcap\ \ b, c := (\mathsf{true}\ {}_{1/2}\oplus \mathsf{false}), (\mathsf{true}\ {}_{1/2}\oplus \mathsf{false})\ . &
\end{array}
$$

Running the assignments "in parallel" avoids the interaction of demonic choice in one with probabilistic choice in the other.

B.2   Loop rule for *demonic* iterations

Here we finish off our proof of the invariant-implies-termination loop rule Thm. 7.3.3, showing that it extends to demonic loops as well. The approach is to use Fact B.3.5 to replace the loop body by an appropriate deterministic refinement of it, as in the following lemma.

Lemma B.2.1 Let the program *dloop* be defined

$$dloop \quad := \quad \mathbf{do} \ G \to det \ \mathbf{od} \ ,$$

for any choice of standard predicate G (the loop guard) and deterministic program *det* (the loop body). Then for any *loop* and post-expectation *postE* there is a *det* such that $body \sqsubseteq det$ and

$$wp.dloop.postE \quad \equiv \quad wp.loop.postE \ . \quad ^9 \qquad (B.4)$$

Proof Define $preE := wp.loop.postE$, and use Fact B.3.5 to choose *det* so that $body \sqsubseteq det$ and

$$wp.body.preE \quad \equiv \quad wp.det.preE \ . \qquad (B.5)$$

Then we have

$$
\begin{array}{lll}
 & wp.det.preE & \mathbf{if} \ G \ \mathbf{else} \quad postE \\
\equiv & wp.body.preE & \mathbf{if} \ G \ \mathbf{else} \quad postE \qquad \text{by construction (B.5)} \\
\equiv & preE \ , & \text{definition } preE; \text{ definition (1.19) of iteration}
\end{array}
$$

so that $preE$ satisfies the (least) fixed-point equation (1.19) given for $wp.dloop.postE$, as well as (by construction) the equation given for $wp.loop.postE$. Hence $wp.dloop.postE \Rrightarrow preE$ and, from $body \sqsubseteq det$ and monotonicity, we have

$$wp.dloop.postE \quad \equiv \quad wp.loop.postE$$

as required.

$\square$

With Lem. B.2.1 we have our theorem easily.

Theorem B.2.2 INVARIANT-IMPLIES-TERMINATION LOOP RULE
 In the terminology of Sec. 7.3, if I is a *wp*-invariant of *loop* and $I \Rrightarrow T$ then

$$I \quad \Rrightarrow \quad wp.loop.([\overline{G}] * I) \ .$$

Proof Use Lem. B.2.1 to choose deterministic refinement *det* of *body* so that

$$wp.dloop.([\overline{G}] * I) \quad \equiv \quad wp.loop.([\overline{G}] * I) \ ,$$

and observe that since $body \sqsubseteq det$ we have I a *wp*-invariant of *dloop* also.
 The result is then immediate from Thm. 7.3.3. $\square$

---

[9]A similar trick is required for the proof of (2.24) for intrinsically unbounded invariants (p. 72). If all deterministic $det \sqsupseteq body$ satisfy Condition (6) in Sec. 2.12 — that the expected value of the invariant I "while still iterating" tends to zero — then in particular the one satisfying (B.4) does, and we proceed as follows.

We suppose $\left[\overline{G}\right] * I \Rrightarrow postE$ for some invariant I and, using the least fixed-point definition (1.19) of iteration and the linearity of $wp.det$, we have

$$wp.dloop.(\left[\overline{G}\right] * I) \quad\equiv\quad (\sqcup n\colon \mathbb{N} \bullet I_n) \,,$$

where we define

$$
\begin{aligned}
I_0 &:= && 0 \\
I_1 &:= && \left[\overline{G}\right] * I \\
I_2 &:= && \left[\overline{G}\right] * I + [G] * wp.det.(\left[\overline{G}\right] * I) \\
I_3 &:= && \left[\overline{G}\right] * I + [G] * wp.det.(\left[\overline{G}\right] * I) + [G] * wp.det.([G] * wp.det.(\left[\overline{G}\right] * I)) \,.
\end{aligned}
$$
$$\vdots$$

Now we formalise "the expected value of the invariant I after n iterations" as

$$E_n \quad:=\quad (\lambda X \bullet [G] * wp.det.X)^n.I \,,$$

and since by Condition (6) we have $(\lim_{n\to\infty} E_n) = 0$, we can add the E_n's into the above limit, giving

$$wp.dloop.(\left[\overline{G}\right] * I) \quad\equiv\quad (\lim_{n\to\infty} I_n + E_n) \,,$$

where we must write "lim" on the right because the terms are no longer necessarily increasing. Then we have for example

$$I_3 + E_3$$

$$
\equiv \quad
\left.
\begin{aligned}
&\; \left[\overline{G}\right] * I \\
&+ \; [G] * wp.det.(\left[\overline{G}\right] * I) \\
&+ \; [G] * wp.det.([G] * wp.det.(\left[\overline{G}\right] * I))
\end{aligned}
\right\} I_3
$$

$$+ \; [G] * wp.det.([G] * wp.det.([G] * wp.det.I)) \quad \} \; E_3$$

$$
\Lleftarrow \quad
\begin{aligned}
&\; \left[\overline{G}\right] * I \\
&+ \; [G] * wp.det.(\left[\overline{G}\right] * I) \\
&+ \; [G] * wp.det.([G] * wp.det.(\left[\overline{G}\right] * I)) \\
&+ \; [G] * wp.det.([G] * wp.det.([G] * I))
\end{aligned}
\qquad\qquad \text{invariance of } I
$$

$$
\equiv \quad
\begin{aligned}
&\; \left[\overline{G}\right] * I \\
&+ \; [G] * wp.det.(\left[\overline{G}\right] * I) \\
&+ \; [G] * wp.det.([G] * wp.det.I)
\end{aligned}
\qquad\qquad \text{linearity of } det
$$

$$\Lleftarrow \quad I \,, \qquad\qquad\qquad\qquad\qquad\qquad \text{above two steps twice more}$$

which inequality is easily verified for all n by induction. Thus we have

$$wp.dloop.(\left[\overline{G}\right] * I) \quad\equiv\quad (\sqcup n\colon \mathbb{N} \bullet I_n) \quad\equiv\quad (\lim_{n\to\infty} I_n + E_n) \quad\Lleftarrow\quad I \,,$$

as required.

B.3 Further facts about probabilistic *wp* and *wlp*

Proofs of these facts can be constructed within a probabilistic *wlp* semantics [MM01b]; they are placed here in the appendix, rather than in the main text, because that theory has not been included in this volume.

Fact B.3.1 For standard program *prog* and standard postcondition *post* we have

$$wlp.prog.post \sqcap wp.prog.1 \quad \Rightarrow \quad wp.prog.post .$$

□

Fact B.3.2 *sub-distributivity of* &   For program *prog* and post-expectations $postE_0, postE_1$ we have

$$wlp.prog.postE_0 \ \& \ wp.prog.postE_1 \quad \Rightarrow \quad wp.prog.(postE_0 \ \& \ postE_1) .$$

□

Fact B.3.3 *sub-distributivity of* +   For any program *prog* and post-expectations $postE_0, postE_1$ we have

$$wp.prog.postE_0 \ + \ wlp.prog.postE_1 \quad \Rightarrow \quad wlp.prog.(postE_0 + postE_1) ,$$

with equality when *prog* is deterministic.   □

Fact B.3.4 For any program *prog* we have $wlp.prog.1 \equiv 1$.   □

Fact B.3.5 For any program *prog* and post-expectation *postE* there is a deterministic refinement of it — a deterministic *det* with $prog \sqsubseteq det$ — such that

$$wp.prog.postE \quad \equiv \quad wp.det.postE .$$

□

Fact B.3.5 is related to continuity, and is most easily understood by working beyond the logic as in the geometric argument given in Sec. 6.9.5.

Fact B.3.6 For any program *prog* and post-expectation *postE* we have

$$\lfloor wp.prog.postE \rfloor \quad \Rightarrow \quad wp.prog.\lfloor postE \rfloor .$$

□

B.4   Details of the extension to infinite state spaces

We now fill in the details behind the ideas introduced in Sec. 8.2 that allow us to extend our semantic models, and the results concerning them, to infinite state spaces. We begin by examining some properties of infinitary discrete probability distributions.

For emphasis, in this section only we use $\mathbb{E}_B S$ for the set of bounded-above expectations over S.

B.4.1   Topological preliminaries; compactness of $\overline{S}$

Our set of discrete sub-distributions $\overline{S}$ is a subset of the function space $S \to \mathbb{R}_{\geq}$, which in turn can be regarded as Euclidean space $\mathbb{R}_{\geq}^S$ but now possibly of infinite dimension. That is, each state s in S corresponds to a dimension: a point in the non-negative hyper-octant is determined by a function from dimension to non-negative co-ordinate.[10]

The *Euclidean topology* for the real line $\mathbb{R}_{\geq}$ corresponds to the usual Euclidean metric, and the topology $\mathcal{E}_S$ we use for $\mathbb{R}_{\geq}^S$ is the product of the Euclidean topologies over $\mathbb{R}_{\geq}$ for each dimension in S. A basis for $\mathcal{E}_S$ is then the collection of sets of the form

$$N \times \mathbb{R}_{\geq}^{S-P} \tag{B.6}$$

(ignoring order in the product), for all finite subsets P of S and open subsets N of $\mathbb{R}_{\geq}^P$. For each dimension s in S we say that s is in the *support* of an open set in $\mathcal{E}_S$ just when the projection of that set onto s is not all of $\mathbb{R}_{\geq}$. Similar notions of finiteness apply to expectations:

Definition B.4.1 FINITARY EXPECTATION   An expectation α in $\mathbb{E}S$ is *finitary* if there is a finite subset P of S such that $\alpha.s \neq 0$ only for states s in P. (Note that such expectations are bounded by construction.)

The *support* of an expectation α is the set of states s such that $\alpha.s \neq 0$; thus an expectation is finitary iff it has finite support.

We write $\mathbb{F}S$ for the set of finite subsets of S, and $\mathbb{E}_f S$ for the set of finitary expectations over S.

For subset P of S, we define the *restriction* of α to P as

$$(\alpha{\downarrow}P).s \; := \quad \begin{array}{ll} \alpha.s & \text{if } s \in P \\ 0 & \text{otherwise.} \end{array}$$

Thus $\alpha{\downarrow}P$ is finitary if P is finite.   □

---

[10] A review of Chap. 6 makes this clear for finite state spaces. In particular note that points in the Euclidean space do not correspond to states of the program: rather they correspond to "tuples" of numbers, one for each state.

Our technique for extending our results will be to show that they continue to hold in infinite S provided we restrict our attention to finitary expectations; then continuity of the transformers involved will carry the equalities through to all expectations, since any expectation can be written as a directed $\sqcup$-limit of its finitary restrictions.

First we must investigate some properties of finitary expectations themselves; we begin by showing that finitary expectations can be used to define "closed half-spaces" in $\mathbb{R}_{\geq}^{S}$.

Lemma B.4.2 CLOSED FINITARY HALF-SPACE   For (finitary) α in $\mathbb{E}_{f}S$ and r in $\mathbb{R}_{\geq}$, sets of the forms

$$\{\Delta : \mathbb{R}_{\geq}^{S} \mid \int_{\Delta} \alpha \leq r\} \quad \text{and} \quad \{\Delta : \mathbb{R}_{\geq}^{S} \mid \int_{\Delta} \alpha \geq r\}$$

are closed in the topology $\mathcal{E}_{S}$ over $\mathbb{R}_{\geq}^{S}$.

Proof   Let the support of α be $\bar{P}$, finite because α is finitary. Then in either case above the projection of the set is all of $\mathbb{R}_{\geq}$ for each dimension outside of P; and its projection into $\mathbb{R}_{\geq}^{P}$ is the complement of an open set there.   □

For infinitary expectations, however, we have closure in one direction only.

Lemma B.4.3 CLOSED INFINITARY HALF-SPACE   For any expectation α in $\mathbb{E}_{B}S$ and r in $\mathbb{R}_{\geq}$, the set of distributions[11]

$$\{\Delta : \mathbb{R}_{\geq}^{S} \mid \int_{\Delta} \alpha \leq r\}$$

is closed in $\mathbb{R}_{\geq}^{S}$.

Proof   We have

$$\begin{array}{ll}
 & \{\Delta : \mathbb{R}_{\geq}^{S} \mid \int_{\Delta} \alpha \leq r\} \\
= & \{\Delta : \mathbb{R}_{\geq}^{S} \mid \int_{\Delta} (\sqcup P : \mathbb{F}S \cdot \alpha {\downarrow} P) \leq r\} \\
\\
= & \text{bounded monotone convergence for } \int_{\Delta} \text{ [Jon90]} \\
 & \{\Delta : \mathbb{R}_{\geq}^{S} \mid (\sqcup P : \mathbb{F}S \cdot \int_{\Delta} \alpha {\downarrow} P) \leq r\} \\
\\
= & \{\Delta : \mathbb{R}_{\geq}^{S} \mid (\forall P : \mathbb{F}S \cdot \int_{\Delta} \alpha {\downarrow} P \leq r)\} \\
= & (\cap P : \mathbb{F}S \cdot \{\Delta : \mathbb{R}_{\geq}^{S} \mid \int_{\Delta} \alpha {\downarrow} P \leq r\}) \,,
\end{array}$$

which is an intersection of a (P-indexed) collection of closed sets (Lem. B.4.2), since $\alpha {\downarrow} P$ is finitary for all P in $\mathbb{F}S$.   □

---

[11]We really do mean "$\mathbb{E}_{B}S$", not "$\mathbb{E}_{f}S$". The former are the bounded expectations over the whole space S; the latter are the expectations of finite support, bounded by construction. We use least upper bounds within $\mathbb{E}_{f}S$ to approximate elements of $\mathbb{E}_{B}S$ in these arguments.

That Lem. B.4.3 works in only one direction is because our expectations (finitary or not) take only non-negative values. To see that its dual does not hold in general, consider the half-space

$$\{\Delta : \mathbb{R}^S_\geq \mid \int_\Delta \underline{1} \geq 1\} .\tag{B.7}$$

The origin (Δ everywhere 0) does not lie within it, yet every open set in the basis (B.6) containing the origin also intersects the half-space — for example, take Δ in $\mathbb{R}^S_\geq$ to be one at some point in $S-P$ and zero elsewhere. Thus the origin is a limit point of (B.7) but is not in it.

With the above we now have the compactness property that we need for our space of distributions.

Lemma B.4.4 PROBABILITY DISTRIBUTION SPACE IS COMPACT
The space $\overline{S}$ of distributions over S is a compact subset of $(\mathbb{R}^S_\geq, \mathcal{E}_S)$.
    Proof   The set $\overline{S}$ may be written

$$[0,1]^S \cap \{\Delta : \mathbb{R}^S_\geq \mid \int_\Delta \underline{1} \leq 1\} .$$

But $[0,1]^S$ is compact by Tychonoff's Theorem,[12] and $\{\Delta : \mathbb{R}^S_\geq \mid \int_\Delta \underline{1} \leq 1\}$ is closed by Lem. B.4.3. Thus $\overline{S}$ is the intersection of a compact set and a closed set, and is therefore compact. (It is also closed.)   □

B.4.2 The Galois functions; continuity

We now return to the connection (Thm. 5.7.7) between the relational space $\mathbb{H}S$ and the expectation-transformer space $\mathbb{T}S$: they are linked by the functions

$$rp\colon\ \mathbb{T}S \to \mathbb{H}S \qquad \text{(Def. 5.7.1)}$$
$$\text{and}\quad wp\colon\ \mathbb{H}S \to \mathbb{T}S \qquad \text{(Def. 5.5.2)} \quad .$$

For well-definedness of rp in the finite case we showed that given any t in $\mathbb{T}S$ (for which rp is defined), and s in S, the set $rp.t.s$ was up-closed, convex and Cauchy closed: and they are just the conditions placed on result sets in $\mathbb{H}S$ by Def. 5.4.4.

In the infinite case rp is not so well behaved, although up closure and convexity follow as before. But Cauchy closure now requires an explicit appeal to the bounded continuity of t (Def. 5.6.6 on p. 147): it can no longer be proved from sublinearity. The following technical lemma shows that when t is boundedly continuous we may restrict ourselves to finitary expectations in the use of Def. 5.7.1:

---

[12]TYCHONOFF'S THEOREM states that a product of compact sets is compact in the product topology.

Lemma B.4.5 For any boundedly-continuous t in $\mathbb{T}S$ and s in S we have

$$rp.t.s \;=\; \{\Delta\colon \overline{S} \cdot (\forall \alpha\colon \mathbb{E}_f S \cdot t.\alpha.s \le \int_\Delta \alpha)\}\,.$$

The only change from Def. 5.7.1 is the type of the universally quantified α — it is now drawn from $\mathbb{E}_f S$ rather than $\mathbb{E}S$.

Proof   For arbitrary α in $\mathbb{E}_B S$ and Δ in $\overline{S}$ we have

$$(\forall P\colon \mathbb{F}S \cdot t.(\alpha{\downarrow}P).s \le \textstyle\int_\Delta \alpha{\downarrow}P)$$
implies $(\sqcup P\colon \mathbb{F}S \cdot t.(\alpha{\downarrow}P).s) \le (\sqcup P\colon \mathbb{F}S \cdot \int_\Delta \alpha{\downarrow}P)$

iff t boundedly continuous; bounded monotone convergence
$$t.(\sqcup P\colon \mathbb{F}S \cdot \alpha{\downarrow}P).s \le \textstyle\int_\Delta(\sqcup P\colon \mathbb{F}S \cdot \alpha{\downarrow}P)$$

iff $t.\alpha.s \le \int_\Delta \alpha\,.$

The result then follows directly from Def. 5.7.1, since any $\alpha{\downarrow}P$ is finitary.   □

With Lem. B.4.5 we can re-establish Cauchy closure in the infinitary case of $rp.t$ for boundedly-continuous t.

Lemma B.4.6 For any boundedly-continuous t in $\mathbb{T}S$ and state s in S, the set of distributions $rp.t.s$ is Cauchy closed in $\overline{S}$.

Proof   We reason

$rp.t.s$
$=\quad \{\Delta\colon \overline{S} \cdot (\forall \alpha\colon \mathbb{E}_f S \cdot t.\alpha.s \le \int_\Delta \alpha)\}$ Lem. B.4.5
$=\quad (\cap \alpha\colon \mathbb{E}_f S \cdot \{F\colon \overline{S} \mid t.\alpha.s \le \int_\Delta \alpha\})\,,$

which by Lem. B.4.2 is an intersection of Cauchy-closed sets.   □

To see that continuity is a necessary condition for Lem. B.4.6, define t in $\mathbb{T}S$ by

$$t.\alpha.s \;:=\; \sqcap \alpha \tag{B.8}$$

for all s in S; it is not continuous when S is infinite,[13] and corresponds to the unboundedly demonic standard program **chaos** that chooses any final state in S. Now for all Δ in $rp.t.s$ we have $\sqcap \alpha \le \int_\Delta \alpha$ by Def. 5.7.1, so that in particular $1 \le \int_F \underline{1}$. But conversely $1 \le \int_F \underline{1}$ implies

$$\sqcap \alpha \;\le\; (\sqcap \alpha)\int_\Delta \underline{1} \;=\; \int_\Delta \underline{\sqcap \alpha} \;\le\; \int_\Delta \alpha\,,$$

so that distribution $\int_\Delta$ is in $rp.t.s$ iff $1 \le \int_\Delta \underline{1}$. Reasoning similar to (B.7) above then shows that $rp.t.s = \{\Delta\colon \overline{S} \mid 1 \le \int_\Delta \underline{1}\}$ is not Cauchy closed.

---

[13]To see that this t is not continuous, consider $\underline{1}$, the limit of the family of (finitary) expectations $\{\underline{1}{\downarrow}P \mid P\colon \mathbb{F}S\}$, and note that $t.\underline{1} \equiv \underline{1}$ whereas $t.(\underline{1}{\downarrow}P) \equiv \underline{0}$ for any P in $\mathbb{F}S$.

In fact the set $\{\Delta \colon \overline{S} \mid 1 \leq \int_{\Delta} 1\}$ is the convex closure of the set of all standard final states $\{s \colon S \cdot \overline{s}\}$ — that is why it represents the program "choose any final state in S." Applying Cauchy closure as well would include the origin (by the argument at (B.7) again), whence up closure would include all of $\overline{S}$. Thus the only probabilistic and continuous program that can choose demonically from all of infinitely many final states must possibly fail to terminate as well. (See also Footnote 33 on p. 289.)

This is how the often-used restriction of "bounded nondeterminism" guarantees continuity in the probabilistic models.

Having shown boundedly-continuous t to yield Cauchy-closed $rp.t$, we turn to the converse: that Cauchy-closed r yields boundedly-continuous $wp.r$ — for infinite S it must be done directly, rather than from sublinearity.

Lemma B.4.7 For any relational program r in $\mathbb{H}S$, the corresponding expectation transformer $wp.r$ is boundedly continuous.

Proof   Let $\mathcal{A}$ be a $\Rightarrow$-directed and bounded subset of expectations in $\mathbb{E}_B S$; we show that for any $c > 0$ and state s in S

$$wp.r.(\sqcup\mathcal{A}).s \quad \leq \quad (\sqcup\alpha \colon \mathcal{A} \cdot wp.r.\alpha.s) + c \ ,$$

which is sufficient for continuity since c may be arbitrarily small. (The other direction is given by monotonicity.)

Define $x \colon = (\sqcup\alpha \colon \mathcal{A} \cdot wp.r.\alpha.s)$; then we have

$$
\begin{array}{lll}
 & (\forall\alpha \colon \mathcal{A} \cdot wp.r.\alpha.s \leq x) & \text{definition of } r \\
\text{iff} & (\forall\alpha \colon \mathcal{A} \cdot (\sqcap\Delta \colon r.s \cdot \int_{\Delta} \alpha) \leq x) & \text{Def. 5.7.1} \\
\text{implies} & (\forall\alpha \colon \mathcal{A} \cdot (\exists\Delta \colon r.s \cdot \int_{\Delta} \alpha \leq x + c)) & c > 0 \\
\text{iff} & (\forall\alpha \colon \mathcal{A} \cdot \{\Delta \colon r.s \cdot \int_{\Delta} \alpha \leq x + c\} \neq \emptyset)
\end{array}
$$

implies   Lem. B.4.3; $r.s$ closed in $\overline{S}$; $\mathcal{A}$ directed; $\overline{S}$ compact — see (†) below
$$(\cap\alpha \colon \mathcal{A} \cdot \{\Delta \colon r.s \cdot \int_{\Delta} \alpha \leq x + c\}) \neq \emptyset$$

$$
\begin{array}{lll}
\text{iff} & (\exists\Delta \colon r.s \cdot (\forall\alpha \colon \mathcal{A} \cdot \int_{\Delta} \alpha \leq x + c)) & \\
\text{iff} & (\exists\Delta \colon r.s \cdot \int_{\Delta}(\sqcup\mathcal{A}) \leq x + c) & \text{bounded monotone convergence} \\
\text{implies} & (\sqcap\Delta \colon r.s \cdot \int_{\Delta}(\sqcup\mathcal{A})) \leq x + c & \\
\text{iff} & wp.r.(\sqcup\mathcal{A}).s \leq x + c \ . & \text{Def. 5.5.2}
\end{array}
$$

†   For the deferred justification note that Lem. B.4.3 and Cauchy closure of $r.s$ imply Cauchy closure of each set $\{\Delta \colon r.s \cdot \int_{\Delta} \alpha \leq x + c\}$, and $\mathcal{A}$'s being directed ensures that they have the finite-intersection property; we thus appeal to the Finite-Intersection Lemma for non-emptiness of their intersection.   □

The geometric-distribution program of Fig. 2.11.1 on p. 69 illustrates continuity for a transformer even where the set of possible final states is infinite. Although that program is "infinitely branching" — every natural number is reachable with nonzero probability — it is still $\sqcup$-continuous since the branching is probabilistic, not demonic or angelic. Recall that in contrast infinitely $\sqcap$-branching programs are generally not $\sqcup$-continuous.

B.4.3 The Galois connection

Now we re-establish the partial Galois connection, and with it our characterisation of demonic programs, for infinite state spaces.

Let $\mathbb{T}_c S$ be the boundedly continuous expectation transformers in $\mathbb{T}S$. The fact that rp and wp form a partial Galois connection between $\mathbb{H}S$ and $\mathbb{T}_c S$ is not difficult to show: the inequalities

$$rp \circ wp \sqsubseteq \mathsf{id} \quad \text{and} \quad \mathsf{id} \sqsubseteq wp \circ rp$$

do not require finiteness of S. However our original proofs of the stronger results, namely the equality $rp \circ wp = \mathsf{id}$ and that sublinearity characterises the wp images in $\mathbb{T}S$ (reported above as Lem. 5.7.2 and Lem. 5.7.6), did use finiteness.

Our new proof for the first is as follows.

Lemma B.4.8 For any r in $\mathbb{H}S$ we have

$$rp.(wp.r) \quad = \quad r \ .$$

Proof The proof for the finite case (p. 150) appealed to the *Separating-Hyperplane Lemma* Lem. B.5.1. Using Lem. B.5.3 instead allows the proof to go through for the infinite case. □

To recover the second result we must strengthen its assumptions to include bounded continuity; then we have

Lemma B.4.9 If (boundedly-continuous) t in $\mathbb{T}_c S$ is sublinear, then

$$wp.(rp.t) \quad = \quad t \ .$$

Proof We are able to replay our earlier proof of Lem. 5.7.6 that established $wp.(rp.t).\alpha \equiv t.\alpha$, but need consider only expectations α with finite support. For by continuity of t, Lem. B.4.6 and Lem. B.4.7 we have continuity of $wp.(rp.t)$ also, and every element of $\mathbb{E}_B S$ is the limit of a directed subset of $\mathbb{E}_f S$.

Thus we proceed as in the proof of Lem. 5.7.4, but (less generally) for finitary $\alpha \colon \mathbb{E}_f S$ and using Lem. B.4.5 instead of Def. 5.7.1; thus we reach that

$$(\cap \alpha' \colon \mathbb{E}_f S \bullet \{\Delta \colon \overline{S} \mid t.\alpha'.s \leq \textstyle\int_\Delta \alpha'\}) \tag{B.9}$$
$$\cap \ \{\Delta \colon \overline{S} \mid -t.\alpha.s \leq (\textstyle\int_\Delta -\alpha)\}$$
$$= \ \emptyset \ .$$

Again we argue by the *Finite-Intersection Lemma* that some finite sub-collection of the sets (B.9) has empty intersection.

Consider the union of the supports of all the finitary predicates α' or α generating that M-collection; it is a finite subset T of the state space S.

The distributions and predicates involved in (B.9) can then be restricted to $\mathbb{R}^T_{\ge}$, a finite dimensional space, and we are effectively considering an empty intersection of M sets based on

$$
\begin{aligned}
& (\cap \alpha' \colon \mathbb{E}_f T \cdot \{\Delta \colon \mathbb{R}^T_{\ge} \mid t.\alpha'.s \le \textstyle\int_\Delta \alpha'\}) \\
\cap \; & \{\Delta \colon \mathbb{R}^T_{\ge} \mid -t.\alpha.s \le (\textstyle\int_\Delta -\beta)\} \\
\cap \; & \{\Delta \colon \mathbb{R}^T_{\ge} \mid -1 \le \textstyle\int_\Delta(\underline{-1})\} \\
= \; & \emptyset .
\end{aligned}
$$

The subsequent contradiction is achieved exactly as in Lem. 5.7.4.   □

B.4.4   Healthiness conditions

The results above show that in our characterisation of wp-images of programs in $\mathbb{H}S$, the already-established healthiness conditions of Fig. 5.6.7 are applicable to the infinite case as well. The first three conditions in Fig. 5.6.7 are consequences of sublinearity as shown earlier and, in the special case where the scalars are $\{0,1\}$-valued, all are generalisations of properties of predicate transformers.

The results of Sections 8.3–8.5 also remain valid within the space of boundedly-continuous expectation transformers. All our constructions preserve continuity, although the extra work in the case of demonic programs (Sec. 8.4) merits some extra explanation, given below.

For the angelic case (Sec. 8.5) we need only ensure that Def. 8.5.4 is applied to finitary expectations, and then Thm. 8.5.8 establishes any boundedly-continuous and semi-sublinear transformer as a supremum of sublinear ones. As summarised in this section, the details of our proofs where we must appeal to theorems from the theory of convex sets are reduced to reasoning over a finite projection of the state space, where those theorems are still valid.

B.4.5   Demonic closure in the infinite case

In Sec. 8.4 we remarked (Footnote 13 on p. 225) that in the infinite case we must take a supremum of finitary infima to construct an arbitrary continuous demonic transformer from pre-deterministic ones. We sketch the details here, to some extent working beyond the logic by relying on geometric arguments. (Refer Chap. 6.)

Since a clump of possible final distributions can have a "smooth" — i.e. not piecewise straight — boundary, we see immediately that a finitary infimum will not do. Instead we approach the clump of distributions via a nested sequence of enclosing polytopes (the supremum),[14] each one formed as the intersection of finitely many closed upwards half-spaces — and each

---

[14] A POLYTOPE is a polygon generalised to dimensions other than two.

such intersection, as we see below, can be expressed as a finitary union (an infimum) of up-closures of a single distribution (the pre-deterministic programs).

Let our "target" demonic transformer be t. Take any finite subset P of the state space S, and fix some integer N. From P, N construct a set $\mathcal{A}_{N,P}$ of expectations (of size $\#P * (N+1)$, the set is therefore finite) by taking for each state $s \in P$ all possible values n/N for $0 \leq n \leq N$, and for each state $s \notin P$ the value zero.

‡   For each initial state $s \in P$ consider each (post-)expectation $\alpha \in \mathcal{A}_{N,P}$ in turn, and construct the half-space $\{\Delta \colon \overline{S} \mid t.\alpha.s \leq \int_\Delta \alpha\}$ — that will be the space above the hyperplane with normal α which is just touching the clump $t.\alpha.s$ from below. The intersection of all these α-hyperplanes (still holding s fixed) clearly contains $t.\alpha.s$.

To express that intersection as a finitary union, we refer to Fig. 6.6.1 where we can see (in two dimensions) a clump whose lower boundary is formed from three straight lines; the clump itself is therefore the intersection of the three half-planes lying above (the extensions of) those lines, restricted of course to the space of sub-distributions lying below the base $x + y = 1$ of the triangle. But that clump is also the *union* of the pre-deterministic programs given by the vertices of that clump, as shown in the smaller illustrations below it, of up-closure.

Thus in general (*i.e.* in any finite number of dimensions) we can express the finitary intersection of up half-spaces as the finitary union (with closure) of the pre-deterministic clumps determined by the intersection's vertices. (Because the polytope has finitely many faces (the half-planes in $\mathcal{A}_{N,P}$), it will have only finitely many vertices.)

Holding N, P (and hence $\mathcal{A}_{N,P}$) fixed, we can do this for each s in P, and so construct an approximation $t_{N,P}$ (from below, or from "outside") of our target program t. (For $s \notin P$ we take the whole space $\overline{S}$, *i.e.* behave as **abort** as we did explicitly with the assertion $\{n \leq N\}$ in the example of Sec. 8.4.) And each approximation is the finitary union of pre-deterministic programs.[15]

To conclude our argument, we must show that t itself is the supremum of all our continuous approximants $t_{N,P}$, *i.e.* that for all α, s we have

$$t.\alpha.s \quad = \quad (\sqcup N \colon \mathbb{N}; \, P \colon \mathbb{F}S \cdot t_{N,P}.\alpha.s) \ . \tag{B.10}$$

Informally we can say that as P increases in size, more and more of S is covered; and as N increases, the polytope encloses the target clump more and more closely. Formally, we proceed as follows.

---

[15]It is difficult to count how many exactly; but it is no more than the product over all $s \in P$ of the number of vertices of the polytope formed from t and $\mathcal{A}_{N,P}$ for each s by the construction above.

Fix *finitary* expectation α and state s in S; it is clear from the construction that $t.\alpha.s \geq t_{N,P}.\alpha.s$ for any integer N and finite subset P of S. To show the limit is attained, pick arbitrary real $x > 0$: we show there is a pair N, P such that $t_{N,P}.\alpha.s \geq t.\alpha.s - x$.

Choose P so that it contains both the support of α and the state s; we assume without loss of generality that α is not everywhere zero. Choose N large enough so that there is an $\alpha' \in \mathcal{A}_{N,P}$ with $\alpha' \ominus x/\sqcup\alpha \Rrightarrow \alpha/\sqcup\alpha \Rrightarrow \alpha'$ — we can do this "bracketing" of $\alpha/\sqcup\alpha$ because we have only finitely many differences $\alpha'.s - \alpha.s/\sqcup\alpha$ to consider, since P is finite and outside of P we know that α is zero. Now we reason as follows:

$$
\begin{array}{lll}
& t_{N,P}.\alpha.s & \\
= & \sqcup\alpha * t_{N,P}.(\alpha/\sqcup\alpha).s & t_{N,P} \text{ scaling} \\
\geq & \sqcup\alpha * t_{N,P}.(\alpha' \ominus x/\sqcup\alpha).s & \text{choice of } \alpha' \\
\geq & \sqcup\alpha * (t_{N,P}.\alpha'.s \ominus x/\sqcup\alpha) & \ominus \text{ sub-distribution} \\
\geq & \sqcup\alpha * t_{N,P}.\alpha'.s \;-\; x & \text{arithmetic} \\
= & \sqcup\alpha * t.\alpha'.s \;-\; x & t \text{ and } t_{N,P} \text{ agree here — see (†) below} \\
\geq & \sqcup\alpha * t.(\alpha/\sqcup\alpha).s \;-\; x & \text{choice of } \alpha' \\
= & t.\alpha.s \;-\; x \;. & t \text{ scaling}
\end{array}
$$

†   The reason that t and $t_{N,P}$ agree at α', s is that, because $s \in P$, the polytope determining $t_{N,P}$ at s used $\alpha' \in \mathcal{A}_{N,P}$ as one of its constructing faces, and that face was positioned precisely so that it touched the clump determined by t and s. (Recall (‡) on p. 339.)

Thus we have shown that (B.10) holds for all finitary α; but since both t (by assumption) and $t_{N,P}$ (by construction) are continuous, the result extends to all α.

B.4.6 Angelic closure in the infinite case

For the infinite angelic case we must show that if semi-sublinear t is continuous then it is the supremum of *continuous* sublinear t_α's. Our first step is to re-work the proof of Thm. 8.5.8 as follows:

$$
\begin{array}{lll}
& t.\beta & \\
\equiv & (\sqcup\alpha\colon \mathbb{E}_f S \mid \alpha \Rrightarrow \beta \cdot t.\alpha) & t \text{ continuous} \\
\equiv & (\sqcup\alpha\colon \mathbb{E}_f S \mid \alpha \Rrightarrow \beta \cdot t_\alpha.\alpha) & \text{Lem. 8.5.5} \\
\Rrightarrow & (\sqcup\alpha\colon \mathbb{E}_f S \mid \alpha \Rrightarrow \beta \cdot t_\alpha.\beta) & \alpha \Rrightarrow \beta \\
\Rrightarrow & (\sqcup\alpha\colon \mathbb{E}_f S \cdot t_\alpha.\beta) & \\
\Rrightarrow & t.\beta \;, & \text{Lem. 8.5.6}
\end{array}
$$

showing that continuous t is attainable as the supremum of t_α's for finitary α. But, as we see now, if α is finitary then t_α is itself continuous.

Let P be the support of α, take any directed set $\mathcal{B}$ of expectations, and for arbitrary $x > 1$ exploit the finiteness of P by choosing $\beta_x \in \mathcal{B}$ so that $(\sqcup\mathcal{B}){\downarrow}P \Rrightarrow x * \beta_x{\downarrow}P$. We then have

$$t_\alpha.(\sqcup\mathcal{B})$$

$$
\begin{aligned}
&\equiv && (\sqcup c, c' \colon \mathbb{R}_{\geq} \mid c\alpha - \underline{c'} \Rrightarrow \sqcup\mathcal{B} \bullet c(t.\alpha.s) - c') && \text{definition } t_\alpha \\
&\Rightarrow && (\sqcup c, c' \colon \mathbb{R}_{\geq} \mid c\alpha - \underline{c'} \Rrightarrow x * \beta_x \bullet c(t.\alpha.s) - c') && \alpha \text{ zero outside of } P \\
&\equiv && x * (\sqcup c, c' \colon \mathbb{R}_{\geq} \mid c\alpha - \underline{c'} \Rrightarrow \beta_x \bullet c(t.\alpha.s) - c') && \text{arithmetic} \\
&\equiv && x * t_\alpha.\beta_x && \text{definition } t_\alpha \\
&\Rightarrow && x * (\sqcup\beta \colon \mathcal{B} \bullet t_\alpha.\beta) \ . && \beta_x \in \mathcal{B}
\end{aligned}
$$

Since x can be taken arbitrarily close to one, we have our result.

B.5 Linear-programming lemmas

The first two of these lemmas are well known in linear programming.

Lemma B.5.1 THE SEPARATING-HYPERPLANE LEMMA   Let $\mathcal{C}$ be a convex and Cauchy-closed subset of $\mathbb{R}^N$, and p a point in $\mathbb{R}^N$ that does not lie in $\mathcal{C}$. Then there is a separating hyperplane S with p on one side of it and all of $\mathcal{C}$ on the other.

Proof   See for example a standard text on Linear Programming [Tru71, p.8] or Game Theory [Kuh03].   □

Lemma B.5.2 FARKAS' LEMMA   Let A be an $M \times N$ matrix, x an $N \times 1$ column-vector and r an $M \times 1$ column-vector, and suppose that A and r are so that the system of equations

$$A \cdot x \geq r \tag{B.11}$$

has no solution in x, where $\cdot$ denotes matrix multiplication. Then there is a $1 \times M$ row-vector C of non-negative values such that

$$C \cdot A = 0 \quad \text{but} \quad C \cdot r > 0 \ . \tag{B.12}$$

Proof   See Schrijver's text for example [Sch86, p.89], taking the contrapositive of Corollary 7.1e there.   □

Lem. B.5.2 can be motivated by considering its converse, also true but trivially so: if there is a C satisfying (B.12) then inequation (B.11) can have no solution—for if it did, we could reason

$$0 \ < \ C \cdot r \ \leq \ C \cdot A \cdot x \ = \ 0 \cdot x \ = \ 0 \ ,$$

a contradiction. Thus the lemma can be read "if (B.11) has no solution in x then there is a witness C to that fact."

The connection between probability and linear programming is reported by Fagin *et al.* also [FHM90].

Lemma B.5.3 THE SEPARATING-HYPERPLANE LEMMA — INFINITE CASE
Let $\mathcal{C}$ be a convex subset of $\mathbb{R}^S$ that is compact (hence closed) in the product $\mathcal{E}_S$ of the Euclidean topologies over its constituent projections $\mathbb{R}$. If some p does not lie in $\mathcal{C}$, then there is a separating hyperplane with p on one side of it and all of $\mathcal{C}$ on the other.

Proof   If $p \notin \mathcal{C}$, then because $\mathcal{C}$ is closed there is some neighbourhood N in the basis of $\mathcal{E}_S$ with $p \in N$ and $N \cap \mathcal{C} = \emptyset$.

Let T in $\mathbb{F}S$ be the support of N. Writing $(\downarrow T)$ for projection onto T, we then have $p{\downarrow}T \in N{\downarrow}T$ and $N{\downarrow}T \cap \mathcal{C}{\downarrow}T = \emptyset$, because for the latter $N{\downarrow}(S-T) = \mathbb{R}^{S-T}$, and thus $p{\downarrow}T \notin \mathcal{C}{\downarrow}T$. Note that $\mathcal{C}{\downarrow}T$ is compact because $\mathcal{C}$ is, hence closed; and it is convex also because $\mathcal{C}$ is.

Applying the standard *Separating Hyperplane Lemma* for the finite dimensional space Lem. B.5.1 in the case of $p{\downarrow}T$ and $\mathcal{C}{\downarrow}T$, within the finite-dimensional $\mathbb{R}^T$, gives us a separating hyperplane there; its extension parallel to the remaining axes $S-T$ is then the hyperplane we seek in $\mathbb{R}^S$.   □

B.6   Further lemmas for *eventually*

In this section we set out the proofs for a number of simple lemmas concerning the quantitative *eventually* operator $\Diamond$.

Lemma B.6.1 PROBABILISTIC DOUBLE-EVENTUALLY   [16]
    For all expectations A we have

$$\Diamond\Diamond A \quad \equiv \quad \Diamond A .$$

Proof   Replace $\subseteq$ by $\Rightarrow$ in the proof of Lem. 10.1.2 (p. 266).   □

Lemma B.6.2 DATA REFINEMENT OF *eventually* $\Diamond$
    If for any monotonic transformer t we have

$$\circ(t.A) \;\Rightarrow\; t.(\circ A)$$

for all expectations A, then we have also

$$\Diamond(t.A) \quad \Rightarrow \quad t.(\Diamond A)$$

for all A.

---

[16]See Footnote 18 on p. 94 for an example of this lemma in use.

Proof   We reason

$$\Diamond(t.A) \;\;\Rightarrow\;\; t.(\Diamond A)$$

if	$t.A \;\sqcup\; \circ t.(\Diamond A) \;\;\Rightarrow\;\; t.(\Diamond A)$		$\Diamond(t.A)$ *least*, Fig. 10.3.1
if	$t.A \;\sqcup\; t.(\circ\Diamond A) \;\;\Rightarrow\;\; t.(\Diamond A)$		assumption
if	$t.(A \sqcup \circ\Diamond A) \;\;\Rightarrow\;\; t.(\Diamond A)$		t monotonic
iff	$t.(\Diamond A) \;\;\Rightarrow\;\; t.(\Diamond A)$.		Def. 9.3.2

$\square$

Lemma B.6.3 $\Diamond$ MONOTONICITY   For all expectations A, B we have

$$A \Rrightarrow B \;\;\text{ implies }\;\; \Diamond A \Rrightarrow \Diamond B\ .$$

Proof   Again use $\Diamond A$ *least* (Fig. 10.3.1), and check that

	$A \;\sqcup\; \circ\Diamond B$	
$\Rrightarrow$	$B \;\sqcup\; \circ\Diamond B$	$A \Rrightarrow B$
$\equiv$	$\Diamond B$,	

as required.

$\square$

Lemma B.6.4 $\Diamond$ EXCLUDED MIRACLE   For all expectations A we have

$$\Diamond A \;\;\Rrightarrow\;\; \underline{\sqcup A}\ .$$

Proof   Use $\Diamond A$ *least*, and check that

	$A \;\sqcup\; \circ\underline{\sqcup A}$	
$\Rrightarrow$	$A \;\sqcup\; \underline{\sqcup\sqcup A}$	$\circ$ *excluded miracle*
$\equiv$	$\underline{\sqcup A}$.	

$\square$

Lemma B.6.5 $\Diamond$ SCALING   For all expectations A and scalars p in $[0, 1]$ we have

$$\Diamond(pA) \;\;\equiv\;\; p(\Diamond A)\ .$$

Proof   We prove $\Diamond(pA) \Rrightarrow p(\Diamond A)$ first, using $\Diamond(pA)$ *least* and checking

	$pA \;\sqcup\; \circ(p(\Diamond A))$	
$\equiv$	$pA \;\sqcup\; p(\circ\Diamond A)$	$\circ$ *scaling*
$\equiv$	$p(A \;\sqcup\; \circ\Diamond A)$	
$\equiv$	$p(\Diamond A)$.	

For $\Diamond(pA) \Lleftarrow p(\Diamond A)$ note first that it is trivial when $p = 0$. For $p > 0$ we prove equivalently $(\Diamond(pA))/p \Lleftarrow \Diamond A$, where Lem. B.6.4 guarantees well-definedness of the left-hand side:

$$(\Diamond(pA))/p \;\;\Rrightarrow\;\; \underline{\sqcup(pA)}/p \;\;\Rrightarrow\;\; \underline{p}/p \;\;\equiv\;\; \underline{1}\ .$$

Then using $\Diamond A$ *least*, we check

$$
\begin{aligned}
& A \ \sqcup \ \circ((\Diamond(pA))/p) \\
\equiv\ & A \ \sqcup \ (p/p)(\circ((\Diamond(pA))/p)) && p \neq 0 \\
\equiv\ & A \ \sqcup \ (\circ(p(\Diamond(pA))/p))/p && \circ\ scaling \\
\equiv\ & A \ \sqcup \ (\circ\Diamond(pA))/p \\
\equiv\ & (pA \ \sqcup \ \circ\Diamond(pA))/p && p \neq 0 \\
\equiv\ & \Diamond(pA)/p \ .
\end{aligned}
$$

$\square$

Bibliography

[ABL96]   J.-R. Abrial, E. Börger, and H. Langmaack, editors. *Formal Methods for Industrial Applications: Specifying and Programming the Steam Boiler Control*, volume 1165 of *LNCS*. Springer-Verlag, 1996.

[Abr96a]   J.-R. Abrial. *The B Book: Assigning Programs to Meanings.* Cambridge University Press, 1996.

[Abr96b]   J.-R. Abrial. Extending *B* without changing it (for developing distributed systems). In H. Habrias, editor, *First Conference on the B Method*, pages 169–190. Laboratoire LIANA, L'Institut Universitaire de Technologie (IUT) de Nantes, November 1996.

[AH90]   James Aspnes and M. Herlihy. Fast randomized consensus using shared memory. *J. Algorithms*, 11(3):441–61, 1990.

[AL96]   Christoph Andriessens and Thomas Lindner. Using FOCUS, LUSTRE, and probability theory for the design of a reliable control program. In Abrial *et al.* [ABL96], pages 35–51.

[ASBSV95]   A. Aziz, V. Singhal, F. Balarinand R.K. Brayton, and A.L. Sangiovanni-Vincentelli. It usually works: The temporal logic of stochastic systems. In *Computer-Aided Verification, 7th Intl. Workshop*, volume 939 of *LNCS*, pages 155–65. Springer-Verlag, 1995.

[AZP03]   T. Arons, L. Zuck, and A. Pnueli. Parameterized verification by probabilistic abstraction. In Andrew D. Gordon, editor, *FOSSACS 2003*, volume 2620 of *LNCS*, pages 87–102. Springer-Verlag, 2003.

[Bac78]   R.-J.R. Back. On the correctness of refinement steps in program development. Report A-1978-4, Department of Computer Science, University of Helsinki, 1978.

[Bac88]   R.-J.R. Back. A calculus of refinements for program derivations. *Acta Informatica*, 25:593–624, 1988.

[BAPM83]   M. Ben-Ari, A. Pnueli, and Z. Manna. The temporal logic of branching time. *Acta Informatica*, 20:207–26, 1983.

[BB96]   G. Brassard and P. Bratley. *Fundamentals of Algorithmics*. Prentice-Hall, 1996.

[BdA95]   Andrea Bianco and Luca de Alfaro. Model checking of probabilistic and nondeterministic systems. In *Foundations of Software Technology and Theoretical Computer Science*, volume 1026 of *LNCS*, pages 499–512, December 1995.

[BFL+99]   B. Bérard, A. Finkel, F. Laroussine, A. Petit, L. Petrucci, Ph. Schoebelen, and P. McKenzie. *Systems and Software Verification: Model-Checking Techniques and Tools*. Springer-Verlag, 1999.

[BKS83]   R.-J.R. Back and R. Kurki-Suonio. Decentralisation of process nets with centralised control. In *2nd ACM SIGACT-SIGOPS Symp. Principles of Distributed Computing*, pages 131–42, 1983.

[Boo82]   H. Boom. A weaker precondition for loops. *ACM Transactions on Programming Languages and Systems*, 4:668–77, 1982.

[BvW90]   R.-J.R. Back and J. von Wright. Duality in specification languages: a lattice theoretical approach. *Acta Informatica*, 27:583–625, 1990.

[BvW93]   R.-J. Back and J. von Wright. Predicate transformers and higher-order logic. In J. W. de Bakker, W.-P. de Roever, and G. Rozenberg, editors, *REX Workshop*, volume 666 of *LNCS*, pages 1–20. Springer-Verlag, 1993.

[BvW96]   R.-J.R. Back and J. von Wright. Interpreting nondeterminism in the refinement calculus. In He Jifeng, John Cooke, and Peter Wallis, editors, *Proceedings of the BCS-FACS 7th Refinement Workshop*, Workshops in Computing. Springer-Verlag, July 1996. www.springer.co.uk/ewic/workshops/7RW.

[BvW98]   R.-J.R. Back and J. von Wright. *Refinement Calculus: A Systematic Introduction*. Springer-Verlag, 1998.

[Chr90]   I. Christoff. Testing equivalences and fully abstract models for probabilistic processes. In *CONCUR '90*, volume 458 of *LNCS*, pages 126–40. Springer-Verlag, 1990.

[CM88]   K.M. Chandy and J. Misra. *Parallel Program Design: A Foundation*. Addison-Wesley, Reading, Mass., 1988.

[CMA02]   D. Cansell, D. Méry, and J.-R. Abrial. A mechanically proved and incremental development of the IEEE 1394 tree-identify protocol. *Formal Aspects of Computing*, 14(3):215–27, 2002.

[Coh00]   E. Cohen. Separation and reduction. In *Mathematics of Program Construction, 5th International Conference*, volume 1837 of *LNCS*, pages 45–59. Springer-Verlag, July 2000.

[CY95]   C. Courcoubetis and Mihalis Yannakakis. The complexity of probabilistic verification. *JACM*, 42(4):857–907, 1995.

[dA99]   Luca de Alfaro. Computing minimum and maximum reachability times in probabilistic systems. In *Proceedings of CONCUR '99*, volume 1664 of *LNCS*, pages 66–81. Springer-Verlag, 1999.

[dAH00]   Luca de Alfaro and T. Henzinger. Concurrent ω-regular games. In *Proc. 15th IEEE Symp. Logic in Computer Science*, pages 141–54. IEEE, 2000.

[dAM01]   Luca de Alfaro and Rupak Majumdar. Quantitative solution of omega-regular games. In *Proc. STOC '01*, pages 675–83. ACM, 2001.

[DFP01]   M. Duflot, L. Fribourg, and C. Picaronny. Randomized finite-state distributed algorithms as Markov chains. In *Int. Conf. on Distributed Computing (DISC'2001)*, volume 2180 of *LNCS*, pages 240–54, October 2001.

[DFP02]   M. Duflot, L. Fribourg, and C. Picaronny. Randomized dining philosophers without fairness assumption. In *Proc. 2nd IFIP Int. Conf. Theoretical Computer Science*, volume 223 of *IFIP Conference Proceedings*, pages 169–80, August 2002.

[DGJP02]   José Desharnais, Vineet Gupta, Radha Jagadeesan, and Prakash Panangaden. The metric analogue of weak bisimulation for probabilistic processes. In *Proc. LICS '02*, pages 413–22. IEEE, 2002.

[DGJP03]   José Desharnais, Vineet Gupta, Radha Jagadeesan, and Prakash Panangaden. Approximating labelled Markov processes. *Information and Computation*, 184:160–200, 2003.

[dHdV02]   J.J. den Hartog and E.P. de Vink. Verifying probabilistic programs using a Hoare-like logic. *Int. J. Found. Comp. Sci.*, 13(3):315–40, 2002.

[Dij71]   E.W. Dijkstra. Hierarchical ordering of sequential processes. *Acta Informatica*, 1(2):576–80, 583, October 1971.

[Dij76]   E.W. Dijkstra. *A Discipline of Programming*. Prentice Hall International, Englewood Cliffs, N.J., 1976.

[DIM83]   S. Dolev, A. Israeli, and S. Moran. Analyzing expected time by scheduler-luck games. *IEEE Transactions on Software Engineering*, 21(5):429–39, 1983.

[DP90]   B.A. Davey and H.A. Priestly. *Introduction to Lattices and Order*. Cambridge Mathematical Textbooks. Cambridge University Press, 1990.

[dRE98]   W.-P. de Roever and K. Engelhardt. *Data Refinement: Model-Oriented Proof Methods and their Comparison*, volume 47 of *Cambridge Tracts in Computer Science*. Cambridge University Press, 1998.

[Eda95]   A. Edalat. Domain theory and integration. *Theoretical Computer Science*, 151(1):163–93, 1995.

[EH86]   E.A. Emerson and J.Y. Halpern. "Sometimes" and "not never" revisited: on branching vs. linear time temporal logic. *Journal of the ACM*, 33(1):151–78, 1986.

[Eme90]   E.A. Emerson. Temporal and modal logic. In Jan van Leeuwen, editor, *Handbook of Theoretical Computer Science, Volume B: Formal Models and Semantics*, pages 995–1072. Elsevier and MIT Press, 1990.

[Eve57]   H. Everett. Recursive games. In *Contributions to the Theory of Games III*, volume 39 of *Ann. Math. Stud.*, pages 47–78. Princeton University Press, 1957.

[Far99]   J. Farkas. Die algebraische Grundlage der Anwendungen des mechanischen Princips von Fourier. *Mathematische und Naturwissenschaftliche Berichte aus Ungarn*, 16:25–40, 1899.

[Fel71]   W. Feller. *An Introduction to Probability Theory and its Applications*, volume 2. Wiley, second edition, 1971.

[FH84]   Yishai A. Feldman and David Harel. A probabilistic dynamic logic. *J. Computing and System Sciences*, 28:193–215, 1984.

[FHM90]   R. Fagin, J.Y. Halpern, and N. Megiddo. A logic for reasoning about probabilities. *Information and Computation*, 87:78–128, 1990.

[FHMV95]   R. Fagin, J. Halpern, Y. Moses, and M. Vardi. *Reasoning about Knowledge*. MIT Press, 1995.

[Flo67]   R.W. Floyd. Assigning meanings to programs. In J.T. Schwartz, editor, *Mathematical Aspects of Computer Science*, number 19 in Proc. Symp. Appl. Math., pages 19–32. American Mathematical Society, 1967.

[Fra86]   N. Francez. *Fairness*. Texts and Monographs in Computer Science. Springer-Verlag, 1986.

[FS03]   Colin Fidge and Carron Shankland. But what if I don't want to wait forever? *Formal Aspects of Computing*, 14(3):281–94, 2003.

[FV96]   J. Filar and O.J. Vrieze. *Competitive Markov Decision Processes: Theory, Algorithms, and Applications*. Springer-Verlag, 1996.

[GM91]   P.H.B. Gardiner and C.C. Morgan. Data refinement of predicate transformers. *Theoretical Computer Science*, 87:143–62, 1991. Reprinted in [MV94].

[GM93]   P.H.B. Gardiner and C.C. Morgan. A single complete rule for data refinement. *Formal Aspects of Computing*, 5(4):367–82, 1993. Reprinted in [MV94].

[Gol03]   R. Goldblatt. Mathematical modal logic: a view of its evolution. *Journal of Applied Logic*, 1(5–6):309–92, 2003.

[GR97]   Lindsay Groves and Steve Reeves, editors. *Formal Methods Pacific '97*, Discrete Mathematics and Computer Science. Springer-Verlag, 1997.

[Gro]   Probabilistic Systems Group. A quantified measure of security 1: a relational model. Available at [MMSS, key QMSRM].

[GS92]   G.R. Grimmett and D. Stirzaker. *Probability and Random Processes*. Oxford Science Publications, second edition, 1992.

[GW86]   G.R. Grimmett and D. Welsh. *Probability: an Introduction*. Oxford Science Publications, 1986.

[Heh89]   E.C.R Hehner. Termination is timing. In J.L.A. van de Snep-
            scheut, editor, *Mathematics of Program Construction*, volume 375
            of *LNCS*, pages 36–47. Springer-Verlag, 1989. Invited presentation.

[Her90]   T. Herman. Probabilistic self-stabilization. *Inf. Proc. Lett.*, 35(2):63–
            7, 1990.

[Hes92]   Wim H. Hesselink. *Programs, Recursion and Unbounded Choice*.
            Number 27 in Cambridge Tracts in Theoretical Computer Science.
            Cambridge University Press, Cambridge, U.K., 1992.

[Hes95]   Wim H. Hesselink. Safety and progress of recursive procedures.
            Formal Aspects of Computing, 7:389–411, 1995.

[HHS87]   C.A.R. Hoare, Jifeng He, and J.W. Sanders. Prespecification in
            data refinement. *Inf. Proc. Lett.*, 25(2):71–6, May 1987.

[HJ94]   H. Hansson and B. Jonsson. A logic for reasoning about time and
            probability. *Formal Aspects of Computing*, 6(5):512–35, 1994.

[HK94]   Joseph Y. Halpern and Bruce M. Kapron. Zero-one laws for modal
            logic. *Annals of Pure and Applied Logic*, 69:157–93, 1994.

[HK97]   Michael Huth and Marta Kwiatkowska. Quantitative analysis and
            model checking. In *Proceedings of 12th Annual IEEE Symposium
            on Logic in Computer Science*, pages 111–22, 1997.

[HMM04]   Joe Hurd, A.K. McIver, and C.C. Morgan. Probabilistic guarded
            commands mechanised in HOL. To appear in *Proc. QAPL '04
            (ETAPS)*, 2004.

[Hoa69]   C.A.R. Hoare. An axiomatic basis for computer programming.
            Communications of the ACM, 12(10):576–80, 583, October 1969.

[Hon03]   Ross Honsberger. *Mathematical Diamonds*, volume 26 of *Dolcani
            Mathematical Expositions*. Mathematical Society of America, 2003.

[HP01]   Oltea Mihaela Herescu and Catuscia Palamidessi. On the gener-
            alized dining philosophers problem. In *Proc. 20th Annual ACM
            Symposium on Principles of Distributed Computing (PODC'01)*,
            pages 81–9, 2001.

[HP02]   Joseph Y. Halpern and Riccardo Pucella. Reasoning about expec-
            tation. In *Proceedings of the Eighteenth Conference on Uncertainty
            in AI*, pages 207–15, 2002.

[HS86]   Sergiu Hart and Micha Sharir. Probabilistic propositional temporal
            logics. *Information and Control*, 70:97–155, 1986.

[HSM97]   Jifeng He, K. Seidel, and A.K. McIver. Probabilistic models for the
            guarded command language. *Science of Computer Programming*,
            28:171–92, 1997. Available at [MMSS, key HSM95].

[HSP83]   S. Hart, M. Sharir, and A. Pnueli. Termination of probabilistic con-
            current programs. *ACM Transactions on Programming Languages
            and Systems*, 5:356–80, 1983.

[Hur02]   Joe Hurd. A formal approach to probabilistic termination. In
            Víctor A. Carreño, César A. Muñoz, and Sofiène Tahar, editors,
            *15th International Conference on Theorem Proving in Higher Or-
            der Logics: TPHOLs 2002*, volume 2410 of *LNCS*, pages 230–45,

Hampton, VA., August 2002. Springer-Verlag.
`www.cl.cam.ac.uk/~jeh1004/research/papers`.

[Hut03]   Michael Huth. An abstraction framework for mixed non-deterministic and probabilistic systems. In *Validation of Stochastic Systems*, Lecture Notes in Computer Science Tutorial Series. Springer-Verlag, 2003.

[JHSY94]   B. Jonsson, C. Ho-Stuart, and W. Yi. Testing and refinement for nondeterministic and probabilistic processes. In Langmaack, de Roever, and Vytopil, editors, *Formal Techniques in Real-Time and Fault-Tolerant Systems*, volume 863 of *LNCS*, pages 418–30. Springer-Verlag, 1994.

[Jon86]   C.B. Jones. *Systematic Software Development using VDM*. Prentice-Hall, 1986.

[Jon90]   C. Jones. Probabilistic nondeterminism. Monograph ECS-LFCS-90-105, Edinburgh University, 1990. (Ph.D. Thesis).

[KB00]   M. Kwiatkowska and C. Baier. Domain equations for probabilistic processes. *Mathematical Structures in Computer Science*, 10(6):665–717, 2000.

[Koz81]   D. Kozen. Semantics of probabilistic programs. *Jnl. Comp. Sys. Sciences*, 22:328–50, 1981.

[Koz83]   D. Kozen. Results on the propositional μ-calculus. *Theoretical Computer Science*, 27:333–54, 1983.

[Koz85]   D. Kozen. A probabilistic PDL. *Jnl. Comp. Sys. Sciences*, 30(2):162–78, 1985.

[Kuh03]   Harold W. Kuhn. *Lectures on the Theory of Games*. Princeton University Press, 2003. First appeared, as lecture notes, in 1952.

[Lam80]   L. Lamport. "Sometimes" is sometimes "not never": on the temporal logic of programs. In *Proc. 7th ACM PoPL*, pages 174–85. ACM, 1980.

[Lam83]   L. Lamport. What good is temporal logic. In R.E.A. Mason, editor, *Proc. IFIP 9th World Congress*, pages 657–68. Elsevier, 1983.

[LPZ85]   O. Lichtenstein, A. Pnueli, and L.D. Zuck. The glory of the past. In *Proc. Logics of Programs Workshop*, volume 193 of *LNCS*, pages 198–218. Springer-Verlag, 1985.

[LR94]   D. Lehmann and M.O. Rabin. On the advantages of free choice: a symmetric and fully-distributed solution to the Dining Philosophers Problem. In Roscoe [Ros94], pages 333–52. An earlier version appeared in *Proc. 8th Ann. Symp. PoPL, 1981*.

[LS82]   D. Lehmann and S. Shelah. Reasoning with time and chance. *Information and Control*, 53(3):165–98, 1982.

[LS89]   Nancy Lynch and Roberto Segala. An introduction to I/O automata. *CWI Quarterly*, 2(3):219–46, 1989.

[LS91]   K.G. Larsen and A. Skou. Bisimulation through probabilistic testing. *Information and Computation*, 94(1):1–28, 1991.

[LSS94]   Nancy Lynch, Isaac Saias, and Roberto Segala. Proving time bounds for randomized distributed algorithms. In *Proc. 13th ACM Symp. Prin. Dist. Comp.*, pages 314–323, 1994.

[LvdS92]   J.J. Lukkien and J.L.A. van de Snepscheut. Weakest preconditions for progress. *Formal Aspects of Computing*, 4:195–236, 1992.

[McI01]   A.K. McIver. A generalisation of stationary distributions, and probabilistic program algebra. In Stephen Brookes and Michael Mislove, editors, *Electronic Notes in Theo. Comp. Sci.*, volume 45. Elsevier, 2001.

[McI02]   A.K. McIver. Quantitative program logic and expected time bounds in probabilistic distributed algorithms. *Theoretical Computer Science*, 282(1):191–219, 2002.

[MG90]   C.C. Morgan and P.H.B. Gardiner. Data refinement by calculation. *Acta Informatica*, 27:481–503, 1990. Reprinted in [MV94].

[MM]   C.C. Morgan and A.K. McIver. Proofs for Chapter 11. Draft presentations of the full proofs can be found at [MMSS, key Games02].

[MM97]   C.C. Morgan and A.K. McIver. A probabilistic temporal calculus based on expectations. In Groves and Reeves [GR97], pages 4–22. Available at [MMSS, key PTL96].

[MM99a]   C.C. Morgan and A.K. McIver. An expectation-based model for probabilistic temporal logic. *Logic Journal of the IGPL*, 7(6):779–804, 1999. Available at [MMSS, key MM97].

[MM99b]   C.C. Morgan and A.K. McIver. *pGCL*: Formal reasoning for random algorithms. *South African Computer Journal*, 22, March 1999. Available at [MMSS, key PGCL].

[MM01a]   A.K. McIver and C.C. Morgan. Demonic, angelic and unbounded probabilistic choices in sequential programs. *Acta Informatica*, 37:329–54, 2001. Available at [MMSS, key PPT2].

[MM01b]   A.K. McIver and C.C. Morgan. Partial correctness for probabilistic programs. *Theoretical Computer Science*, 266(1–2):513–41, 2001. Available at [MMSS, key PCFPDP].

[MM01c]   C.C. Morgan and A.K. McIver. Cost analysis of games using program logic. In *Proc. of the 8th Asia-Pacific Software Engineering Conference (APSEC 2001)*, December 2001. Abstract only: full text available at [MMSS, key MDP01].

[MM02]   A.K McIver and C.C. Morgan. Games, probability and the quantitative μ-calculus qMu. In *Proc. LPAR*, volume 2514 of *LNAI*, pages 292–310. Springer-Verlag, 2002. Revised and expanded at [MM04b].

[MM03]   C.C. Morgan and A.K. McIver. Almost-certain eventualities and abstract probabilities in the quantitative temporal logic qTL. *Theoretical Computer Science*, 293(3):507–34, 2003. Available at [MMSS, key PROB-1]; earlier version appeared in CATS '01.

[MM04a]   A.K. McIver and C.C. Morgan. An elementary proof that Herman's Ring has complexity $\Theta(N^2)$. Available at [MMSS, key HR04], 2004.

[MM04b]   A.K. McIver and C.C. Morgan. Results on the quantitative
             μ-calculus $qM\mu$. To appear in *ACM TOCL*, 2004.

[MMH03]   A.K. McIver, C.C. Morgan, and Thai Son Hoang. Probabilistic ter-
             mination in B. In D. Bert, J.P. Bowen, S. King, and M. Waldén,
             editors, *ZB 2003: Formal Specification and Development in Z and
             B*, volume 2651 of *LNCS*, pages 216–39. Springer-Verlag, 2003.

[MMS96]   C.C. Morgan, A.K. McIver, and K. Seidel. Probabilistic predicate
             transformers. *ACM Transactions on Programming Languages and
             Systems*, 18(3):325–53, May 1996.
             doi.acm.org/10.1145/229542.229547.

[MMS00]   A.K. McIver, C.C. Morgan, and J.W. Sanders. Probably Hoare?
             Hoare probably! In J.W. Davies, A.W. Roscoe, and J.C.P. Wood-
             cock, editors, *Millennial Perspectives in Computer Science*, Corner-
             stones of Computing, pages 271–82. Palgrave, 2000.

[MMSS]   A.K. McIver, C.C. Morgan, J.W. Sanders, and K. Seidel. Proba-
             bilistic Systems Group: Collected reports.
             web.comlab.ox.ac.uk/oucl/research/areas/probs.

[MMSS96]   C.C. Morgan, A.K. McIver, K. Seidel, and J.W. Sanders. Refinement-
             oriented probability for CSP. *Formal Aspects of Computing*,
             8(6):617–47, 1996.

[MMT98]   A.K. McIver, C.C. Morgan, and E. Troubitsyna. The probabilis-
             tic steam boiler: a case study in probabilistic data refinement. In
             J. Grundy, M. Schwenke, and T. Vickers, editors, *Proc. Interna-
             tional Refinement Workshop, ANU, Canberra*, Discrete Mathemat-
             ics and Computer Science, pages 250–65. Springer-Verlag, 1998.
             Available at [MMSS, key STEAM96].

[Mon01]   David Monniaux. *Analyse de programmes probabilistes par in-
             terprètation abstraite*. Thèse de doctorat, Université Paris IX
             Dauphine, 2001. Résumé étendu en français. Contents in English.

[Mor87]   J.M. Morris. A theoretical basis for stepwise refinement and
             the programming calculus. *Science of Computer Programming*,
             9(3):287–306, December 1987.

[Mor88a]   C.C. Morgan. Auxiliary variables in data refinement. *Inf. Proc.
             Lett.*, 29(6):293–6, December 1988. Reprinted in [MV94].

[Mor88b]   C.C. Morgan. The specification statement. *ACM Transactions on
             Programming Languages and Systems*, 10(3):403–19, July 1988.
             Reprinted in [MV94].

[Mor90]   J.M. Morris. Temporal predicate transformers and fair termination.
             Acta Informatica, 27:287–313, 1990.

[Mor94a]   C.C. Morgan. The cuppest capjunctive capping, and Galois. In
             Roscoe [Ros94], pages 317–32.

[Mor94b]   C.C. Morgan. *Programming from Specifications*. Prentice-Hall,
             second edition, 1994.

[Mor96]   C.C. Morgan. Proof rules for probabilistic loops. In He Jifeng, John
             Cooke, and Peter Wallis, editors, *Proceedings of the BCS-FACS 7th*

Refinement Workshop, Workshops in Computing. Springer-Verlag, July 1996.
ewic.bcs.org/conferences/1996/...
...refinement/papers/paper10.htm.

[Mor97] J.M. Morris. Nondeterministic expressions and predicate transformers. *Inf. Proc. Lett.*, 61(5):241–6, 1997.

[MR95] Rajeev Motwani and Prabhakar Raghavan. *Randomized Algorithms*. Cambridge University Press, 1995.

[MV94] C.C. Morgan and T.N. Vickers, editors. *On the Refinement Calculus*. FACIT Series in Computer Science. Springer-Verlag, Berlin, 1994.

[NCI99] N. Narasimha, R. Cleaveland, and P. Iyer. Probabilistic temporal logics via the modal mu-calculus. In *Proceedings of the Foundation of Software Sciences and Computation Structures, Amsterdam*, number 1578 in LNCS, pages 288–305, 1999.

[Nel89] G. Nelson. A generalization of Dijkstra's calculus. *ACM Transactions on Programming Languages and Systems*, 11(4):517–61, October 1989.

[PRI] PRISM. Probabilistic symbolic model checker.
www.cs.bham.ac.uk/~dxp/prism.

[PZ93] A. Pnueli and L. Zuck. Probabilistic verification. *Information and Computation*, 103(1):1–29, March 1993.

[Rab76] M. O. Rabin. Probabilistic algorithms. In J. F. Traub, editor, *Algorithms and Complexity: New Directions and Recent Results*, pages 21–39. Academic Press, 1976.

[Rab82] M.O. Rabin. The choice-coordination problem. *Acta Informatica*, 17(2):121–34, June 1982.

[Rao94] J.R. Rao. Reasoning about probabilistic parallel programs. *ACM Transactions on Programming Languages and Systems*, 16(3):798–842, May 1994.

[Ros94] A.W. Roscoe, editor. *A Classical Mind: Essays in Honour of C.A.R. Hoare*. Prentice-Hall, 1994.

[Roy68] H.L. Royden. *Real Analysis*. MacMillan, second edition, 1968.

[Sch86] A. Schrijver. *Theory of Integer and Linear Programming*. Wiley, New York, 1986.

[SD80] N. Saheb-Djahromi. CPO's of measures for nondeterminism. *Theoretical Computer Science*, 12(1):19–37, 1980.

[Seg95] Roberto Segala. *Modeling and Verification of Randomized Distributed Real-Time Systems*. PhD thesis, MIT, 1995.

[Sha53] L. S. Shapley. Stochastic games. In *Proc. National Academy of Sciences USA*, volume 39, pages 1095–1100, 1953.

[SHRT04] S. Schneider, T.S. Hoang, K.A. Robinson, and H. Treharne. Tank monitoring: a case study in *pAMN*. Technical Report CSD-TR-03-17, Royal Holloway College, 2004. In draft.

[SL95]   Roberto Segala and Nancy Lynch. Probabilistic simulations for
             probabilistic processes. *Nordic Journal of Computing*, 2(2):250–73,
             1995.

[SMM]   K. Seidel, C.C. Morgan, and A.K. McIver. Probabilistic imperative
             programming: a rigorous approach. Revises [SMM96]; available at
             [MMSS, key SMM96]; extended abstract appears in [GR97].

[SMM96]   K. Seidel, C.C. Morgan, and A.K. McIver. An introduction to
             probabilistic predicate transformers. Technical Report PRG-TR-
             6-96, Programming Research Group, February 1996. Available at
             [MMSS, key SMM96].

[Smy78]   M.B. Smyth. Power domains. *Jnl. Comp. Sys. Sciences*, 16:23–36,
             1978.

[Smy89]   M.B. Smyth. Power domains and predicate transformers: a topo-
             logical view. In *Automata, Languages and Programming 10th
             Colloquium, Barcelona, Spain*, volume 298 of *LNCS*, pages 662–75.
             Springer-Verlag, 1989.

[SPH84]   M. Sharir, A. Pnueli, and S. Hart. Verification of probabilistic
             programs. *SIAM Journal on Computing*, 13(2):292–314, May 1984.

[Spi88]   J.M. Spivey. *Understanding Z: a Specification Language and its
             Formal Semantics*. Cambridge University Press, 1988.

[Sti92]   C. Stirling. Modal and temporal logics. In S. Abramsky, D. Gabbay,
             and T. Maibaum, editors, *Handbook of Logic in Computer Science,
             Volume 2: Computational Structures*, pages 478–551. Clarendon
             Press, 1992.

[Sti94]   David Stirzaker. *Elementary Probability*. Cambridge University
             Press, 1994.

[Sti95]   C. Stirling. Local model checking games. In *CONCUR '95*, vol-
             ume 962 of *LNCS*, pages 1–11. Springer-Verlag, 1995. Extended
             abstract.

[Sto88]   Allen Stoughton. *Fully Abstract Models of Programming Languages*.
             Research Notes in Theoretical Computer Science. Pitman/Wiley,
             1988.

[Sto96]   N. Storey. *Safety-Critical Computer Systems*. Addison-Wesley,
             1996.

[Sut75]   W. Sutherland. *An Introduction to Metric and Topological Spaces*.
             Oxford University Press, 1975.

[SV03]   M.I.A. Stoelinga and F.W. Vaandrager. A testing scenario for
             probabilistic automata. In *Proceedings ICALP '03*, volume 2719
             of *LNCS*, pages 407–18. Springer-Verlag, 2003.

[Tar55]   A. Tarski. A lattice-theoretic fixpoint theorem and its applications.
             Pacific Journal of Mathematics, 5:285–309, 1955.

[Tru71]   K. Trustrum. *Linear Programming*. Library of Mathematics. Rout-
             ledge and Kegan Paul, London, 1971.

[Tur84]   R. Turner. *Logics for AI*. Artificial Intelligence. Ellis Horwood,
             1984.

[Van98]   B. Van Roy. *Learning and Value Function Approximation in Complex Decision Processes*. PhD thesis, EECS, MIT, 1998.

[Var85]   M.Y. Vardi. Automatic verification of probabilistic concurrent finite-state programs. In *Proc. 26th IEEE Symp. on Foundations of Computer Science*, pages 327–38, Portland, October 1985.

[Var01]   M.Y. Vardi. Branching vs. linear time: Final showdown. In T. Margaria and Wang Yi, editors, *Seventh International Conference on Tools and Analysis of Systems, Genova*, number 2031 in LNCS, pages 1–22, April 2001.

[vBMOW03]   Franck van Breugel, Michael W. Mislove, Joel Ouaknine, and James Worrell. An intrinsic characterization of approximate probabilistic bisimilarity. In *Proceedings of FOSSACS 2003*, number 2620 in LNCS, pages 200–15. Springer-Verlag, 2003.

[vGB98]   A.J.G van Gasteren and A. Bijlsma. An extension of the program derivation format. In D. Gries and W.P. de Roever, editors, *PROCOMET*, volume 125 of *IFIP Conference Proceedings*, pages 167–85. Chapman & Hall, 1998.

[vNM47]   J. von Neumann and O. Morgenstern. *Theory of Games and Economic Behavior*. Princeton University Press, second edition, 1947.

[War89]   M. Ward. *Proving Program Refinements and Transformations*. PhD thesis, Programming Research Group, 1989.

[Yin02]   Mingsheng Ying. Bisimulation indexes and their applications. *Theoretical Computer Science*, 275(1–2):1–68, 2002.

[Yin03]   Minsheng Ying. Reasoning about probabilistic sequential programs in a probabilistic logic. *Acta Informatica*, 39(5):315–89, 2003.

[YL92]   W. Yi and K.G. Larsen. Testing probabilistic and nondeterministic processes. In R.J Linn and M. Ümit Uyar, editors, *Proceedings of 12th IFIP International Symposium on Protocol Specification, Testing and Verification, Florida, USA*, pages 47–61, 1992.

[YW00]   Mingsheng Ying and Martin Wirsing. Approximate bisimilarity. In Teodor Rus, editor, *AMAST*, volume 1816 of *LNCS*, pages 309–22. Springer-Verlag, 2000.

Index of citations

General index

Symbols are indexed in order of occurrence, and alphabetically if appropriate, referring to a spelled-out entry whose name may be sufficient to jog the memory.

Hierarchical topics are cross-referenced both up and down, with page-number references occurring at the tips. Thus "semantics" leads down to "expectation-transformer semantics" and then to "assignment," where the expectation-transformer semantics of assignment is located on p. 7.

In the reverse direction, starting from "assignment" a reference leads back to "expectation-transformer semantics" (showing what other constructs also have that kind of semantics) and then up to "semantics" (showing what other kinds of semantics there are).

Bold page numbers indicate definitions; underlined numbers are named figures, definitions *etc*. A "*qv*" after a sub-item is a cross-reference to the main item beginning with those words.

A —

abort

    is bottom (⊥) among programs, 131, 133

    is (pre-)deterministic, 131

    "jail" metaphor for, 12

probabilistic transformer semantics,
    see expectation transformer...
probabilistically guarded iteration,
    syntax, **196**
probability
  defiance, 89
  theory, 16ff
probability distribution (Pr), **16**
  continuous, 16, 220, 238, 297
  discrete, **16**, 166, 219, 220, 238
  discrete over infinite state space,
      219, 220, 332
  explicit in formulae, 316ff
  generalised to *evaluation*, 18, 36
  geometric, 70, 208 *see also*
      Programs
  infinitary, 332
  not confused with expectation, 137
  order, **131**
  point ($\overline{s}$), 132
  Pr$_n$, in proof of loop rules, **76**
  space ($\overline{S}$), **130**
  space as (hyper-) pyramid,
      166ff, 239
  space is compact, 334
  stationary, 291
  sub-, **130**
  "sub-" omitted, 131
  uniform over [0, 1], 220
  written Δ, 130, 219
  see also sets of distributions
product ($\prod$), **62**
program
  atomic, **160**
  complexity *qv*
  depicted as a diamond, 169
  depicted as a line, 169
  least, *see* **abort**
  maximal, *see* deterministic program
  miraculous, 138 *see also* feasibility
  relational semantics for, *see*
      probabilistic relational...,
      standard relational...
  transformer semantics for, *see*
      expectation transformer...,
      standard transformer...
program algebra, 6, 28ff, 161–3, 170
  collection of laws, 321–8

example, 10, 11, 112, 114ff,
    120, 325–6
iteration, 38, 187
justified by healthiness conditions,
    161
of *wp/wlp*, 331
of recursion, 38
program superposition, 64
Programs
  card-and-dice game, 11ff
  choice coordination
      (Rabin's tourists), 79–87
  coin-flip implementation of $_p\oplus$, 210
  control system, 325–6
  counter-example to loop rules, 71
  demonic, 8, 226
  demonic not additive, 226
  demonic/probabilistic, 21ff
  deterministic, 7
  dining philosophers, 88–98
  Duelling Cowboys, 210–14
  faulty factorial, 51–3
  faulty skipper, 111, 115
  flip a coin, 6
  geometric distribution, 69, 196, 208
  Herman's (token) Ring, 56–61
  Las-Vegas, 44–6
  martingale gambling, 44–6, 64ff
  Monte-Carlo, 46–51
  Monty-Hall game, 22, 27ff
  not continuous, 335
  primality testing (Miller-Rabin),
      46–51
  probabilistic amplification, 46–51
  probabilistic termination example,
      40
  random walk, bounded
      two-dimensional, *see* Three-up
  random walk, general, 99–105
  random walk, symmetric, 71
  refinement example, 10
  self-stabilisation, 56–61
  software publishing, 301–4
  square root, 5
  standard variant fails, 55, 56
  steam boiler (safety-critical,
      fault-tolerant), 117–23
  Three-up game, 72ff
  unboundedly nondeterministic, 335